S0-ENV-204

Ocean Reverberation

Ocean Reverberation

edited by

DALE D. ELLIS,
JOHN R. PRESTON
and
HEINZ G. URBAN

SACLANT Undersea Research Centre,
La Spezia, Italy

KLUWER ACADEMIC PUBLISHERS
DORDRECHT / BOSTON / LONDON

Library of Congress Cataloging-in-Publication Data

```
Ocean reverberation / edited by Dale D. Ellis, John R. Preston, and
   Heinz G. Urban.
       p.    cm.
    Includes index.
    ISBN 0-7923-2420-X
    1. Underwater acoustics.   I. Ellis, Dale D.   II. Preston, John R.
  III. Urban, Heinz G., 1937-      .
  QC225.15.O28  1993
  620.2'5--dc20                                                  93-27852
```

ISBN 0-7923-2420-X

Published by Kluwer Academic Publishers,
P.O. Box 17, 3300 AA Dordrecht, The Netherlands.

Kluwer Academic Publishers incorporates
the publishing programmes of
D. Reidel, Martinus Nijhoff, Dr W. Junk and MTP Press.

Sold and distributed in the U.S.A. and Canada
by Kluwer Academic Publishers,
101 Philip Drive, Norwell, MA 02061, U.S.A.

In all other countries, sold and distributed
by Kluwer Academic Publishers Group,
P.O. Box 322, 3300 AH Dordrecht, The Netherlands.

Printed on acid-free paper

All Rights Reserved
© 1993 Kluwer Academic Publishers
No part of the material protected by this copyright notice may be reproduced or
utilized in any form or by any means, electronic or mechanical,
including photocopying, recording or by any information storage and
retrieval system, without written permission from the copyright owner.

Printed in the Netherlands

CONTENTS

PREFACE xi

Section 1 – Scattering Mechanisms 1

Sea surface reverberation (Invited paper)
S.T. McDaniel 3

Properties of bubble distributions relevant to surface reverberation
(Invited paper)
D.M. Farmer 11

Sound scattering from microbubble distributions near the sea surface
(Invited paper)
W.M. Carey and R.A. Roy 25

Theoretical modeling of low frequency acoustic scattering from
subsurface bubble structures
R.F. Gragg and D. Wurmser 45

Discrete backscatter can be dominant in rough bottom reverberation
I. Dyer, A.B. Baggeroer, H. Schmidt, J.R. Fricke, N. Ozluer and D. Giannoni 51

Sea-bottom reverberation: the role of volume inhomogeneities of the
sediment
M. Gensane 59

The perturbation characterization of ocean reverberations
E.Y.T. Kuo 65

The reduction of surface backscatter and ambient noise by monomolecular films on the ocean surface
J. Rohr, R. Kolesar and F. Ryan 71

Section 2 – High Frequency Measurements and Mechanisms 77

High frequency acoustic bottom backscatter mechanisms at shallow
grazing angles
N.P. Chotiros 79

Measurements of high frequency reverberation in shallow water
J.G. Kelly, R.N. Carpenter, M. Buffman and E.R. Levine 85

Frequency response measurements on backscattering from a shallow sea floor, using a parametric source
T.G. Muir, L.A. Thompson, J.A. Shooter, T.E. DeMary and R.J. Wyber 91

Measurements of acoustic backscattering of the near-surface layer
B. Nützel and H. Herwig 97

Section 3 – Reverberation Modelling 103

Numerical modeling of three-dimensional reverberation from bottom facets (Invited paper)
H. Schmidt 105

Numerical simulations of lower-frequency acoustic propagation and backscatter from solitary internal waves in a shallow water environment
S.A. Chin-Bing, D.B. King and J.E. Murphy 113

Reverberation modeling with the two-way parabolic equation
M.D. Collins, G.J. Orris and W.A. Kuperman 119

Calculations of ocean bottom and sub-bottom backscattering using a time-domain finite-difference code
D.D. Ellis, N.A. Kampanis and R.A. Stephen 125

Time and angle spreading from rough sediments
D.F. McCammon 131

Rough surface scattering as seen through the Renormalization Group
G.J. Orris and R.F. Dashen 139

Long range 3-D reverberation and scattering modeling methodology
R.B. Williams, H.P. Bucker, A. D'Amico and D.F. Gordon 145

Range-dependent, normal-mode reverberation model for bistatic geometries
D.M. Fromm, B.J. Orchard and S.N. Wolf 155

Measurement, characterization, and modeling of ice and bottom backscattering and reverberation in the Arctic Ocean
T.C. Yang and T.J. Hayward 161

A reverberant-undersea model for bottom-limited environments
H.P. Bucker, J.A. Rice and N.O. Booth 167

Shallow water reverberation modeling
J.K. Fulford 175

Section 4 – ARSRP Mid-Atlantic Ridge Experiment 181

An overview of the 1991 Reconnaissance Cruise of the Acoustic
   Reverberation Special Research Program
A.B. Baggeroer and J.A. Orcutt 183

Long-range measurements of seafloor reverberation in the Mid-Atlantic
   Ridge area
J.M. Berkson, N.C. Makris, R. Menis, T.L. Krout and G.L. Gibian 189

Ocean-basin scale inversion of reverberation data
N.C. Makris, J.M. Berkson, W.A. Kuperman and J.S. Perkins 195

ARSRP reconn results and BISSM modeling of direct path backscatter
J.W. Caruthers and E.J. Yoerger 203

Directional processing of the simulated direct path, broadband
   reverberation data and its interpretation
A.K. Kalra and J.K. Fulford 209

Numerical scattering results for a rough, unsedimented seafloor
A. Levander, A. Harding and J.A. Orcutt 215

The effects of seafloor roughness on reverberation: finite difference and
   Kirchhoff simulations
J.A. Orcutt, A. Harding and A. Levander 221

A numerical scattering chamber for studying reverberation in the seafloor
R.A. Stephen 227

Section 5   Low Frequency Measurements 233

Low-frequency direct-path surface and volume scattering measured
   using narrowband and broadband pulses
R.C. Gauss, J.M. Fialkowski and R.J. Soukup 235

Low-frequency surface and bottom scattering strengths measured using
   SUS charges
P.M. Ogden and F.T. Erskine 241

Upslope propagation data versus two-way PE
G.J. Heard, D.J. Thomson and G.H. Brooke 247

Section 6 – Volume Scattering 253

Low-frequency volume reverberation measurements
T. Akal, R.K. Dullea, G. Guidi and J.H. Stockhausen 255

Low-frequency volume scatter and distant reverberation: 200-1500 Hz
J.M. Monti and C.S. Hayek 263

GINRUNS – A 1991 volume scattering and convergence zone
   experiment in the Norwegian Sea
J.A. Doutt and J.R. Preston 271

Volume reverberation at mid frequencies in the Norwegian Sea
C.H. Thompson and R.H. Love 279

Volume reverberation in the marginal ice zone of Fram Strait
M.A. Wilson and R.H. Love 285

Section 7 – Signal Processing Issues 291

Analytical description and experimental results for reverberation-resistant
   acoustic tracking signals
T.J. Curry and T.A. Casey 293

Broadband adaptive beamformer for linear-frequency-modulation active
   signals
J.C. Lockwood 299

Broadband pulse distortion: waveform design issues for active systems
F.J. Ryan and E.F. Rynne 305

Active matched field processing for clutter rejection
L.B. Dozier and H.A. Freese 313

Reverberation suppression and modeling
D.W. Tufts, D.H. Kil and R.R. Slater 319

Sub-bottom scattering at low frequencies
J.H. Wilson, M. Bradley and M. Wagstaff 331

Broadband detection and classification of underwater sources
A.H. Quazi and A.H. Nuttall 343

Section 8 – Applications 349

A geometrical approach to medium low frequency reverberation
   modeling (Invited paper)
A. Plaisant 351

The reverberation array heading surface
R.A. Wagstaff, M.R. Bradley and M.A. Hebert 361

Bistatic ocean reverberation effects
S. Sutherland-Pietrzak 367

Modeling of wide area ocean reverberation and noise
A.I. Eller, L. Haines, W. Renner, R. Cavanagh and E.D. Chaika 373

Rapid environmental acoustic survey and modeling system
D. Rubenstein and D.S. Hansen 379

LIST OF PARTICIPANTS 385

AUTHOR INDEX 391

SUBJECT INDEX 397

PREFACE

During the past decade there has been a renewed interest in active sonar systems at both low and medium frequencies. More recently this interest has been extended to very high frequencies in shallow water. Reverberation often limits the detection performance of these systems, and there is a need to understand the underlying mechanisms that cause the scattering. With more emphasis being given to reverberation phenomena in the Scientific Program of Work at the SACLANT Undersea Research Centre, it was considered an opportune time to host a meeting, bringing together scientists from NATO countries to foster cross-disciplinary dialogue and generate ideas for new research directions.

Consequently the Ocean Reverberation Symposium was held 25–29 May 1992 in La Spezia, Italy. Over 60 presentations were made on a diverse selection of topics, of which ten papers will be published as a SACLANTCEN Conference Proceedings. The papers in this volume are grouped into 8 sections, usually in the same order as presented at the corresponding session of the Symposium:

Section 1 – Scattering Mechanisms
Section 2 – High Frequency Measurements and Mechanisms
Section 3 – Reverberation Modelling
Section 4 – ARSRP Mid-Atlantic Ridge Experiment
Section 5 – Low Frequency Measurements
Section 6 – Volume Scattering
Section 7 – Signal Processing Issues
Section 8 – Applications

Taken together the papers show some emerging trends in the research.

The availability of high-power low-frequency sources and highly directional arrays has brought with it the tools, and the need, to study long-range reverberation. The use of projector sources and various waveforms rather than explosives allows the use of signal processing techniques to enhance the extraction of information about the reverberation and scattering processes.

Two projects that are having a major impact on the understanding of reverberation are the US Office of Naval Research Special Research Project (SRP) on surface reverberation, and a similar project to study bottom scattering – the Acoustic Reverberation Special Research Project (ARSRP). The experiments have shown that perturbation theory can be applied to surface scatter in low wind conditions. The scattering from near-surface bubbles has been recognized as an important mechanism contributing to surface backscatter, especially at higher wind conditions. The ARSRP experiment near the Mid-Atlantic Ridge will hopefully lead to an appreciation of the effects of feature scattering. One possibility is the investigation of the conjecture that bottom scattering is dominated by glints from facets approximately one wavelength in size. Another hypothesis, suggested by numerical modelling, is that the elastic properties of hard bottoms result in enhanced scattering. These programs and other initiatives are very important in obtaining high-quality acoustic and environmental data, to guide the research, and provide a focus for the modelling work.

High levels of reverberation observed in some northern waters have been associated with volume scattering. Subsequent experiments have shown that resonant scattering from air-filled swim bladders of various fish can be a dominant contributor to reverberation, even at frequencies below 1 kHz.

There were a number of papers on high-frequency reverberation in shallow water. With the changing emphasis from deep water to shallow water, we expect high-frequency scattering to be the subject of increased study in the future.

The increase in computing power has enabled significant advances in both data processing and modelling. Traditional sonar modelling techniques are now being applied to bistatic and 3-D range-dependent environments. The advances in full-wave propagation techniques over the past two decades are now being applied to scattering, though only two-dimensional problems are presently tractable. Scattering is intrinsically a difficult mathematical problem and analytical techniques have had limited success even for pressure-release surfaces; the need to handle the subbottom and subsurface scattering from bubbles complicates the issue even more. The application of numerical techniques to scattering should lead to further understanding of the processes in the future.

New signal processing techniques, as well as older ones traditionally applied to passive sonar processing are showing great promise as tools to "de reverberate" received signals. Proper understanding of reverberation as a non-stationary process will be essential to improving these techniques.

Though often initially performed in support of military objectives, the research is leading to an increased understanding of our ocean environment, and the development of remote sensing techniques for other applications.

The editors express their gratitude to all contributors for their excellent papers and presentations. Special thanks is given to the invited speakers. We also wish to thank the Chairmen of the various sessions, and the members of the Advisory Committee. Many thanks are also given to the SACLANTCEN staff members for their help in organizing the Symposium, and managing the many details that were so important to its success. In particular, thanks to the Conference Secretary, Anna Bizzarri, and to Adolf Legner who took care of all the manuscripts and was responsible for the final layout of the papers. Finally, we thank our wives and families for their help and patience as we prepared for the Symposium and assembled the proceedings.

Dale D. Ellis
John R. Preston
Heinz G. Urban

Section 1

Scattering Mechanisms

SEA SURFACE REVERBERATION

Suzanne T. McDaniel
Pennsylvania State University
University Park, PA 16804

ABSTRACT

The past decade has seen a significant improvement of the understanding of acoustic reverberation from the sea surface and of the basic physical mechanisms driving this process: scattering due to surface roughness and to resonant microbubbles or bubble clouds. This paper reviews recent experimental work as well as related studies to characterize sea surface roughness and the density of near surface microbubbles. Theoretically predicted sea surface scatter is compared with experimental data over a broad frequency range to determine the conditions for which additional scattering mechanisms must be invoked to explain the experimental results.

1. INTRODUCTION

The scattering of acoustic energy from the sea surface has long been studied because it constitutes an important source of interference for active sonars. More recent applications include the sensing of upper ocean dynamical processes. In this paper, measurements of acoustic reverberation are compared with theoretical predictions of scattering due to sea surface roughness to establish the relative importance of scattering from resonant microbubbles or from clouds of these bubbles. Theories for acoustic backscatter due to surface roughness and subsurface bubbles are introduced in Section 2, and the environmental inputs to these theories are discussed in Section 3. The theoretical predictions are compared with acoustic data in Section 4 where data are presented which show that resonant scattering from subsurface bubbles may constitute an important contribution to the reverberant field for frequencies as low as 3 kHz. At lower frequencies, where resonant bubbles are no longer a viable scattering mechanism, the disparity between low grazing angle measurements and the predicted backscattering strength due to rough surface scattering indicates the presence of an additional scattering mechanism at high wind speeds. The findings of this study are summarized and discussed in Section 5.

2. SCATTERING THEORIES

The geometry for surface backscatter is shown in Fig. 1, which also illustrates the two major sources of reverberation at high acoustic frequencies. Scattering due to surface roughness may be treated using composite roughness theory[1] where the surface roughness is partitioned into two regimes: a large scale surface to which the Kirchhoff approximation is applied, and a small scale surface that is treated using the Rayleigh-Rice approach. For the low grazing angles of interest in this study, the backscattering strength S obeys

$$S = (k^4/\pi) < \sin^4(\theta + \epsilon) > W(K), \qquad (1)$$

where k is the acoustic wavenumber, $K = 2k/\cos\theta$, and θ and ϵ are respectively the grazing angle and large scale surface slope. In Eq. 1, $W(K)$ is the wave number spectrum of the ocean surface elevation. This result has the very simple physical interpretation that the scatter is governed by Bragg diffraction from small scale elements of the surface, and the only effect of the large scale surface is a modulation of the local grazing angle. One disadvantage of this theory is that its predictions are dependent on the surface partitioning. Hence, in many of the examples that follow, the predicted backscatter is shown for a range of partitioning choices.

Scattering from resonant microbubbles may be treated using a theory due to Crowther[2]. For sparse microbubble populations of the form $N(a) \sim 1/a^m$; $a_{min} < a < a_{max}$, the backscattering strength at a frequency for which the resonant bubble radius is a_r, where $a_{min} \ll a_r \ll a_{max}$, is given by

$$S = (C\pi/\delta)a_r^3 \int N(a_r)dz \qquad (2)$$

where C is a constant that is weakly dependent on the coefficient m, δ is the bubble damping coefficient at resonance, and it has been assumed that multiple paths by which the scattering may occur add incoherently. Because acoustic energy is strongly attenuated as it traverses a dense layer of subsurface bubbles, for sufficiently high bubble populations, this effect dominates and a saturation level is reached where further increases in bubble density produce no further increase in backscattering strength. The predicted backscatter in this case takes a particularly simple form

$$S = (\delta_r/\delta)\sin\theta/(8\pi), \qquad (3)$$

where δ_r is the reradiation damping constant. Experimental observation of the level

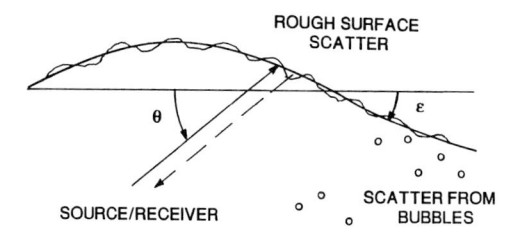

Figure 1. Geometry for monostatic sea surface backscatter.

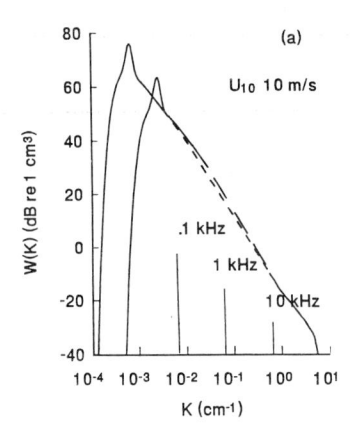

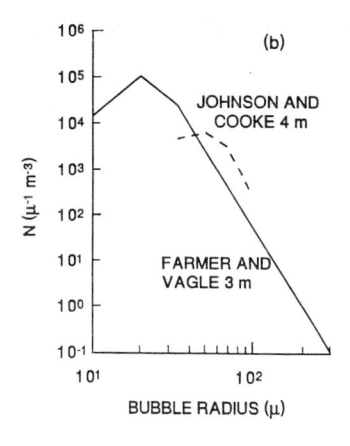

Figure 2. Environmental inputs: (a) ocean wave number spectrum; (b) subsurface bubble populations.

and angular dependence of this saturation limit provides a means of determining if the scatter is due to resonant microbubbles or some other mechanism.

Acoustic backscattering from bubble clouds or plumes has recently been treated by McDonald[3] and Henyey[4]. They model the plumes of bubbles generated by breaking waves as cylinders having an air void fraction that diminishes exponentially with depth, or as finite length cylinders of constant void fraction, respectively. Although these models provide agreement with some of the anomalous low-frequency data, a more complete description is beyond the scope of this paper.

3. ENVIRONMENTAL INPUTS

The theories addressed in the previous sections require environmental inputs: in the first, the wave number spectrum $W(K)$ of the sea surface is needed, and in the latter two, subsurface bubble populations. The ocean wave number spectrum that will be used in this paper combines the recent low wave number model of Donelan et al.[5] with Pierson's[6] high wave number spectrum, which yields good agreement with high-frequency acoustic data when scattering from resonant bubbles is unimportant. Following Pierson, these two regions are connected by a power law as shown in Fig. 2 (a), where the spectrum shown for a fetch of 50 km is 3 dB higher than that for a fully developed sea when K is 0.02 cm^{-1}. In this figure, the wave numbers responsible for Bragg scattering at low grazing angles are shown for frequencies of .1, 1 and 10 kHz.

There are disparities between the bubble populations reported by different researchers. An example is presented in Fig. 2 (b) in which the acoustical measurements of Farmer and Vagel[7] are compared with the optical measurements of Johnson and Cooke[8] for wind speeds of 12 to 14 m/s and 11 to 13 m/s, respectively. The differences between the two data sets which may be a consequence of optical resolution, water temperature, or other factors, severely limit our ability to verify theories for scattering from microbubbles and bubble clouds.

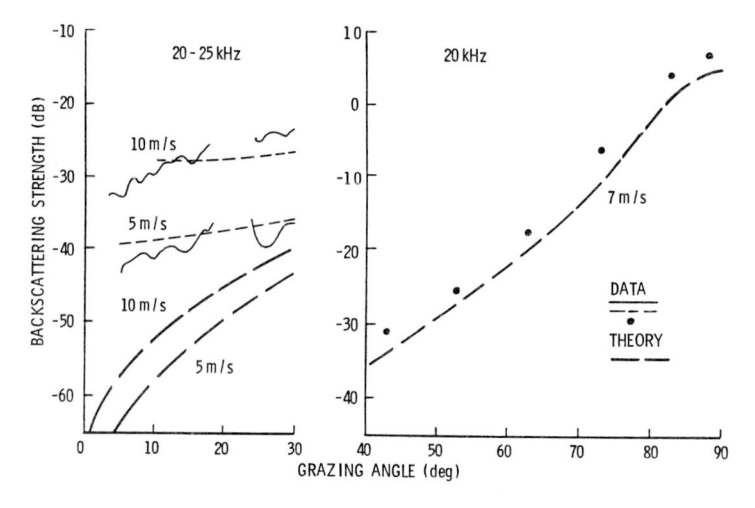

Figure 3. Grazing angle dependence of backscatter for frequencies of 18-25 kHz.

4. COMPARISON OF THEORY WITH EXPERIMENT

Backscatter data for frequencies of 18 to 25 kHz is compared in Fig. 3 with the predictions of composite roughness theory. In addition to the data of Ref. 9, more recent data acquired by Nützel, et al.[10] have been included. As in Ref. 9, a fully developed omnidirectional sea will be assumed for scattering at frequencies above 3 kHz. It is evident in this comparison, that good agreement is obtained with composite roughness theory at grazing angles above 40°, but that at lower grazing angles, the measured backscattering strengths are significantly higher than those predicted by the theory. It is also evident, that the disparity between theory and experiment increases with increasing wind speed and decreasing grazing angle.

It can be shown that scattering from subsurface bubbles is responsible for this disparity by examining acoustic data at high wind speeds where the saturation effect predicted by the bubble scattering theory has been observed. Figure 4 compares acoustic backscatter data at high wind speeds with the predicted scatter due to res-

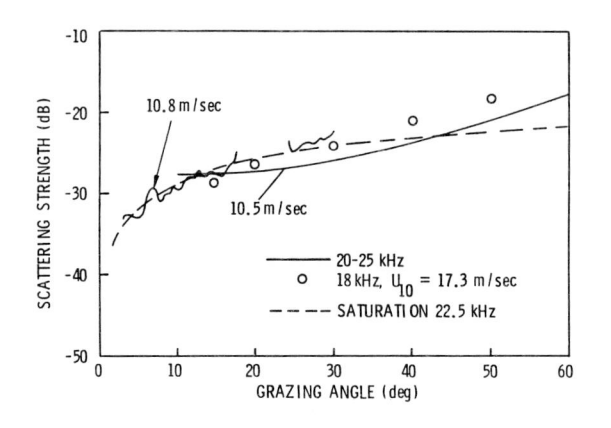

Figure 4. Bubble saturation limit for frequencies of 18-25 kHz.

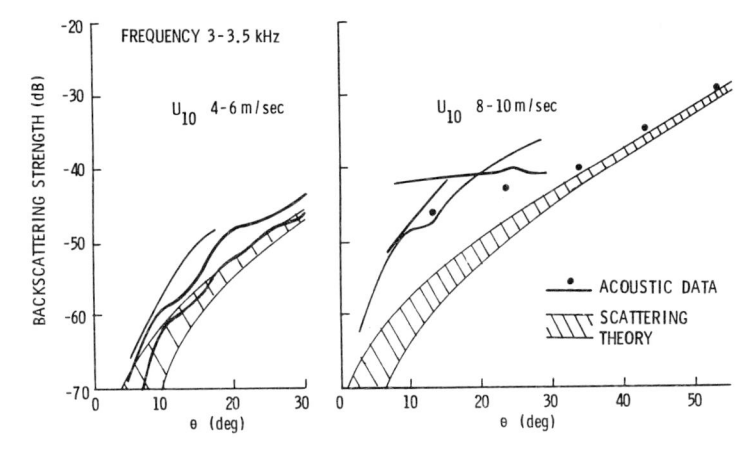

Figure 5. Grazing angle dependence of backscatter for frequencies of 3 and 3.5 kHz.

onant microbubbles in the saturation limit. Both the angular dependence and level of the data are in accord with the predictions.

Let us now consider backscatter at a lower frequency. Figure 5 presents backscatter measurements at frequencies of 3 and 3.5 kHz[2,10,11] acquired using narrow beam sources and receivers. The data are in fair agreement with the predictions for wind speeds of 4 to 6 m/s, as shown on the left. However, with increasing wind speeds of 8 to 10 m/s, an effect similar to that which was observed at higher frequencies is apparent: good agreement is obtained at the higher grazing angles, with generally poor agreement below a grazing angle of 30°.

It is of interest to examine the wind speed dependence of backscatter in this frequency range to determine if a saturation effect exists. Figure 6 shows the dependence of the narrow beam data from Fig. 5 on wind speed for a grazing angle of 15°. Also shown in this figure are data[12-14] acquired using explosive charges that were selected because of their agreement with the narrow beam data. It is clear in this figure that

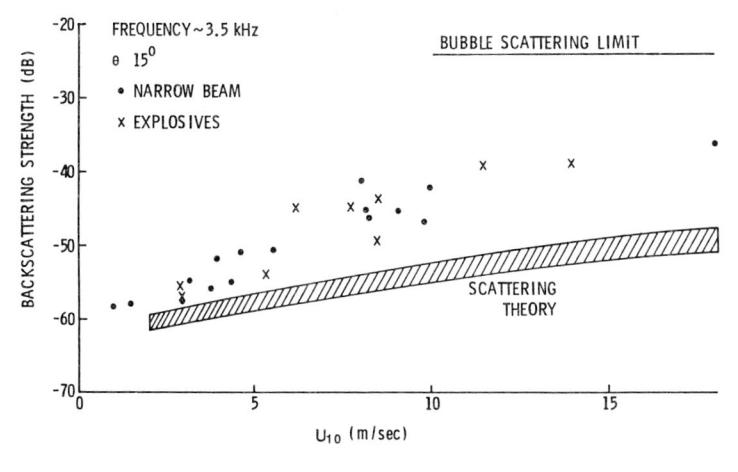

Figure 6. Dependence of backscatter on wind speed for frequencies of 3 and 3.5 kHz.

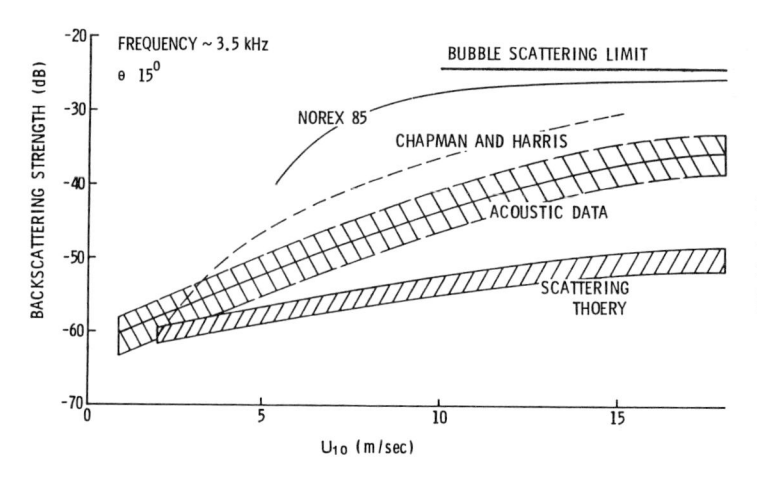

Figure 7. Data at frequencies of 3-3.5 kHz acquired in duration or fetch limited seas.

although the data increase much more rapidly with wind speed than the theory predicts, the bubble saturation limit is not reached, even at a wind speed of 18 m/s. For comparison with other explosive charge data and more recently acquired narrow beam data, a least squares curve, with the standard deviation indicated, will be passed through the data.

Figure 7 shows additional narrow beam data acquired in the North Sea[10] (NOREX 85) for duration and fetch limited seas. These data clearly exhibit a saturation effect, reaching the theoretical limit at a wind speed slightly above 10 m/s. It thus appears that scattering from resonant bubbles is an important contribution at this frequency. Also shown in this figure are the predictions of the Chapman and Harris empirical model[15] derived from explosive charge measurements in duration limited seas. The Chapman and Harris model displays a stronger dependence on wind speed than the acoustic data of Fig. 6 at low wind speeds: for wind speeds of 5 to 15 m/s, this model's predictions are uniformly high by 8 dB.

Backscatter measurements at still lower frequencies are shown in Fig. 8. This figure shows acoustic data and scattering theory predictions as a function of frequency for a fixed grazing angle of 20° and wind speeds of 7.5 to 12.5 m/s. The narrow beam data are from the previously cited sources, and the explosive charge data from the same selected data sets that provided good agreement with the 3.5 kHz narrow beam data. Because all of the measurements below a frequency of 3 kHz are omnidirectional, an omnidirectional wave number spectrum and a fully developed sea has been assumed for the scattering predictions. At frequencies below 1 kHz, the effect of the large scale surface slope of Eq. 1 on the predicted reverberation levels is less than 1 dB. The data clearly depart from the theory at a frequency of about 1 kHz. As in Fig. 7, the Chapman and Harris predictions lie uniformly 8 dB above the other acoustic data.

Additional low-frequency acoustic data are shown in Fig. 9 for higher wind speeds, 10 to 14 m/s, and a lower grazing angle, 10°. In this case, all of the acoustic data lie clearly above the theory, even at frequencies below 1 kHz.

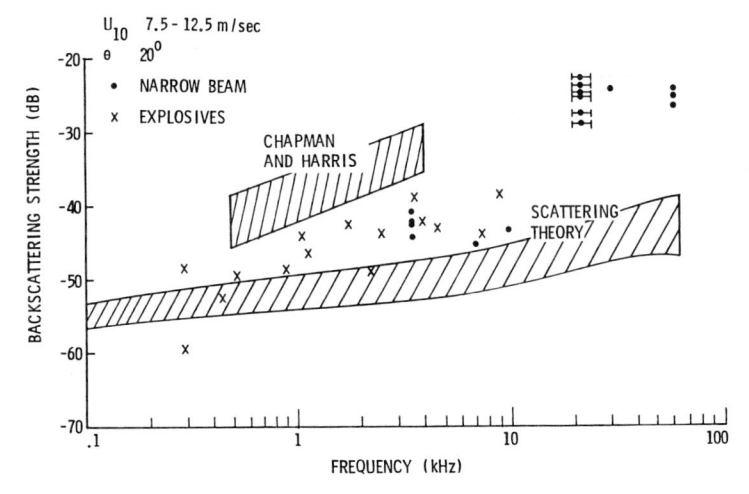

Figure 8. Dependence of backscatter on frequency for a grazing angle of 20°.

5. DISCUSSION

At frequencies above 3 kHz, measured reverberation levels and their observed saturation with wind speed are consistent with the view that two basic mechanisms contribute: scattering due to resonant microbubbles and ocean surface roughness. In this frequency range, there is a wide disparity in the backscattering strengths reported at a given wind speed for conditions in which the sea is duration or fetch limited. This disparity is comparable in magnitude to the disparity in measured microbubble populations.

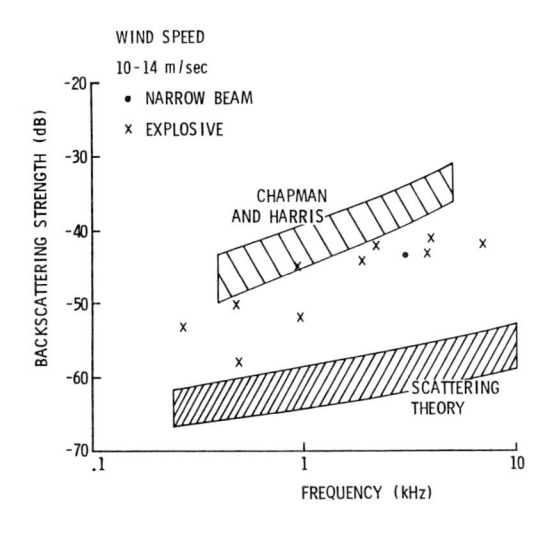

Figure 9. Dependence of backscatter on frequency for a grazing angle of 10°.

At frequencies below 1 kHz, only at very high wind speeds and low grazing angles, is there a clear disparity between the predictions of composite roughness theory and the measured backscattering levels. Although models of scattering from clouds of subsurface microbubbles can account for this disparity, it is premature to rule out the effects of surface roughness.

REFERENCES

1. Kur'ynov, B. F., "The scattering of sound waves at a rough surface with two types of irregularity," Sov. Phys. Acoust. **8**, 252 (1963).
2. Crowther, P. A., "Acoustic scattering from near-surface bubble layers," in *Cavitation and Inhomogeneities in Underwater Acoustics*, ed. W. Lauterborn (Springer, New York, 1979).
3. McDonald, B. E., "Echoes from vertically striated subresonant bubble clouds: A model for ocean surface reverberation," J. Acoust. Soc. Am. **89**, 617 (1991).
4. Henyey, F. S., "Acoustic scattering from ocean microbubble plumes in the 100 Hz to 2 kHz region," J. Acoust. Soc. Am. **90**, 399 (1991).
5. Donelan, M. A., J. Hamilton and W. H. Hui, "Directional spectra of wind generated waves," Phil. Trans. Roy. Soc. Lond. **A 315**, 509 (1985).
6. Pierson, W. J., "The theory and applications of ocean wave measuring systems at and below the sea surface, on the land, and from aircraft and spacecraft," NASA Contract. Rep. CR-2646, NASA, Washington, DC (1976).
7. Farmer, D. and S. Vagle, "Waveguide propagation of ambient sound in the ocean surface bubble layer," J. Acoust. Soc. Am. **86**, 1897 (1989).
8. Johnson, B. D., and R. C. Cooke, "Bubble populations and spectra in coastal waters: A photographic approach," J. Geophys. Res. **84**, 3761 (1979).
9. McDaniel, S. T., and A. D. Gorman, "Acoustic and radar sea surface backscatter," J. Geophys. Res. **87**, 4127 (1982).
10. Nützel, B., H. Herwig, J. M. Monti, and P. D. Koenigs, "The influence of surface roughness and bubbles on sea surface acoustic back-scattering," NUSC Tech. Rep. 7955, Naval Underwater Systems Center, New London, CT (1987).
11. Bachmann, W., "A theoretical model for the backscattering strength of a composite roughness sea," J. Acoust. Soc. Am. **54**, 712 (1973).
12. Richter, R. M., "Measurements of backscattering from the sea surface," J. Acoust. Soc. Am. **36**, 864 (1964).
13. Brown, J. R., J. A. Scrimger, and R. G. Turner, "Reverberation from the ocean surface," Tech. Mem. 66-8, Pacific Naval Laboratory, Canada (1966).
14. Percy, J. L., private communication (1970).
15. Chapman, R. P., and J. H. Harris, "Surface backscattering strengths measured with explosive charges," J. Acoust. Soc. Am. **34**, 1592 (1962).

PROPERTIES OF BUBBLE DISTRIBUTIONS RELEVANT TO SURFACE REVERBERATION

David M. Farmer
Institute of Ocean Sciences
Sidney, B.C., Canada

ABSTRACT Recent interest in near-surface bubble distributions stems both from an environmental interest in the role bubbles play in the upper ocean layer and from the recognition that they may contribute significantly to surface reverberation. Bubble clouds are injected by breaking waves, but their subsequent organization depends on the bubble sizes, subsurface flow field, and other factors. Recent work carried out in the ONR Special Research Project has addressed several aspects of bubble distribution and behaviour. A brief summary of recent results is presented, identifying some of the factors that are relevant to understanding bubble distributions and their potential significance to surface scattering.

1. INTRODUCTION

The speed of sound in water is sensitive to the inclusion of a small fraction of air (Wood, 1941; Clay & Medwin, 1977). For acoustic frequencies well below the resonance of microbubbles, the dominant effect arises from an increase in the bulk compressibility of water. Thus, clouds of microbubbles result in patches of water with anomalously low sound-speed, which will influence the scattering of low frequency sound from the sea surface. Attempts to explain low frequency reverberation levels, in particular the observations of Chapman and Harris (1962) and Ogden and Erskine (1992), have recently stimulated efforts to combine the measurement of bubble distributions with low frequency scattering studies. A joint US-Canadian experiment (CST-7) was conducted in the Gulf of Alaska in February-March of 1992. Further experiments are planned. Here we identify some of the features of ocean surface bubble fields that appear to be relevant to surface scattering and comment on some of the work that needs to be done and the experiments that are planned.

It is apparent that an understanding of the bubble fields near the ocean surface cannot be achieved without a broader understanding of their origin, subsurface behaviour and ultimate fate. A comprehensive study of bubble fields must therefore include the study of other aspects of air-sea interaction and of circulation in the upper ocean boundary layer. Although bubbles may originate from biological processes, the dominant source at higher wind speeds is certainly breaking waves. (From the point of view of surface reverberation studies, anomalously high backscatter only occurs at higher wind speed.) Once formed, bubbles respond to buoyancy, subsurface currents, differences in gas partial pressures, the influence of surface films and other factors. Thus a predictive capability ultimately depends on our understanding of several aspects of air-sea interaction and near-surface dynamics. Studies of bubbles relevant to the surface reverberation problem can therefore be expected both to benefit from and contribute to ongoing research into the upper ocean boundary layer.

2. WAVE-BREAKING

When a surface wave breaks, much of the air that is incorporated rises rapidly to the surface. The air fraction (or void fraction) in these events is of order 0.1 to 0.6 and is to be distinguished from the longer-lived clouds of microbubbles that can last minutes, but which are orders of magnitude less dense. Figure 1a shows a fairly representative measurement of void fraction at different depths that we obtained during the CST-7 cruise. The instrument used a set of four conductivity sensors at depths of 13, 30, 59 and 98cm, suspended in such a way as to remain at a fixed depth beneath the local water surface. As is well known (cf. Lamarre & Melville, 1992), the conductivity depends on the relative proportion of air in the neighbourhood of the sensor and can thus be used to calculate void fraction for known salinity and temperature. The sensors used in the CST-7 experiment have been adjusted to be sensitive to low levels of void fraction and saturate at 0.4. The wind speed during this observation was 14ms^{-1}, the significant wave height 3.5m and wave period 8s. Most of the air injections are relatively short. of order 1s or less, but occasionally (i.e., at 32s) there is a longer-lived event. Most breaking events involve high void fraction penetration to less than 1m. However, smaller fractions of air, of order 0.004, penetrate to the deepest sensor in nearly every case; this is seen in Figure 1a where the scale is greatly expanded for the two deeper sensors. Although these void fractions are small relative to those detected nearer the surface, they are still acoustically dense. These measurements emphasize the need for void fraction measurements with high sensitivity at greater depth.

A statistical summary of air penetration over a 3h period is shown in Figure 1b. This shows that most of the events lasted 0.6s or less. These observations should be contrasted with the duration of the wave-breaking event, which can last for two or more seconds. This suggests a description of the breaking event as a rapidly moving curtain of air that continues travelling forward for up to two or more seconds, but lasts only for some fraction of a second at any one location.

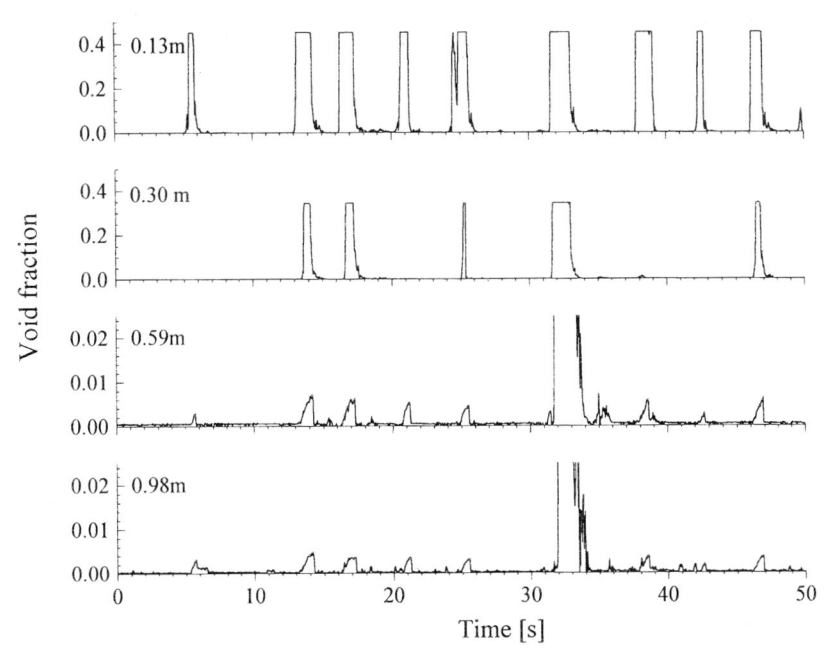

Figure 1a. Time series of void fraction (i.e., fraction of air in water) measured with conductivity sensors in the surface 1m during the CST-7 cruise, February 1992. Wind speed was 14ms^{-1}. Note expanded scale for two deepest sensors.

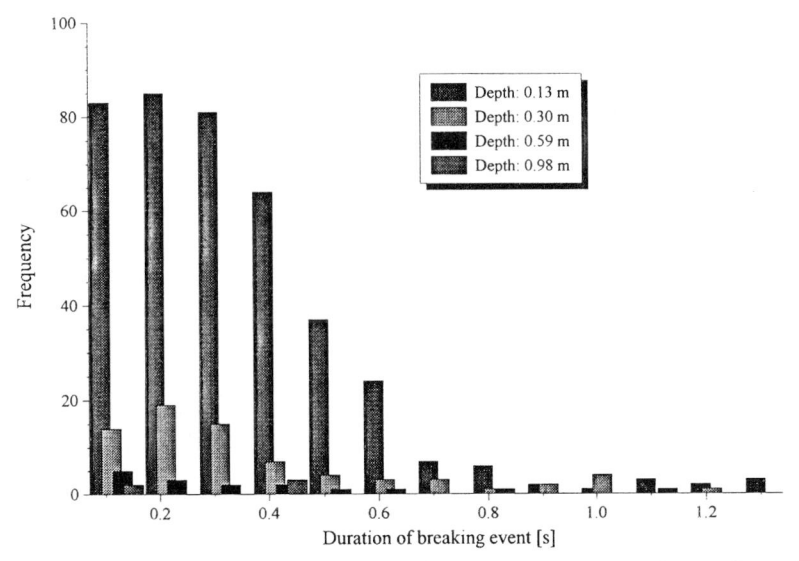

Figure 1b. Histogram showing distribution of event duration for a 3h period with wind speeds of 12-14ms^{-1}. Most events are less than 0.6s.

If these high void fraction injections play a role in surface reverberation, it is desirable to know more about their distribution and properties. (This is true whether they impact the scattering process directly, as discussed here, or by virtue of being a source of clouds of microbubbles.) We have recently developed passive acoustic methods involving the use of a small hydrophone array for tracking wave-breaking events in space and time as they move across the ocean surface (Farmer & Ding, 1992; Ding and Farmer, 1992). This allows the acquisition of statistics of wave-breaking properties. Figure 2a shows observed relationships between the speed of breaking events and the wind speed; Figure 2b shows the same dependence for the duration of breaking events. In general, both speed and duration increase with wind speed. These results are scattered, but there is no *a priori* reason to expect a simple relationship. In fact, wave-breaking is dependent on the wave-field, which is not a simple function of wind speed. Present work is oriented towards a more direct analysis of this problem by combining simultaneous observations of the directional wave spectrum and the wave-breaking field, and using simulation models of the sea surface to reconcile the observations with wave-breaking criteria.

The spectrum and coherence of sound radiated by a breaking wave can also provide some clues as to its properties. For example, Figure 3 shows the temporal evolution of the ambient sound spectrum of an event in which the dominant frequency rises during its passage. This would be consistent with a diminishing volume of entrained air due to outgassing, resulting in a progressively decreasing resonant frequency of the air-water mixture. Farmer & Ding (1992) show how strong anisotropies in the coherence of radiated sound can be related to the shape of the sound source.

The relevance of these results to surface reverberation is that they imply constraints on models of the Doppler properties of scattering from these dense but brief injections of air. The short, instantaneous life of the air injection implies a corresponding spectral spread; the propagation of the whitecap similarly implies a spreading of the Doppler spectrum (and an anisotropy in Doppler shift for directional reverberation measurements). Measurements of the void fraction (Figure 1) also emphasize the strong vertical gradients which need to be considered in resonant oscillation calculations. Results from the CST-7 scattering study described elsewhere in this volume should provide an opportunity for directly testing the hypothesis that high scattering levels result from high void fraction injections.

3. BUBBLE DISTRIBUTIONS

Although the high void fraction component of a breaking event may last only for some fraction of a second, clouds of microbubbles can persist much longer and penetrate much deeper. From the standpoint of reverberation analysis the greater penetration depth is important because of reduced interference with the surface reflection.

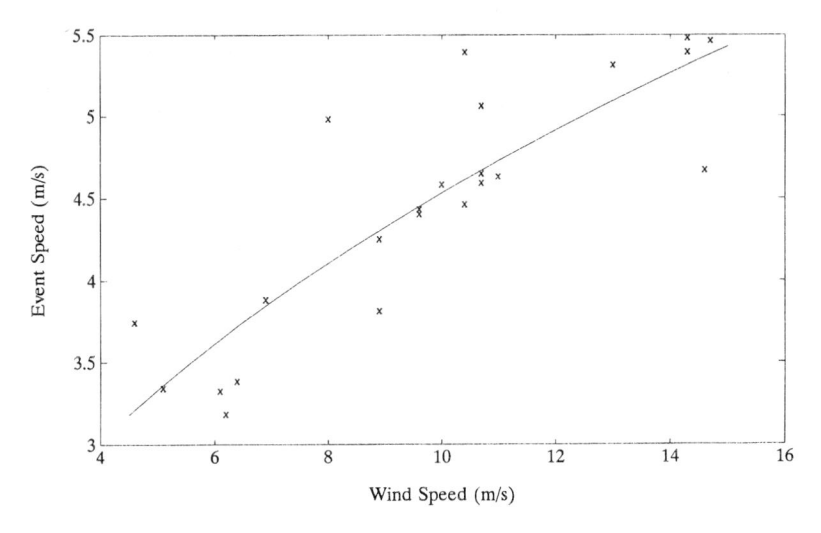

Figure 2a. Measured breaking event speed as a function of wind speed measured during the Surface Wave Process Program, using passive sensing with a hydrophone array.

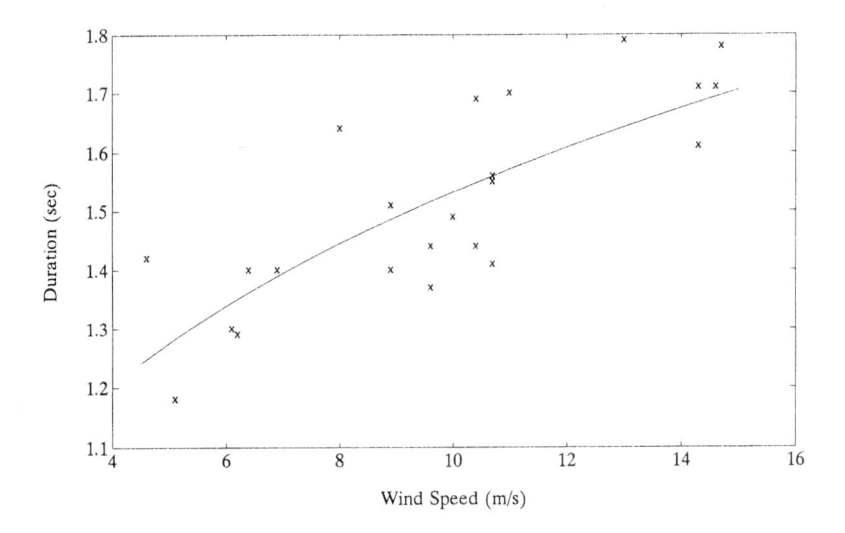

Figure 2b. Measured event duration as a function of wind speed corresponding to

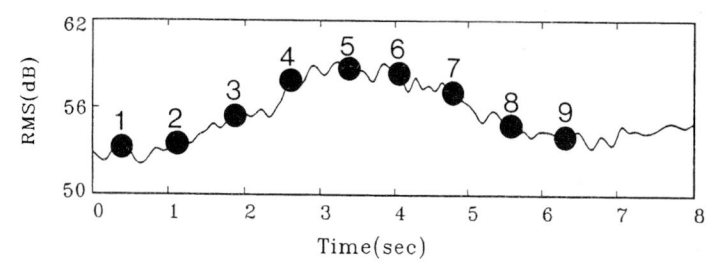

Figure 3a. Time series of rms sound pressure level measured by a hydrophone 28m beneath a breaking wave. Filled circles (●) correspond to sound spectra in Figure 3b.

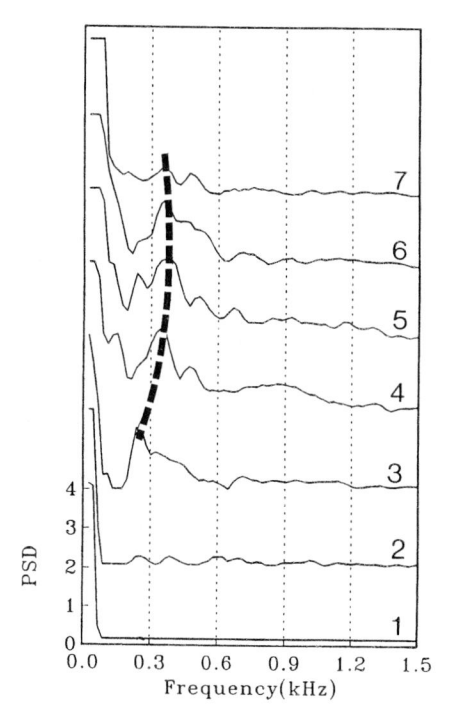

Figure 3b. Sequence of sound spectra corresponding to successive portions of the time series in Figure 3a. As the breaking event evolves, the peak frequency of the radiated sound increases from 245 to 375Hz, possibly due to degassing and consequent decrease in resonant frequency of the air-water mixture.

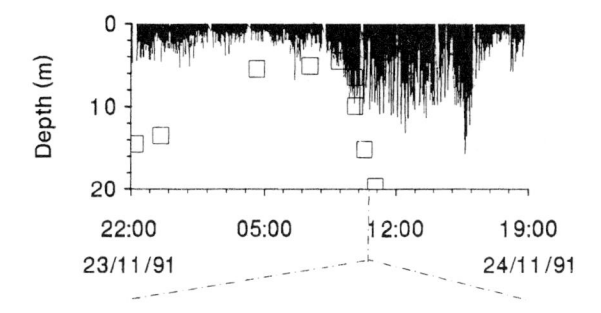

Figure 4a. Measurements of bubble penetration depth as successive bubble clouds pass over a drifting sensor during a storm (wind speed 13-14ms^{-1}. The concentration as a function of depth can be derived from target strength but is not shown in this highly compressed record. Square symbols (□) indicate the depth of the pycnocline, which deepens under the influence of wind mixing.

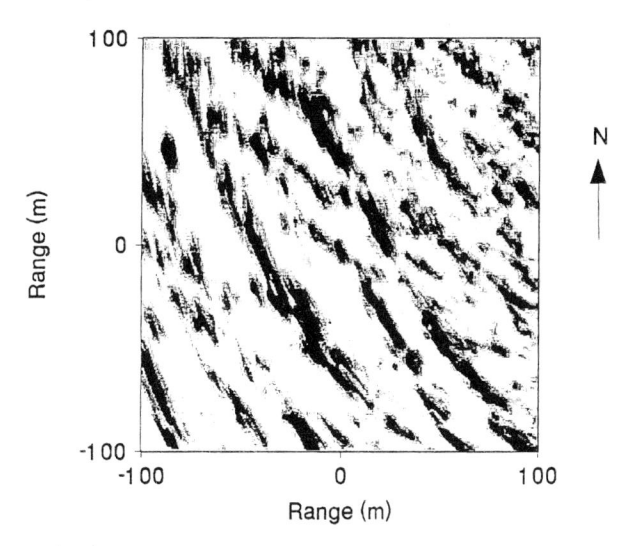

Figure 4b. Two-dimensional image obtained with a mechanically scanned sonar, during the same period represented in Figure 4a, showing horizontal organization of bubble clouds by Langmuir circulation.

Bubbles are organized by subsurface circulation and are subject to compression and dissolution. A bubble is acted upon by buoyancy and thus has a rise speed dependent on its volume. In the ocean bubbles can be expected to acquire a surface coating which has the effect of stiffening the bubble skin and causing it to behave like a rigid sphere (Levich, 1962). If the effects of turbulent diffusion and vertical advection can overcome the natural tendency to rise, the bubble descends and is compressed. Surface tension adds to the internal pressure. Gases will diffuse across the bubble surface at a rate dependent on the difference in partial pressures. In supersaturated water near the surface, bubbles will grow, but deeper bubbles will tend to dissolve. Models have been developed by Thorpe (1982) and Woolf & Thorpe (1991) to describe this process.

The vertical distribution of bubbles is most effectively measured with echo-sounders, and numerous (single-frequency) measurements have been described (cf. Thorpe, 1986). Time series measurements show the passage of successive clouds over our freely drifting instrument (Figure 4a). During the CST-7 cruise bubble clouds penetrated to the depth of the instrument (24m) for extended periods.

These vertical measurements can most readily be put in perspective with the help of observations of the horizontal distribution. Imaging sonar measurements made with a mechanically scanning 100kHz acoustic beam reveal a highly organized pattern of elongated structures aligned approximately with the wind direction (Figure 4b). These are two-dimensional images of structures previously observed by Thorpe and Hall (1983) and others with fixed orientation sonars, and attributed to advection and downwelling of bubbles in Langmuir convergence zones. The two-dimensional imaging sonar we have developed provides unambiguous information about their two-dimensional properties including such behaviour as the merging of adjacent convergence zones. Langmuir circulation is at present the subject of active theoretical and observational study.

Since the rate at which gas goes into solution depends upon the ambient pressure of dissolved gas in the water, this factor can also be expected to be relevant to the resulting bubble distribution. A number of processes are involved and differences in diffusivities and solubilities of the two primary species (N_2, O_2) need to be considered (cf. Woolf & Thorpe, 1991). Figure 5 shows a time series of wind speed, surface layer temperature and gas tension, that is, the sum of partial pressures including water vapour, measured in a storm during the CST-7 cruise. Before the wind speed picks up, the gas tension decreases while the temperature remains constant. This is consistent with the effect of biological respiration depleting the dissolved oxygen. During the period of increasing wind speed, gas tension initially continues to decrease, but later rapidly increases, while the temperature drops by 0.1°C. Evidently neither the drop in temperature, which contributes to an increase in the solubility and thus a decrease in gas tension, nor the biological respiration is sufficient to offset the effect of the injection of air by breaking waves and bubble

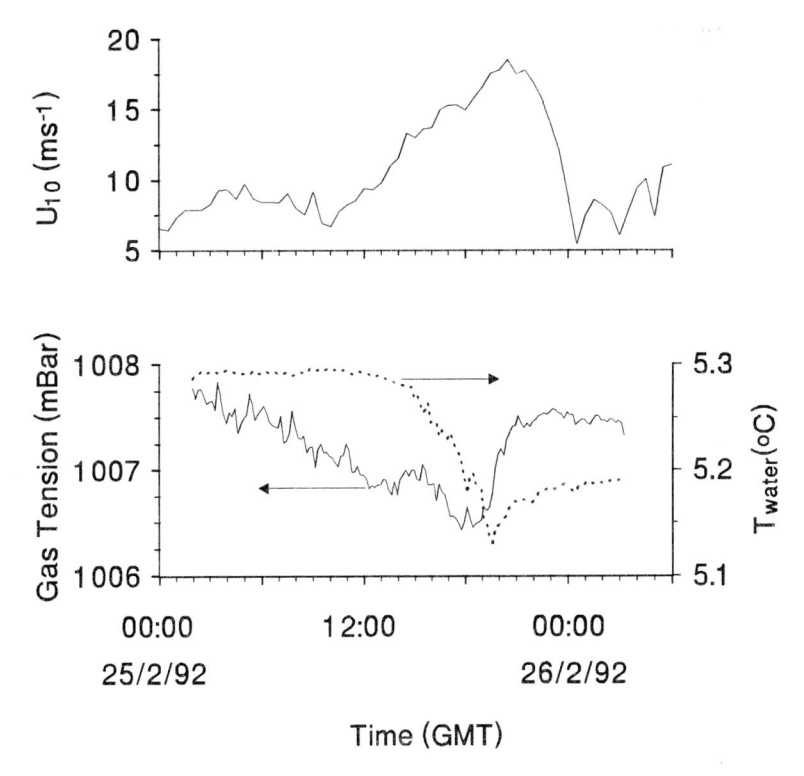

Figure 5. *Above:* Wind speed during the CST-7 Gulf of Alaska cruise.
Below: Mixed layer temperature (dotted curve) and gas tension (solid curve), obtained with a sensor provided by Dr. B. Johnson.

dissolution once the wind speed is high enough. Such observations give some indication of the complexity of processes affecting bubble distributions.

4. BUBBLE SIZE DISTRIBUTIONS

Bubble size distributions in the ocean have been measured both with *in situ* devices and with a remote technique. *In situ* measurements of bubble size distributions include the bubble trap (Kolovayev, 1976; Blanchard & Woodcock, 1957), optical sensing (Johnson & Cooke, 1979; Ling & Pao, 1988; Su, Ling & Cartmill, 1988; Walsh & Mulhearn, 1987), and acoustic resonator (Medwin & Breitz, 1989; Breitz & Medwin, 1989). Akulichev and Bulanov (1987) among others have used non-linear acoustic techniques. Remote sensing of ocean bubble distributions has been carried out by multiple frequency acoustic backscatter (Løvik, 1980; Szczucka, 1989; Farmer & Vagle, 1989; Vagle & Farmer, 1992). Wu (1981, 1988) has reviewed many of these measurements and given empirical formulae for bubble concentration as a function of wind speed and depth.

There is an important qualitative distinction between the *in situ* and remote measurements. *In situ* observations provide the opportunity for size distributions to be resolved with fine detail; remote measurements have thus far lacked this resolution, but instead give a much more comprehensive picture of spatial variability. Significant discrepancies continue to exist in the measured distributions; Akulichev and Bulanov (1987) attribute these differences to meteorological and oceanographic forcing, and it is certain that many of the reported differences are due to environmental conditions. However, systematic discrepancies remain to be explained.

Uncertainties exist with respect to the distribution of the smallest bubbles, which are hard to measure optically. For the larger bubbles, discrepancies show up especially in the slope of the size distribution.

Author	Method	Size Range (μm)	Slope	Comment
Akulichev & Bulanov, 1987	Parametric sonar	90 - 700	-3.7	2.5m depth, "slight sea"
Blanchard & Woodcock, 1957	Bubble trap	>38	-4.7	Wind 8-10ms^{-1}
Farmer & Lemon, 1984	Ambient sound inversion	20 - 400	-4.5 to -3.8	Slope decreases with increasing wind speed
Farmer & Vagle, 1989	4-freq. inverted echo-sounder	40 - 120	-4 to -6	10-12ms^{-1} winds
Johnson & Cooke, 1979	Photographic	51 - 316	-4.5	11-13ms^{-1} winds, slope increases with depth
Kolovayev, 1976	Bubble trap	100 - 400	-4.0	11-13ms^{-1} winds
Lovik, 1980	3-freq. echo-sounder	49 - 380	-3.8 avg.	Slope is -2.6 betw.120 and 380, wind 35-40ms^{-1}
Medwin & Breitz, 1989	Acoustic resonator	60 - 120	-2.6	25cm beneath breaking wave
Su et al., 1988	Optical counter	150 - 400	-5 to -6	9.5-13ms^{-1} winds
Szczucka, 1989	4-freq. echo-sounder	40 - 100	-3.64	Shallow water, wind speed not given
Walsh & Mulhearn, 1987	Photographic	102 - 300	-4.0	8-14ms^{-1} wind speed

Table 1. Bubble size spectrum slope for larger bubbles (>50 μm) measured at sea.

Most of the published results indicate a relatively uniform slope in size spectra for bubble radii greater than about 50μm, but with some variation in reported values; a uniform slope of -4 is included in Wu's (1988) functional approximation based on results in the literature. Some of these size spectrum slopes are listed in Table 1. Most of the measurements indicate a uniform slope of -3.6 or greater with many of the measurements, including the most widely quoted values of Johnson and Cooke (1979), in the range -4 to -6. Variability is certainly to be expected, depending on the balance between buoyancy and turbulent diffusion. An important exception to the range of most of these distributions is the *in situ* acoustical measurement of Medwin and Breitz (1989), in which a change is seen in the size slope of the bubble size spectrum from -4 at radii less than 50μm to -2.6 for bubble radii greater than about 60μm. Medwin and Breitz obtained their results close (25cm) to the sea surface where a higher population of larger bubbles is to be expected, and they note the similarity with Baldy's (1988) laboratory results. In fact, there are some reasons to expect a -3 slope close to the surface, based on dimensional analysis (Crowther, 1980, 1988). However, the slope should steepen, due to buoyancy effects, at greater depths. A steeper slope at greater depths is in fact clearly apparent in the Johnson and Cooke (1979) data. Increased turbulence at higher wind speed can be expected to reduce the slope of the distribution at a given depth, and this effect was observed by Farmer and Lemon (1984).

On the other hand, Su (Air-Sea Acoustics Workshop, 19-21, May, personal communication) reports preliminary analysis of observations obtained during the CST-7 cruise with an acoustic resonator similar to that described by Medwin and Breitz (1989), in which the slope is approximately -3 out to the maximum detectable radius of 1200μm, even at a depth of 5.75m. This is a surprising result, the more so because it extends to quite large bubbles with high rise velocities (~20cms^{-1} for 1200μm) and significant depths. A slope of -3 implies a void fraction that increases linearly with respect to the upper bound of integration. The implication of these results would be that the void fraction cannot be determined with confidence, even by integration out to 1200μm, and that still larger bubbles must be measured. Clearly, there is an urgent need to reconcile these results with the much steeper size spectrum slopes found by other means. The difference between the sound speed anomaly inferred by the resonator and that found by optical and other measurement techniques is so great that it should be possible to resolve the issue unambiguously using an independent measure of low frequency sound speed. Such a test is now planned.

5. SUMMARY

Several aspects of air-sea interaction are relevant to the scattering of sound by bubbles. Recent observations have been described, including several results from the Surface Reverberation Program and Critical Sea Test. Wave-breaking is both the source of large but short-lived injections of air and also the source of more persistent but tenuous microbubble clouds. The dense mass of bubbles that constitutes a whitecap lasts only some fraction of a second at any single location, although the whitecap itself may last significantly longer (two or more seconds) as it moves forward. Whitecaps travel in the wind direction at speeds of order 5ms^{-1} in a 12ms^{-1} wind.

Clouds of microbubbles are much longer-lived and can penetrate to depths of over 20m. Langmuir circulation organizes the bubbles into long rows aligned with the wind. Langmuir cell convergence draws the bubbles downwards. Gas diffusion across the bubble surface affects bubble size and composition, and can cause the deeper bubbles to go into solution completely. Uncertainties remain with respect to the size distribution; more recent acoustic observations suggest a peak in the spectrum at a radius of 10-20μm, although there is a need to verify this result with additional measurements using different techniques. Discrepancies exist also for the spectral slope of the larger bubbles; the recent acoustic resonator results imply much smaller slopes (-3) for the larger bubbles than generally determined using other techniques. Development of scattering models requires as input the sound speed anomaly associated with bubble clouds. While this can in principle be achieved by integration over observed bubble size distributions, uncertainties about spectral slope and the outer bounds of the integration lead to uncertainties about the magnitude of the anomaly. Thus an independent check on the accuracy of the estimate based on bubble measurements is especially desirable and is presently being planned for future experiments.

New measurement techniques are contributing to our understanding of these phenomena. Hydrophone arrays have been used to track the distribution and properties of breaking waves. Imaging sonars can reveal the horizontal distribution of bubbles and vertical sonars probe their vertical structure. Theoretical models are now being developed to account for the two- and three-dimensional distribution of bubbles as they move under the influence of subsurface circulation and are subject to the effects of gas dissolution. From the standpoint of surface reverberation there is a pressing need for deterministic measurements of the bubble field, together with related environmental conditions, coincident with measurements of the low frequency acoustic backscatter using a narrow beam sonar. The ONR-funded Surface Reverberation Project is directed towards such experiments, the first of which will be carried out in 1993.

Acknowledgement: The observations presented here were obtained with support from the US Office of Naval Research, under the direction of Drs M. Orr and M. Briscoe, and the Institute of Ocean Sciences. The author is indebted to his students C. McNeil, L. Ding and J. Gemmrich for providing the figures. The gas tension measurements (Figure 5) were obtained with a sensor provided by Dr. B. Johnson.

REFERENCES

1. Akulichev, V.A., and Bulanov, V.A. (1987) 'The study of sound backscattering from micro-inhomogeneities in sea water', in Progress in Underwater Acoustics, ed. H.M. Merklinger, New York, pp. 86-92.
2. Baldy, S. (1988) 'Bubbles in the close vicinity of breaking waves: statistical characteristics of the generation and dispersion mechanism', Journal of Geophysical Research 93, 8239-8248.
3. Blanchard, D.C., and Woodcock, A.H. (1957) 'Bubble formation and modification in the sea and its meteorological significance', Tellus 9, 145-158.
4. Breitz N. and Medwin, H. (1989) 'Instrumentation for in situ acoustical measurements of bubble spectra under breaking waves', Journal of the Acoustical Society of America 86, 739-743.
5. Chapman, R., and Harris, J. (1962) 'Surface backscattering strength measured with explosive sound sources', Journal of the Acoustical Society of America 34, 1592-1597.
6. Clay, C.S., and Medwin, H (1977) Acoustic Oceanography, John Wiley and Sons, New York.

7. Crowther, P.A. (1980) 'Acoustical scattering from near-surface bubble layers', in Cavitation and Inhomogeneities in Underwater Acoustics, ed. W. Lauterborn, Springer Verlag, New York, pp. 194-204.
8. Crowther, P.A. (1988) 'Bubble noise creation mechanisms', in Sea Surface Sound: Natural Mechanisms of Surface Generated Noise in the Ocean, ed. B.R. Kerman, Kluwer Academic Publishers, Dordrecht, pp. 131-150.
9. Ding, L., and Farmer, D.M. (1992) 'A signal-processing scheme for passive acoustical mapping of breaking surface waves', Journal of Atmospheric and Oceanic Technology 9, 484-494.
10. Farmer, D.M., and Ding, L. (1992) 'Coherent acoustical radiation from breaking waves', Journal of the Acoustical Society of America 92, 397-402.
11. Farmer, D.M., and Lemon, D.D. (1984) 'The influence of bubbles on ambient noise in the ocean at high wind speeds', Journal of Physical Oceanography 14, 1762-1778.
12. Farmer, D.M., and Vagle, S. (1989) 'Waveguide propagation of ambient sound in the ocean surface bubble layer, Journal of the Acoustical Society of America 86, 1897-1908.
13. Johnson, B.D., and Cooke, R.C. (1979) 'Bubble populations and spectra in coastal water: a photographic approach', Journal of Geophysical Research 84, 3761-3766.
14. Kolovayev, P.A. (1976) 'Investigation of the concentration and size distribution of wind-produced bubbles in the near-surface ocean layer', Oceanology (English version) 15, 659-661.
15. Lamarre, E., and Melville, W.K. (1992) 'Instrumentation for the measurement of void-fraction in breaking waves: laboratory and field results', IEEE Journal of Oceanic Engineering 17, 204-215.
16. Levich, V.G. (1962) Physicochemical Hydrodynamics, Prentice Hall, Inc., Englewood Cliffs, NJ.
17. Ling, S.C. and Pao, H.P. (1988) 'Study of micro-bubbles in the North Sea', in Sea Surface Sound: Natural Mechanisms of Surface Generated Noise in the Ocean, ed. B.R. Kerman, Kluwer Academic Publishers, Dordrecht, pp. 197-210.
18. Løvik, A. (1980) 'Acoustic measurements of the gas bubble spectrum in water', in Cavitation and Inhomogeneities in Underwater Acoustics, ed. W. Lauterborn, Springer Verlag, New York, pp. 211-218.
19. Medwin, H., and Breitz, N.D. (1989) 'Ambient and transient bubble spectral densities in quiescent seas and under spilling breakers', Journal of Geophysical Research 94C, 12,751-12,759.
20. Ogden, P.M., and Erskine, F.T. (1992) An empirical prediction algorithm for low-frequency acoustic surface scattering strengths, April 28, 1992, NRL Report No. NRL/FR/5160-92-9377.
21. Su, M.-Y., Ling, S.C., and Cartmill, J., (1988) 'Optical microbubble measurements in the North Sea', in Sea Surface Sound: Natural Mechanisms of Surface Generated Noise in the Ocean, ed. B.R. Kerman, Kluwer

Academic Publishers, Dordrecht, pp. 211-224.
22. Szczucka, J. (1989) 'Acoustic detection of gas bubbles in the sea', Oceanologia 28, 103-113.
23. Thorpe, S.A. (1982) 'On the clouds of bubbles formed by breaking waves in deep water and their role in air-sea gas transfer', Philosophical Transactions of the Royal Society of London A304, 155-210.
24. Thorpe, S.A. (1986) 'Measurements with an automatically recording inverted echo sounder: ARIES and the bubble clouds', Journal of Physical Oceanography 16, 1462-1478.
25. Thorpe, S.A., and Hall, A. (1983) The characteristics of breaking waves, bubble clouds, and near-surface currents observed using side-scan sonar', Continental Shelf Research 1, 353-384.
26. Vagle S., and Farmer, D.M. (1992) 'The measurement of bubble size distributions by acoustical backscatter', Journal of Atmospheric and Oceanic Technology 49, 630-644.
27. Walsh, A.L. and Mulhearn, P.J. (1987) 'Photographic measurements of bubble populations from breaking wind waves at sea', Journal of Geophysical Research 92C, 14,553-14,565.
28. Wood, A.B. (1941) A Textbook of Sound, G. Bell and Sons, Ltd., London.
29. Woolf, D.K. and Thorpe, S.A. (1991) 'Bubbles and the air-sea exchange of gases in near-saturation conditions', Journal of Marine Research 49, 435-466.
30. Wu, J. (1981) 'Bubble population and spectra in near-surface ocean: summary and review of field measurements', Journal of Geophysical Research 86, 457-463.
31. Wu, J. (1988) 'Bubbles in the near-surface ocean', Journal of Geophysical Research 93C, 587-590.

SOUND SCATTERING FROM MICROBUBBLE DISTRIBUTIONS NEAR THE SEA SURFACE

William M. Carey
Defense Advanced Research Projects Agency
Arlington, VA 22203-1714

Ronald A. Roy
Applied Physics Laboratory
University of Washington
Seattle, WA 98105

ABSTRACT. Backscatter from the sea surface is thought to be governed by the roughness of the surface and subsurface bubble distributions. At low frequencies, due to the paucity of large bubbles, scattering results primarily from coherent and/or collective scatter from bubbles entrained by the subsurface vorticity or carried to depth by the Langmuir circulation and thermal convection. It is shown that scattering from compact regions is a function of the volume fraction of air and to first order can be described by a Minnaert formula modified with the volume fraction. Measurement of sound scattering from a submerged cloud of bubbles produces low-frequency peaks with large low-frequency target strength consistent with this theory.

1. INTRODUCTION

Recent experimental evidence shows that when waves break, microbubble layers and plumes are produced, and these bubbly assemblages are convected to depth by the Langmuir circulation as well as the thermal convection found in the oceanic mixed layer [1,2,3,4]. The fundamental question to be answered is what role, if any, do microbubble layers and plumes play in the near-surface radiation and scattering of sound in the low- to mid-frequency range (20 Hz to 2 kHz). There exists ample evidence that sound scattering at these frequencies has a different characteristic than expected from Bragg scattering from gravity waves [5,6]. It is our hypothesis that if these microbubble clouds and plumes have structures with volume fractions exceeding 10^{-4}, then these structures can radiate as well as scatter sound through a dampened resonant oscillation [6,9,10,11,

12,13]. Recent analytical studies of low-frequency sound scattering from geometrically shaped bubble distribution beneath the sea surface [7,8] indicate that these types of bubble distributions may also coherently scatter sound.

This paper discusses these issues in terms of the classical theories of sound scattering from complex but compact distributions of bubbles. In particular, our intention is to show, with the Morse and Ingard formulation [19], the relationship between the Born approximation [19], continuum theory, and single-bubble incoherent scattering. We show that when the bubbles are treated as scatters in the Lighthill method that one proceeds from single-bubble scattering, to coherent scattering, and finally to the scattering from the distribution as if it were a continuum. In fact, the continuum theory is shown to provide a bound on the region of validity of the Born approximation. The problem is also scaled by the volume fraction and its effect on the liquid and gas mixture compressibility. Finally, we present the results of a free bubble cloud scattering experiment conducted at Lake Seneca. These measurements clearly illustrate the complex nature of the scattering cross section of a compact bubble cloud.

2. A REVIEW OF SCATTERING OF SOUND FROM COMPACT BUBBLE DISTRIBUTIONS

The classical theory of sound radiation and scattering from compact regions containing bubbly liquids is a mature subject and several texts [14-21] contain very useful overviews. In recent years, the scattering of sound from bubble distributions near the sea surface has received renewed interest[6]. This is in part due to the work of Thorpe [1] and his observation of subsurface bubble plumes and clouds. McDaniel [5] has shown that high frequency (>1 kHz) sea surface scattering can be explained using bubble distributions near the sea surface coupled with composite rough surface scattering. However, the bubble size distributions [4,5] measured at sea are too small to explain scattering at frequencies less than 1 kHz using a single bubble scattering formalism. The purpose here is to discuss this low-frequency bubble scattering from bubble plumes and clouds in light of classical scattering theories as well as to place current analytical work [7,8,9,10,11,12,13] in perspective. Consequently, we closely follow and use material found in these texts, most notably the Morse and Ingard text [19].

Fig. 1 shows the geometry for a compact bubble cloud. The primary focus is on distributions of salt water bubbles with radii less than 1 mm, resonant frequencies greater than 3 kHz, and excitation frequencies less than 1 kHz. Since we are treating a compact source, the sea surface will be incorporated using the method of images.

The Fourier transform of the wave equation is

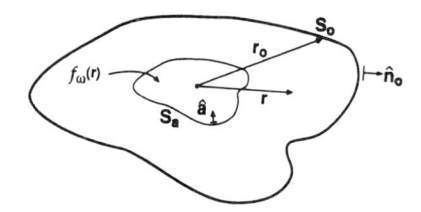

Figure 1. The geometry of the general scattering problem

$$\nabla^2 P_\omega(r) + k^2 P_\omega(r) = -F_\omega(r), \tag{1}$$

where $F_\omega(r)$ represents a general source distribution. Since the problem is a farfield radiation one and since our source region is compact, the more general retardation notation has not been used. The Green's equation is

$$\nabla^2 g_\omega(r|r_o) + k^2 g_\omega(r|r_o) = -\delta(\vec{r}-\vec{r}_o). \tag{2}$$

where $g_\omega(r|r_o) = \exp(i\vec{k} \cdot (\vec{r}-\vec{r}_o))/4\pi|\vec{r}-\vec{r}_o|$ and will be used in this form throughout this paper. The standard procedure is to multiply 1) by g_ω, 2) by P_ω, subtract, use reciprocity to interchange r and r_o, and then use the divergence theorem to yield:

$$P_\omega(r) = \iiint g_\omega(\vec{r}|\vec{r}_o) F_\omega(r_o) dV_o + \iint \left[g_\omega \frac{\partial P_\omega}{\partial n_o} - P_\omega \frac{\partial g_\omega}{\partial n_o} \right] ds_o. \tag{3}$$

This equation describes both the radiation and scattering of sound from a region V_o. The surface integral in the radiation problem goes to zero as r_o increases due to imposition of the radiation condition. This integral in the case of the scattering problem becomes the incident wave coming in from infinity. Thus both the radiation and scattering of sound require the evaluation of the volume integral, the difference being the source of excitation. In this paper we will simply use a plane wave, consequently

$$P_\omega(r) = P_{i\omega}(r) + \iiint g_\omega(\vec{r}|\vec{r}_o) P_{i\omega}(r_o) f_\omega(r_o) dV_o$$

$$F_\omega(r) = P_{i\omega}(r_o) f_\omega(r_o), \quad P_{i\omega}(r) = P_o e^{i\vec{k}_o \cdot \vec{r}}. \tag{4}$$

The various approaches to scattering can be considered as consisting of the method by which $f_\omega(r_o)$ is determined coupled with the means by which the volume integral can be solved. Traditional treatments employed at higher frequencies consider bubble densities, the number of bubbles per wavelength volume, are small and multiple scattering is therefore not a problem. However, individual bubble resonance is important, thus:

$$f_\omega(r_o) = f_n(\omega_o, \theta, r_n) \delta(\vec{r}_o - \vec{r}_n), \tag{5}$$

where $f_n(\omega_o, \theta, r_n)$ represents the response function of the nth bubble. Substitution of this expression for f_ω, and after some algebraic manipulation, we obtain for a uniform distribution of N bubbles the following expression for the scattered intensity

$$I_s(r) = \frac{|P_{s\omega}|^2}{2\rho c} = \frac{P_o^2}{2\rho c r^2}|f(\omega_o,\theta)|^2\left[N+\sum_{m\neq n}\exp\left(i\vec{\mu}\cdot\vec{R}_{mn}\right)\right], \tag{6}$$

where $\vec{\mu} = \vec{k}_o - \vec{k}$, $|\mu| = 2k\sin(\theta/2)$. When the summation within the brackets is zero, the scattered intensity is seen to be proportional to the number of bubbles N; i.e., the incoherent sum of scattering from individual bubbles. For the case of an acoustically compact spherical volume of radius a, the bracketed terms can be shown to be [18]

$$I_s(r) \rightarrow N^2\left[1 - \frac{2}{5}\frac{(N-1)}{N}k^2a^2\sin^2(\theta/2)\cdots\right]. \tag{7}$$

This is the coherent scattering case where the scattered intensity is directional and proportional to N^2. The distinguishing factor between these two expressions is the number of bubbles within V_o or the volume fraction.

The incoherent, proportional to N, scattering particularly applies to non-compact distributions of bubbles which are sparse. This is the case treated by McDaniel [5] and has been shown to describe several sea surface scattering measurements at frequencies >1 kHz and at low grazing angles.

The work of Foldy [14,16] was based on the above case of $f_\omega(r)$, except it included the effects of scattering from neighboring bubbles (here it is assumed that they are monodispersed in radius with resonance frequency ω_o):

$$f_{\omega_n}(r_o) = P_{\omega n}(r_n)f_n(\omega_o,\theta_n,r_n)\delta(\vec{r}_o - \vec{r}_n) \text{ and} \tag{8}$$

$$P_\omega(r) = P_{i\omega}(r) + \sum_{n=1}^{N}\frac{e^{ikr_n}}{r_n}P_{\omega n}(r_n)f_n. \tag{9}$$

The problem when the interaction between bubbles must be considered is the choice of $P_{\omega n}(r_n)$. Foldy [16] recognized that this quantity depends on the incident field and the sum of the scattered field from all the bubbles and could be written as

$$P_{\omega n} = P_{i\omega}(r_n) + \sum_{m\neq n}^{N}\frac{e^{ikr_{mn}}}{r_{mn}}f_m P_{\omega m}(r_n) = P_\omega(r_n) - \frac{f_n}{r_n}P_{\omega n}(r_n)e^{ikr_n}. \tag{10}$$

This result when substituted in equation 9 yields

$$P_\omega(r) \approx P_{i\omega}(r) + \sum_{n=1}^{N}P_\omega(r_n)f_n\frac{e^{ikr_n}}{r_n}, \tag{11}$$

$$\xrightarrow{N\rightarrow\infty} P_{i\omega}(r) + \int P_\omega(r_n)f_n\frac{e^{ikr_n}}{r_n}dr_n.$$

Foldy then recognized that the configurational average of equation (9) is equivalent to the solution of

$$\nabla^2 \langle P_\omega(r) \rangle + k_s^2 \langle P_\omega(r) \rangle = 0, \quad k_s^2 = k^2 + 4\pi \langle f_n \rangle, \tag{12}$$

which is a wave equation with a different wave number or index of refraction. This index of refraction is determined by the continuum properties of the bubbly volume [11,22,23].

The concept that sound scattering from a cloud of air bubbles in water is described as a progression from incoherent, coherent to continuum scattering was fully developed in references [14-21] prior to the mid-nineteen sixties. Morse and Ingard [19] have provided an extension of these results based on the Lighthill method.

The use of Lighthill's approach requires the determination of $f_\omega(r_o)$ from the two-component conservation equations. Retaining only second order terms and neglecting the Lighthill stress tensor yields

$$f_\omega(r_o) = k^2 \gamma_k P_\omega(r_o) + \vec{\nabla}_o \cdot (\gamma_\rho \vec{\nabla}_o P_\omega(r_o)), \tag{13}$$

where γ_k and γ_ρ are the compressibility and density contrasts given by

$$\gamma_k = \frac{\kappa_e - \kappa}{\kappa}, \gamma_\rho = \frac{\rho_e - \rho}{\rho} \text{ for } r_o < a \text{ and } \gamma_k = \gamma_\rho = 0 \text{ for } r_o > a. \tag{14}$$

Substitution of this form into equation (4) yields

$$P_\omega(r) = P_{i\omega}(r) + \iiint \left[k^2 \gamma_k P_\omega g_\omega - \vec{\nabla}_o \cdot (\gamma_\rho \vec{\nabla}_o P_\omega) g_\omega \right] dV_o. \tag{15}$$

This equation is extremely interesting, because substitution of the mean compressibility contrast $(\bar{\gamma}_k)$ and mean density contrast $(\bar{\gamma}_\rho)$ results in a wave equation that is similar to Foldy's with an effective wavenumber given by $[1 + \bar{\gamma}_\rho]\nabla^2 P_\omega(r_o) + k^2(1-\bar{\gamma}_k)P_\omega(r_o)$ where $k^2(1 - \bar{\gamma}_k)/(1 + \bar{\gamma}_\rho)$ is the effective wavenumber squared. This is another way of showing that the region may be described by its mixture density and compressibility. This also provides a means by which the scattering regimes may be distinguished. To show this we integrate equation 15 by parts and use our free space Green's function:

$$P_\omega(r) = P_o e^{i\vec{k}\cdot\vec{r}} + \frac{P_o e^{ikr}}{r} \phi(k_s), \text{ where} \tag{16}$$

$$\phi(k_s) = \frac{k^2}{4\pi P_o} \iiint \left[\gamma_k P_\omega - i\gamma_\rho \frac{\vec{k}_s}{k^2} \cdot \nabla P_\omega \right] e^{-i\vec{k}_s \cdot \vec{r}_o} dV_o$$

This quantity is referred to as the angle distribution factor and is recognized as a three-dimensional transform of the compressibility and density contrasts. Since Fourier methods apply, the function $\phi(\vec{k}_s) = F(\phi(\vec{r}_o))$ can also be determined by the convolution of the spatial transforms of P_ω and the γ's. Since the scattered intensity is proportional to $|\phi(k_s)|^2$, the full statistical description of the power spectral density and correlation properties can thus be used. This is an important observation, for the description of non-compact near-surface bubble layers can be treated in this manner. Gilbert [24] has employed these techniques to characterize low-frequency sound scattering from subsurface bubble distributions. The case of a compact distribution is presented here.

The problem is still to specify P_ω within the volume integral [15]. The method of successive approximations has been used to solve this equation. This method approximates the term P_ω by a sum of successively higher order terms $P_\omega(r_o) = P_{i\omega}(r_o) + P_1 + P_2 + \cdots$. When $P_{i\omega}(r_o) > P_1$ throughout the source volume, we simply allow $P_\omega(r_o) = P_{i\omega}(r_o)$. This approximation is referred to as the Born approximation. This approach has been applied to bubble scattering in underwater acoustics for over 30 years. When we assume the internal field to be given by the incident field, the function $\phi_s(k_s,k_o)$ reduces to (Morse and Ingard [19]):

$$\phi(k_s,k_o) = \frac{k^2}{4\pi}\iiint [\gamma_\kappa + \gamma_\rho \cos\theta] e^{-i\vec{\mu}\cdot\vec{r}} dV_o \qquad (17)$$
$$= 2\pi^2 k^2 [\Gamma_\kappa(\vec{\mu}) + \Gamma_\rho(\vec{\mu})\cos\theta].$$

This expression, when applied to a sphere of radius a, yields the well known result [19]:

$$\phi_s(\vec{\mu}) = 1/3 k^2 a^3 \left(<\frac{\kappa_e - \kappa}{\kappa}> + <\frac{\rho_e - \rho}{\rho}> \cos\theta \right) \qquad (18)$$

This result yields a scattered intensity proportional to the number of bubbles squared, equivalent to our previous result of coherent scattering.

This result from the use of the classical Born approximation can be shown to be limited to the treatment of the bubbly liquid in V_o as a continuum with different ρc^2. To show this result we return to our schematic (Fig. 1) and treat the source volume as being bounded by a surface integral, S_a, the region of r would thus be between S_a and the surface at distance S_o. For the inner surface S_a, $\vec{a}$ is the outward directed normal into volume V_a. The resulting pressure field is

$$P_\omega(r) = P_{i\omega}(r) + \iint \left[P_\omega \frac{\partial g_\omega}{\partial a} - g_\omega \frac{\partial p}{\partial a} \right] dS_a. \qquad (19)$$

The radiation and scattering problems are now reduced to specifying $P_\omega/(\partial P/\partial a)$ on the surface S_a bounding the medium with properties ρ_m, C_m. The material in the sphere will determine this ratio, as this is equivalent of requiring continuity of velocity and pressure at the boundary. The procedure is to use spherical harmonics to expand the field within the sphere, the field external to the sphere, and the incident plane wave (Anderson [21], also Carey [12]). The mathematical manipulation is straightforward but tedious. Morse and Ingard show that to the lowest order in ka the result is

$$\phi_s = 1/3 k^2 a^3 \left(\frac{\kappa_m - \kappa}{\kappa} + \frac{3\rho_m - 3\rho}{2\rho_m - \rho} \cos\theta \right). \tag{20}$$

This expression allows us to define the region of applicability of the Born approximation and the beginning of the region determined by the volume fraction in which the bubbly liquid should be treated as a continuum. The case of air bubbles in water when $\kappa_e > \kappa$ and $\rho_e \sim \rho$ we find

$$\phi_s \rightarrow (1/3) k^2 a^3 \left(\frac{\rho c^2}{\rho_m c_m^2} - 1 \right) \sim 1/3 k^2 a^3 \left(\frac{c^2}{c_m^2} - 1 \right). \tag{21}$$

This theoretical result, which, we stress, is taken from the classical literature, has attempted to place the question of sound scattering from compact bubble distributions in context. In particular it was our intention to show the relationship between the Born approximation, continuum theory, and single incoherent bubble scattering. The validity of these approximations was shown to depend on the quantity $(c^2/c_m^2 - 1)$, which is proportional to χ, the volume fraction of the free air in the water volume.

The discussion has focused on a sphere of radius a. We have used this geometry because of its simplicity, but our conclusions for both the radiation and scattering from a complex region do not depend on the geometry. For an arbitrary complex yet compact distribution of air bubbles in water the radiated field can be shown to consist of monopole, dipole, quadrupole, and higher order terms. In our application the dipole and higher terms are not important. The monopole term corresponds to the volume pulsation; hence this treatment of a spherical source should also apply to any complex distribution radiating or scattering in the volumetric mode. This is similar to the observations made by Strasberg concerning single-bubble radiation [25].

3. SOUND PROPAGATION IN BUBBLY MIXTURES

In a bubbly mixture, in which the of bubbles possess resonance frequencies which are much greater than the frequencies of interest, then the propagation of sound in the mixture can be described by an effective wave equation with mixture density and sound speed. These effective wave equations have been studied by several investigators (i.e., Foldy [14,16], Crighton [26], Van Wijngaarden [24]) and confirmed by Karplus (1958) at frequencies <1 kHz (See Carey [11] as well as [12,13] and Prosperetti [9]; also Ruggles [29]).

A. B. Wood [22] showed that at low frequencies in an air-water mixture that when the size of the bubbles and spacing between bubbles were small compared to a wavelength, that for a given volume fraction (χ) the mixture density (ρ_m) and compressibility (κ_m) are under equilibrium conditions given by

$$\rho_m = (1 - \chi)\rho_l + \chi\rho_g \text{ and } \kappa_m = (1 - \chi)\kappa_l + \chi\kappa_g, \tag{22}$$

where l represents the liquid and g the gaseous component. The low-frequency limiting sonic speed, C_{mlf}, is seen to be

$$C_{mlf}^2 = \frac{dP}{d\rho} = (\rho_m \kappa_m)^{-1}. \tag{23}$$

In the case of the air bubble-water mixture, the process can be considered either adiabatic or isothermal. Since the controlling factor is the rate of heat transfer during bubble compression to the surrounding fluid which is rapid in water due to its large thermal capacity, the bubble oscillations at low frequencies may be considered isothermal. The following equations result:

$$C_{mlf}^{-2} = \frac{(1-\chi)^2}{C^2} + \frac{\chi^2}{C_g^2} + (\chi)(1-\chi)\frac{\rho_g^2 C_g^2 + \rho^2 C^2}{\rho_g C_g^2 \rho C^2} \tag{24}$$

$$\chi \to 0, C_m^{-2} \to C^{-2} \text{ and } \chi \to 1, C_m^{-2} \to C_g^{-2}$$

and when $0.002<\chi<0.94$, $C_{mlf}^2 = P/\rho\chi(1-\chi)$.

These equations show the large effect on sonic speed as a function of void fraction and relative bulk modulus. In the low-frequency range (<1 kHz) Karplus used an acoustic tube to determine the standing wave pattern as a function of air volume fraction and verified these results. Ruggles (1987) [29] has extended Karplus's technique to study a wide range of volume fractions. This behavior is due to the mixture's mass being primarily due to the liquid while its compressibility is due to the gas. Since we are dealing with the low-frequency response, much less than any individual bubble resonant frequency, only the volume fraction is important and not the bubble size distribution.

This equation clearly shows the dispersive character of the mixture speed of sound. In Fig. 2, we show several curves for a cloud of bubbles with individual resonant frequencies near 9.18 kHz. Below the resonance range we see the low-frequency minimum predicted by Wood's equation and above the resonance range we see the curves approach the speed of sound determined by the liquid compressibility. The phase velocity reaches a minimum at the resonance frequency and goes supersonic for a range of frequencies slightly above f_o. This is due to the fact that when driven above resonance bubbles pulsate out of phase with the driving force. Thus the bubbles "appear" to be stiff and the phase velocity subsequently increases dramatically. These strong dispersive effects need to be considered when describing sea surface scattering.

Equation (24) does not show the dispersive nature of this sonic speed variation. If one considers a bubbly liquid with a uniform bubble size whose corresponding resonant frequency is ω_o and dampening $\delta(\omega)$, then one [23,26] can solve the component conservation equations to obtain the following frequency-dependent sonic speed:

$$\frac{1}{C_m^2} = \frac{(1-\chi)^2}{C^2} + \frac{1}{C_{mlf}^2(1 - \omega^2/\omega_o^2 + 2i\delta\omega/\omega_o)}. \tag{25}$$

This expression reduces to our previous equation for C_{mlf}^2 when $\omega << \omega_o$. The equation does show a dispersive character as illustrated in Fig. 2, where we have used the explicit expression from Prosperetti [23]. The dispersive character of the phase velocity is the reason that volume fraction measurements are necessary to determine the low-frequency sound speed changes near the surface of the sea.

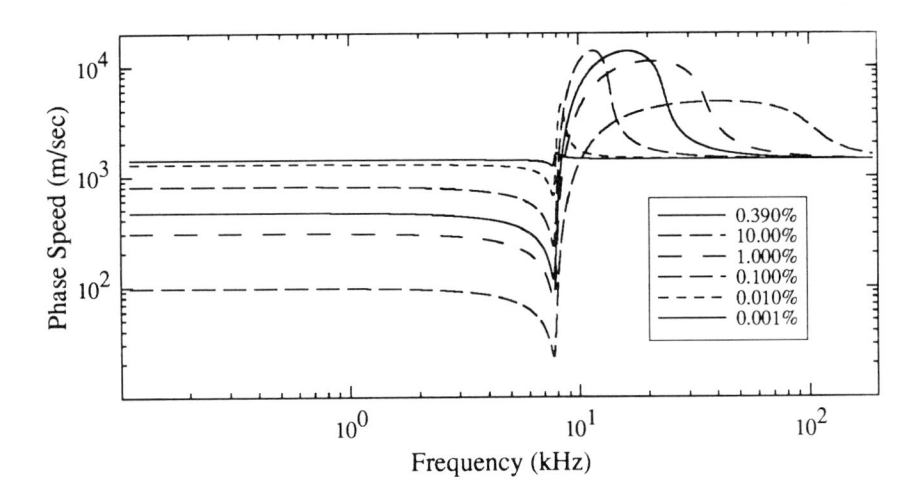

Figure 2. Dispersion curves for a cloud of 9.18 kHz bubbles

4. THE OSCILLATION OF A BUBBLE SPHERICAL VOLUME

The generation and scattering of sound from a compliant sphere immersed in a fluid can be found in classical texts on the theory of acoustics (see Reschevkin, or Morse). In this particular case, the bubbly sphere has no well-defined boundary, but nonetheless is localized by perhaps a vortex or some other circulatory feature beneath the breaking wave. Since this analysis is somewhat standard [20,12,29], only a brief outline will be presented.

Here we assume that the bubbly region (Fig. 3) is compact with an arbitrary radius r_o and the region is composed of microbubbles with resonance frequencies far above the frequency of excitation. Buoyancy forces and restoring forces such as surface tension are not important. The properties of the bubbly region are described by the mixture speed c_m and density ρ_m with a resulting compressibility $1/\rho c^2$. The microbubbles supply the compressibility and the liquid supplies the inertia.

Fig. 3 shows the random collection of microbubbles within a radii r_o from the origin, where (P_i = incoming wave or excitation, ρ_m, c_m = the properties of the mixtures, ρ, c = the properties of water, and P_s = the scattered sound.)

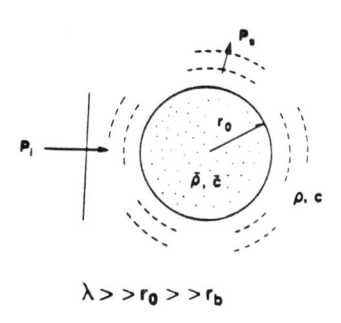

Figure 3. The random collection of micro bubbles.

The source of excitation is assumed to be global compared to the dimensions of the compact sphere. This assumption means we can consider a plane wave expanded in terms of spherical harmonics. The physical reasoning is that the properties of the bubbly region determine its ability to radiate provided there is a source of excitation; i.e.,

$$P_i = P_o \exp(i\omega t - ikr) = P_o \exp(i\omega t) \sum_{m=0}^{\infty} i^m (2m+1) P_m(\theta) J_m(kr). \qquad (26)$$

We require continuity of velocity and pressure at the generalized radius, r_o. Furthermore, a radiation condition is imposed at large r and the field is required to remain finite within the bubbly region.

The particle velocity $V_i = (-1/i\omega\rho) \, \partial P_i/\partial r$, (27)

Continuity of Pressure $P_i(a) + P_s(a) = \overline{P}(a)$, (28)

Continuity of Velocity $V_i(a) + V_s(a) = \overline{V}_n(a)$. (29)

The procedure is to assume the scattered wave or radiated wave is a sum of outward propagating spherical waves.

$$P_s = \exp(i\omega t) \sum_{m=0}^{\infty} A_m \overline{P}_m(\theta) H_m(kr) \exp(-i\epsilon_m kr). \qquad (30)$$

The pressure field inside the volume is expanded in terms of spherical Bessel function of the first kind; i.e.,

$$\bar{P} = \exp(i\omega t) \sum_{m=0}^{\infty} \bar{A}_m P_m(\theta) J_m(kr). \tag{31}$$

When the region is compact $\lambda > 2\pi r_o$, we can solve for the coefficients A_m and $\bar{A}_m$ by using the boundary conditions (27, 28, and 29) and equating each m order term. To $O(k^3r^3)$, we find for A_0 and A_1 the following:

$$A_0 = \frac{iP_o(k_o r_o)^3 (1 - y/\bar{y})/3}{\left[(1 - y/\bar{y})\frac{(kr_o)^2}{3}\right] - i(kr_o)\left[y/\bar{y}\frac{(kr_o)^2}{3}\right]} \quad \text{and} \tag{32}$$

$$A_1 = \frac{P_o(kr_o)^3 (\rho_m - \rho)}{(2\rho_m + \rho)}, \tag{33}$$

where $y = \rho c^2$ and $\bar{y} = \rho_m c_m^2$. The term A_0 becomes interesting when the real part of the denominator is equal to zero, we may say a monopole resonance has occurred. This occurs when

$$(kr_o)^2 = \left(\frac{2\pi f_o r_o}{C}\right)^2 = \frac{3\rho_m C_m^2}{\rho C^2} \quad \text{and} \quad f_o = \frac{C}{2\pi r_o}\sqrt{\frac{3\rho_m C_m^2}{\rho C^2}}, \tag{34}$$

since $\rho_m \sim \rho_e$ and $C_m^2 \sim \gamma P/\rho_l \chi(1-\chi)$, $f_o = \frac{1}{2\pi r_o}\sqrt{\frac{3\gamma P}{\rho \chi}}, \tag{35}$

which we recognize as a modified Minnaert formula for the resonance frequency of a volume oscillation of a gas bubble (Carey and Fitzgerald [12]). (Note that we have inserted a factor γ, the ratio of specific heats. This factor applies to individual bubble pulsations and for the cloud we set $\gamma = 1$ corresponding to isothermal conditions.) This result follows because the resonance angular frequency is proportional to the compressibility of the bubbly region, characterized by the stiffness $(4\pi r_o \rho C^2)$ and the inertia $(4\pi r_o^3 \rho/3)$.

5. THE MEASUREMENT OF BACKSCATTER FROM A SUBMERGED BUBBLE CLOUD

We report results from an experiment designed to measure low-frequency acoustic backscattering from a bubble cloud in fresh water under known propagation conditions. At issue is the scattering cross section for a bubble cloud. Is it determined by the bubble sizes and number density (as in the case of coherent or incoherent resonance and off-resonance scattering from individual bubbles) and discussed previously, or is it related to the free-gas volume fraction and the length scales for the cloud proper? In this effort we did not set out to duplicate "realistic" saltwater bubble plumes, but rather to obtain data to test bubble cloud scattering theories.

5.1 Experimental Procedures

The Naval Undersea Warfare Center (NUWC) Seneca Lake, New York test facility consists of two moored barges in 130 m of water, with the smaller barge (10.7 m x 42.7 m) serving as the platform for our test range. Equipment was deployed with the use of davits, a cable meter, and the edge of the barge, with the resulting vertical geometry shown in Fig. 4. Parametric and conventional transmitters and conventional receivers were orientated colinearly, with the axis of the range intersecting the path of a rising bubble cloud. Though the actual test consisted of measurements for frequencies ranging from 300 Hz to 14 kHz, only the low-frequency (<1.6 kHz) results will be presented here.

The parametric source (PS) was driven with a 22-kHz carrier signal that was up- and down-shifted by one half the difference frequency. The repetition frequency was fixed at either 1 Hz or 2 Hz and the pulse length varied from 5 to 15 msec. Calibration of the source was performed *in situ*. At 500 Hz, the beamwidth and source level was 8.5° and 167 dB *re* 1 µPa respectively. Although parametric sources have the advantage of low-frequency directionality, bubbly liquids are highly non-linear and the possibility of enhanced parametric interaction within the cloud needed to be addressed. Therefore, a conventional source (CS) (Honeywell HX-29) was also utilized. The working depth of 87.6 m was chosen to minimize reverberation.

The receivers consisted of four spherical hydrophones (ITC, 15.2 cm), situated as indicated in Fig. 4 (see H1 through H4). All hydrophone outputs were band-pass filtered, preamplified, and recorded on analog tape. Sound speeds were measured daily. At depths greater than 70 m, isovelocity conditions prevailed, with speeds of 1421.7 m/sec and 1420.8 m/sec at 70 m and 91.4 m respectively. For the calculations that follow, a speed of 1421.5 m/sec was used. We aligned the range using the 22-kHz signal to measure time-of-flight and verify alignment and slant ranges. We then inserted a 1.12-m diameter, hollow, steel spherical target in order to test our ability to measure target strength (TS). Assuming a perfectly reflected surface, this sphere possessed a theoretical

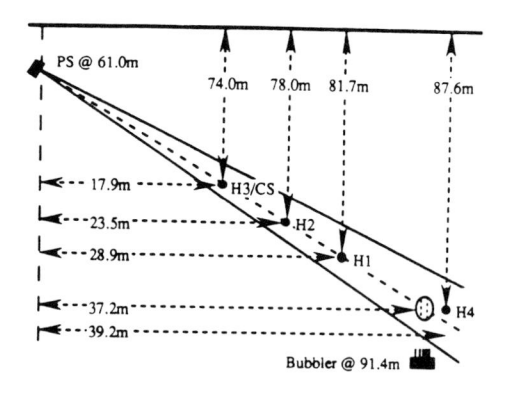

Figure 4. Test range for the backscattering experiment.

TS of -11.1 dB at 5 kHz. Our measured TS was -12.3 dB.

Bubble clouds were produced at 91.4-m depth using a pressurized steel enclosure, vented to the lake via a concentric circular array of 48 22-gauge hypodermic needles connected to a bank of 24 solenoid valves. A 3.5-sec burst of air at 68.95 kPa (10 psi) overpressure yielded a roughly cylindrically shaped cloud length ≈ 1.7 ± 0.3 m, radius ≈ 0.24 ± 0.01 m, with a void fraction of ≈ 0.40 ± 0.007%. The bubble sizes, which were measured with a video camera at depth as well as in a pressurized tube, were normally distributed about a peak at 1.1 ± 0.2 mm radius, which corresponds to $f_o ≈ 9.2 ± 1.7$ kHz at a depth of 87.6 m.

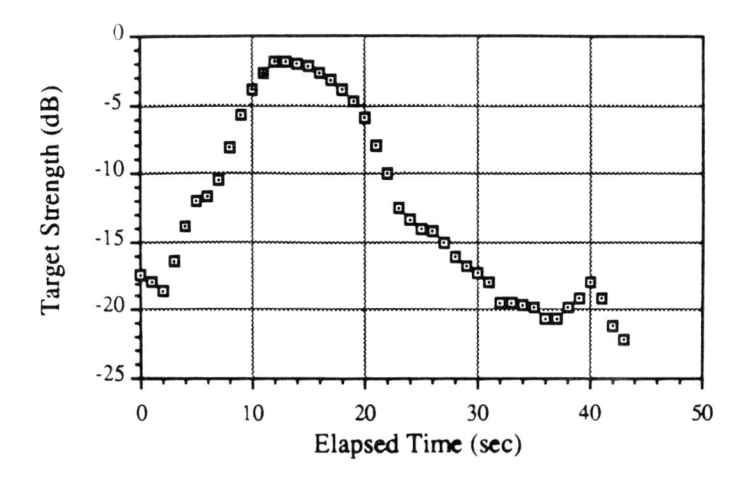

Figure 5. The evolution of the TS as the cloud traverses the beam of the parametric source operating at a pulse repetition frequency of 1 Hz

The clouds rose at ≈ 0.3 m/sec, yielding a time-varying echo that reached a maximum level as the cloud crossed the range axis. Fig. 5 depicts the evolution of the backscattered target strength as the cloud rises through the beam. This was obtained by time-gating the sequence of band-pass filtered echo returns (the filter bandwidth was set to equal the signal bandwidth; i.e., 200 Hz). Rms voltages were computed for both the main bang signal (mv) and for each echo return (ev). These values where inserted, along with the transmission loss factor (TLF), into the following expression for the target strength:

$$TS = 20\log\left[\frac{ev}{mv}\right] - TLF; \quad TLF = TL_{sr} - TL_{st} - TL_{tr}, \qquad (36)$$

where the TLF accounts for source-receiver, source-target, and target-receiver transmission losses, assuming spherical spreading of the propagating waves.

5.2 Experimental Results

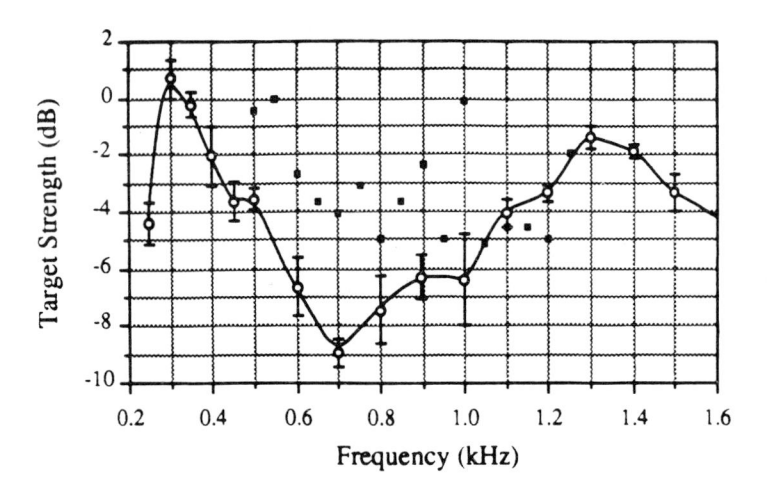

Figure 6. Backscatter TS measurements for conventional (squares) and parametric (circles) sources.

Peak target strength measurements obtained using conventional and parametric sources are plotted in Fig. 6. The target strengths are very high; at 300 Hz, the TS is about 12 dB greater than one would expect from a 1-m diameter perfectly reflecting sphere at 5 kHz. Each point corresponds to a mean value computed over a population of 4-5 clouds run at each frequency. The uncertainty corresponds to the standard error of the mean. There are apparent resonance peaks at 300 Hz and 1.3 kHz. If these peaks were due to single-bubble resonance scattering, the bubbles would possess radii of 3.4 cm and 7.8 mm respectively, which is highly unlikely in light of our revised observations of the bubbles generated at the needles and of the cloud as a whole.

It is probable that these peaks are the result of resonance scattering due to collective oscillations of the bubble cloud. The resonance frequency for a spherical bubble cloud is given by equation 35, where χ, γ, P_o, ρ, and a are, respectively, the void fraction, the polytropic exponent, the ambient pressure, the water density, and the cloud radius. We assume that the spherical cloud which has the same volume as the cylindrical cloud will describe the resonance frequency. Such a cloud, with $\chi = 0.40 \pm 0.07\%$ and

$a = 0.42 \pm 0.04$ m, has an isothermal resonance frequency of 324 ± 53 Hz. Moreover, the target strength is given by:

$$TS = 10 \log\left[\frac{I_s}{I_i}\right]_{r=1m}, \qquad (37)$$

where I_s and I_i are the scattered and incident intensities respectively. From equations 30 and 32 this is given by

$$\frac{I_s}{I_i} \approx \frac{k^4 a^6}{9} \frac{\left[1-\frac{y}{\bar{y}}\right]^2}{\left[\left(1-\left(\frac{y}{\bar{y}}\right)\frac{z_o^2}{3}\right)^2 + \left(\left(\frac{y}{\bar{y}}\right)\frac{z_o^3}{3}\right)^2\right]}, \quad r = 1m, \qquad (38)$$

where $z_o = ka$, $y = \rho c^2$, and $\bar{y} = \rho_m c_m^2$; the subscript denotes the effective properties of the air/water mixture. For the cloud in question, Equations 37 and 38 yield a calculated TS of -3.9 ± 0.8 dB at approximately 309 Hz. Although the measurements and calculations of the fundamental resonance frequency of the bubble cloud agree to within the stated uncertainty, the calculated target strengths are about 4 dB too low. All uncertainties on these results are based on either statistical analysis or *ad hoc* assumption and are propagated through the calculations by Taylor expanding relevant quantities to first order in the perturbed parameters. One such assumption was that the incident wave from the parametric source decreased as 1/r between the hydrophone and the target. This could lead us to underestimate the true P_i and subsequently overestimate the target strength. Consequently, the difference between our measurements and calculations may be attributed to the assumption of propagating spherical wavefronts for the parametric source and/or an underestimate of the damping within the cloud.

Parametric and conventional measurements exhibit "reasonably" good agreement despite the fact that reverberation made it difficult to obtain CS data for frequencies less than 1 kHz. It appears that enhanced parametric excitation in the region occupied by the bubbly fluid did not serve to appreciably bias the TS measurements.

6. SUMMARY AND CONCLUSIONS

We have reviewed the classical theory of sound radiation and scattering from compact bubble clouds. The theory shows that the important factor is the compressibility of the mixture which is characterized by the volume fraction at the lower frequencies. The volume pulsation of the cloud is the important mechanism whereby the cloud radiates or scatters sound; Strasberg has shown for the single bubble that shape may not be important. Since any compact radiator may be expanded in terms of a distribution of simple sources such as monopoles, dipoles, and quadrupoles. An expansion of the radiation field in terms of spherical harmonics identifies the monopole as the oscillation of a spherical source. We have shown that in this limit the cloud of bubbles, radiates with a resonance frequency described by a Minnaert equation modified to include the volume fraction of the gas.

Secondly, we have presented experimental results on the backscatter from a cloud of bubbles in the absence of boundaries. The results were interesting because of the high target strengths and the low-frequency resonance behavior. The experimental and theoretical results were shown to agree in a qualitative manner, we estimated the resonance frequency correctly, but were several dB off with respect to target strength. These results should be considered preliminary as analysis of the data is continuing. Nevertheless, they clearly show the effect of low-frequency resonant scatter.

The presence of the sea surface at these low frequencies may to first order be treated by the method of images. Hence a compact source at a depth below the pressure release surface will have a dipole characteristic and a magnitude which scales as $(z/\lambda)^2$, whereas a scatter below this surface will have a dipole pattern squared and a $(z/\lambda)^4$ dependence. Therefore, the large target strengths observed here may be offset by the proximity of the rough sea surface; the deeper plumes are the most important.

7. ACKNOWLEDGEMENTS

This paper represents the work of a large group of individuals. Jim Fitzgerald, The Kildare Corporation, contributed to this work through his observation of the resonant behavior of bubble clouds. Michael Nicholas of the NCPA was instrumental in all phases of our experiment. Larry Crum, NCPA, Andrea Prosperetti of The Johns Hopkins University, and Murray Korman of the U.S. Naval Academy provided critical insight, advice and criticisms. The experimental team of Jeff Schindall, Bill Konrad, Bill Marshall, Martin Wilson, Lynn Carlton, and the Lake Seneca Crew made this work possible. Ralph Goodman, The Pennsylvania State University/Applied Research Laboratory, has also contributed by his interesting discussions, technical reviews, and moderating influence. Finally, we acknowledge the sponsorship of Marshall Orr, ONR (11250A), Ken Lima, NUWC IR, and additional support from AEAS, ONT, and DARPA.

NOMENCLATURE

$A_m, \bar{A}_m$	series coefficients	P_o	source pressure
C	speed of sound	$P_m(\theta)$	Legendre polynomial
$H_m(kr)$	Hankel function	R	radial distance
I	intensity	S, S_o, S_a	surfaces
$J(kr)$	Bessel function	TS	target strength (dB)
N	number	TL	transmission loss (dB)
$P_\omega, P_{i\omega}, P_\omega$	pressure transform	V, V_o, V_a	volumes
P	ambient pressure		

a	radial distance	t	time
â	normal to S_a	v, v_i, v_o	normal velocities
$c_m, c_{m/p} c_g,$		$y, \bar{y}$	bulk modulii
$c_e, \bar{c}_e$	sonic speeds		
ev	echo voltage	γ	gas constant
f_o	natural frequency	$\gamma_\kappa, \gamma_\rho$	compressibility, density ratio
$f_n(\omega_o, \theta, r_n)$	bubble response		
$f_\omega(r)$	source function	δ	dampening
$g_\omega(r\|r_o)$	Green's function	ε	small quantity
g	subscript for gas	θ	angular measure
i	subscript incident	κ	compressibility
k	wave number	λ	wave length
l	liquid	μ	wave number
m	subscript mixture or index	π	pi
mv	main bang voltage	ρ	density
$\hat{n}_o$	normal to S_o	χ	void fraction
r	radial distance	ω	angular frequency
s	subscript, scatter	Γ	gamma function

REFERENCES

1. Thorpe, S. A. (1982), "On the clouds of bubbles formed by breaking wind-waves in deep water and their role in air-sea gas transfer," Phil. Trans, Roy. Soc. A 304, 155-185.

2. Su, M. Y., Green, A.W., and Bergin, N.T. (1984), "Experimental studies of surface wave breaking and air entrainment," in W. Brutsaert and G. Jirka (eds.), Gas Transfer at Water Surfaces, Reidel Press, pp. 211-219.

3. Monahan, E. C. and MacNiocaill (eds.) (1986), Oceanic Whitecaps and Their Role in Air-Sea Exchange Processes, D. Reidel, Boston.

4. Monahan, E. C. and Lu, M. (1990), "Acoustically relevant bubble assemblages and their dependence on meteorological parameters," IEEE Jour. Ocean Engr. 15(4), 340-349.

5. McDaniel, S. (1988), "High-frequency sea surface scattering: Recent Progress," J. Acoust. Soc. Am 84(S1), S121 (also: ARL/PSU T.M. 88-134-21, July 1988)

6. Carey, W., Goodman, R., McDaniel, S., Jackson, D., Deferrare, H., and Briscoe, M. (August 1985), "Sea surface scatter experiments, a review," Memo to M. Orr, ONR (11250A).

7. McDonald, B. E. (1991), "Echoes from vertically striated subresonant bubble clouds: A model for ocean surface reverberation," J. Acoust. Soc. Am 89(2), 617-622.

8. Henyey, F. (1991), "Acoustic scattering from ocean microbubble plumes in the 100 Hz to 2 kHz region," J. Acoust. Soc. Am 90(1), 399-405.

9. Prosperetti, A. (1988), "Bubble related ambient noise in the ocean," J. Acoust. Soc. Am. 84(3), 1042-1054.

10. Prosperetti, A. (1988), "Bubble dynamics in ocean ambient noise," in B. R. Kerman (ed.), Sea Surface Sound, Kulwer Acad., Boston, MA, pp. 171-171.

11. Carey, W. M. and Browning, D. G. (1988), "Low frequency ocean ambient noise: Measurements and theory," in B. R. Kerman (ed.), Sea Surface Sound, Kulwer Acad., Boston, MA.

12. Carey, W. M. and Fitzgerald, J. W. (1987), "Low frequency noise and bubble plume oscillations," J. Acoustic soc. Am. 82(51), p. 362 (paper 001) (also available NUSC TD 8495, 24 February 1989; DTIC AD206537)

13. Carey, W. M., Fitzgerald, J. W., and Browning, D. G. (in press 1990), "Low-frequency noise from breaking waves," in B. R. Kerman (ed.), Natural Physical Sources of Underwater Sound, Kluwer Acad., Dordrecht, Netherlands, (also: NUSC Tech. Rep. 8783, 5 oct., 1990, avail. DTIC AD227969).

14. Foldy, L. (1944), "Propagation of sound through a liquid containing bubbles," paper, Columbia Univ. NRDC DIV 6.1. (Available DTIC).

15. Spitzer, L. (1943), "Acoustic properties of gas bubbles in a liquid," paper, NRDC DIV 6.1, (available DTIC).

16. Foldy, L. (February 1945), "Multiple scattering of waves," Phys. Rev. 67(3&4) 107-119.

17. Carstensen, E. L. and Foldy, L. (May 1947), "propagation of sound through a liquid containing bubbles," J. Acous. Soc. 19(3), 481-501.

18. Morse, P. M. and Feshback, H. (1943), Methods of Theoretical Physics, Part II, McGraw Hill Book Co., N.Y.

19. Morse, P. M. and Ingard, K. (1968), Theoretical Acoustics, McGraw Hill Book Co., N.Y., pp. 413-414.

20. Rschevkin, S. N. (1963), The Theory of Sound, Pergamon Press Ltd., N.Y.

21. Anderson, V. (1950), "Sound scattering from a fluid sphere," J. Acous. Soc. Am. 22(4), 426-431.

22. Wood, A. B. (1932/1955), A Textbook of Sound, G. Bell and Sons Ltd., London, pp. 360-364.

23. Commander, K. and Prosperretti, A. (1989), "Linear pressure waves in bubbly liquids: Comparison between theory and experiments," J. Acoust. Soc. Am. 85(2), 732-746.

24. Gilbert, K., Wang, L., and Goodman, R. (1991), "A stochastic model for scattering from near-surface oceanic bubble layers," J. Acoust. Soc. Am. 90(s4), 2301.

25. Strasberg, M. (1956), "Gas Bubbles as Sources of Sound in Liquids," J. Acoust. Soc. Am. 28(1), 20-26.

26. Crighton, D. G. and Williams, J.E. (1969), "Sound generation by turbulent two-phase flow," J. Fluid Mech. 36, 585-603.

27. Van Wijngaarden, L. (1958), "On the equation of motions for mixtures of liquid and gas bubbles," J. Fluid Mech. 133(P3), 465-474.

28. Karplus, H. B. (1958), "The velocity of sound in a liquid containing gas bubbles," paper, Armour Research Foundation, University of Illinois, Proj. c00-248, TID-4500.

29. Ruggles, A. (1987), "The Propagation of Pressure Perturbations in Bubbly Air/Water Flows," Ph.D. Thesis, Dept. of Mech. Eng., Rensselaer Polytechnic Institute, Troy, NY.

THEORETICAL MODELING OF LOW FREQUENCY ACOUSTIC SCATTERING FROM SUBSURFACE BUBBLE STRUCTURES

R. F. Gragg and D. Wurmser
Naval Research Laboratory
Washington, D. C. 20375-5000

We have produced a software package (BIRPS — Boundary Integral Resonant Plume Scatter) that simulates scattering from general near-surface bubble structures. It solves the field-matching boundary integral equations and computes complex scattering amplitudes for structures with user-specified depths, shapes and interior sound speed functions. It incorporates surface-image and resonance effects and is currently restricted to fixed plumes under flat, smooth sea surfaces. We describe the operation of BIRPS and present simulations for "intermediate" plumes.

INTRODUCTION

We have extended the boundary integral theory of Kittappa and Kleinman [1] to CW scattering from a body with a *non*-uniform interior. We compute the scattering amplitude as follows. First, for source direction $\hat{s}$, we solve the *continuity equation*

$$\begin{pmatrix} (\bar{\rho}/\rho_\mathrm{I})U(\mathbf{y}|\hat{s}) \\ (\bar{\rho}/\rho_\mathrm{E})V(\mathbf{y}|\hat{s}) \end{pmatrix} + \int_\mathcal{S} dS(\mathbf{x}) \begin{pmatrix} -A(\mathbf{x},\mathbf{y}) & B(\mathbf{x},\mathbf{y}) \\ -C(\mathbf{x},\mathbf{y}) & D(\mathbf{x},\mathbf{y}) \end{pmatrix} \begin{pmatrix} U(\mathbf{x}|\hat{s}) \\ V(\mathbf{x}|\hat{s}) \end{pmatrix} = \begin{pmatrix} P(\hat{s},\mathbf{y}) \\ Q(\hat{s},\mathbf{y}) \end{pmatrix}$$

for the velocity potential $U(\mathbf{x}|\hat{s})$ and its normal derivative $V(\mathbf{x}|\hat{s})$ at points $\mathbf{x}$ throughout the body's surface $\mathcal{S}$. Then we employ these as equivalent sources to compute the scattering amplitude in receiver direction $\hat{r}$ via the *far-field scattering equation*

$$F(\hat{r}|-\hat{s}) = \frac{1}{4\pi} \int_\mathcal{S} dS(\mathbf{x}) \left(Q(\hat{r},\mathbf{x}), -P(\hat{r},\mathbf{x}) \right) \begin{pmatrix} U(\mathbf{x}|\hat{s}) \\ V(\mathbf{x}|\hat{s}) \end{pmatrix}.$$

$P(\hat{s},\mathbf{x})$ and $Q(\hat{s},\mathbf{x})$ are essentially the exterior Green's function $G_\mathrm{E}(\mathbf{x},\mathbf{s})$ and its normal derivative with spherical spreading factors removed, while the kernels A–D are spatial derivatives involving the interior and exterior Green's functions [2]:

$$\begin{aligned}
A(\mathbf{x},\mathbf{y}) &= \frac{\partial}{\partial n(\mathbf{x})}\{G_\mathrm{E}(\mathbf{x},\mathbf{y}) - (\rho_\mathrm{E}/\rho_\mathrm{I})\, G_\mathrm{I}(\mathbf{x},\mathbf{y})\} \\
B(\mathbf{x},\mathbf{y}) &= \{G_\mathrm{E}(\mathbf{x},\mathbf{y}) - G_\mathrm{I}(\mathbf{x},\mathbf{y})\} \\
C(\mathbf{x},\mathbf{y}) &= \frac{\partial^2}{\partial n(\mathbf{y})\partial n(\mathbf{x})}\{G_\mathrm{E}(\mathbf{x},\mathbf{y}) - G_\mathrm{I}(\mathbf{x},\mathbf{y})\} \\
D(\mathbf{x},\mathbf{y}) &= \frac{\partial}{\partial n(\mathbf{y})}\{G_\mathrm{E}(\mathbf{x},\mathbf{y}) - (\rho_\mathrm{I}/\rho_\mathrm{E})\, G_\mathrm{I}(\mathbf{x},\mathbf{y})\}
\end{aligned}$$

where '$\partial/\partial n$' indicates an outward normal derivative.

D.D. Ellis et al. (eds.), Ocean Reverberation, 45–50.
© 1993 Kluwer Academic Publishers. Printed in the Netherlands.

MODELING METHODS

For the exterior Green's function we begin with the free space "$\exp\{ikr\}/r$" form and then incorporate the pressure-release sea surface by including an image contribution. In the interior we use the WKB form $G_I(\mathbf{x},\mathbf{y}) = \exp\{i\varphi(\mathbf{x},\mathbf{y})\}/4\pi|\mathbf{x}-\mathbf{y}|$, computing the phase by a line integral of the local wave number. We find A, B, C, D, P, Q by differentiation of these forms. To solve the continuity condition, we first fit S with a "wireframe" — a mesh of N plane facets or "tiles" that approximate uniform equilateral triangles. We then approximate the surface integral as a sum over this wireframe, converting the continuity equation to the algebraic form

$$\begin{pmatrix} (\bar{\rho}/\rho_I)I - A & B \\ -C & (\bar{\rho}/\rho_E)I + D \end{pmatrix} \begin{pmatrix} U(\hat{s}) \\ V(\hat{s}) \end{pmatrix} = \begin{pmatrix} P(\hat{s}) \\ Q(\hat{s}) \end{pmatrix}$$

in which A, B, C, D, I are the appropriate $N \times N$ matrices and U, V, P, Q are $N \times 1$ column vectors. We solve this using LU decomposition with iterated improvement. In this wireframe approximation, the far-field scattering equation is

$$F(\hat{r}| - \hat{s}) = \bar{Q}^T(\hat{r})U(\hat{s}) - \bar{P}^T(\hat{r})V(\hat{s})$$

where $\bar{Q}$ and $\bar{P}$ correspond to Q and P with components weighted by the tile areas and 'T' indicates a matrix transpose. We initially construct the wireframes to fit spheres and then adjust them for non-spherical plumes by a process of projection followed by annealing [2] as indicated in Fig. (1).

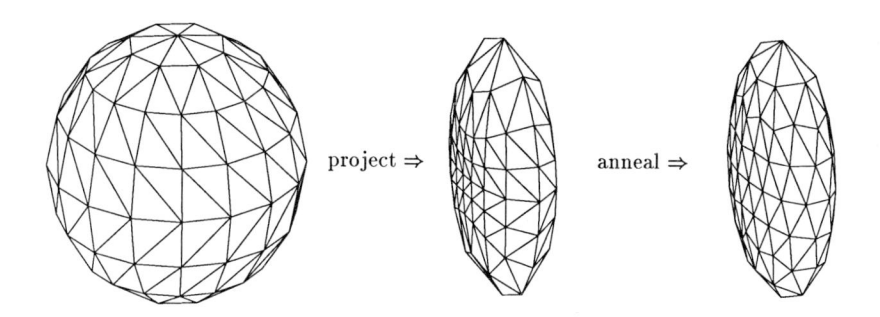

Fig. 1 – Production of an ellipsoidal wireframe.

PLUME RESULTS

To verify the accuracy of the model, we ran the following test case for a spherical plume with a one-meter radius and a uniform interior sound speed of 500 m/s. This was done *without* the pressure-release condition at the air/sea interface so that the model results could be compared to partial wave calculations. The comparison, in Fig. (2), shows excellent agreement for the 192-tile wireframe at low frequencies (through the $\ell = 0, 1, 2$ or s,p,d-wave resonance peaks). The computation fails in the vicinity of 450 Hz because 192 tiles do not provide adequate spatial sampling to model the complicated radiation pattern of the f-wave resonance [2]. Next, we

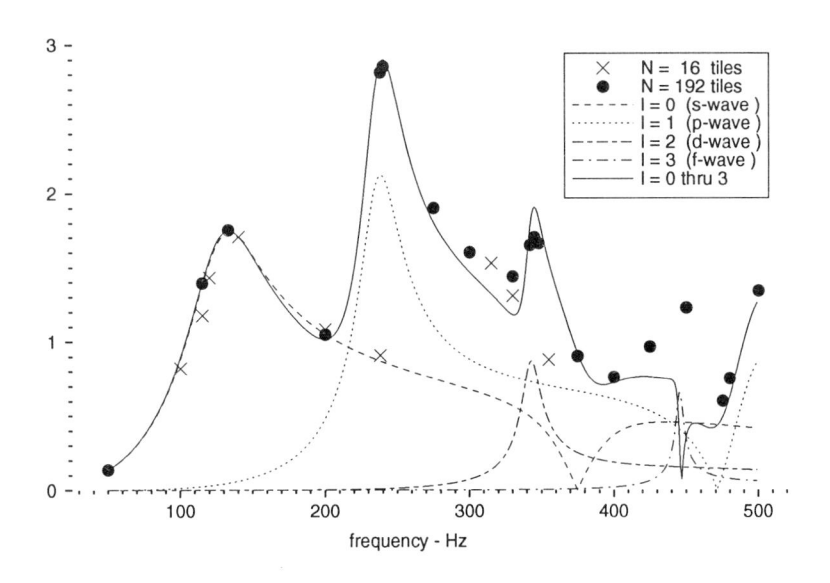

Fig. 2 – $|F|$ at 45° scattering angle *without* the air/sea boundary condition for: 16-tile wireframe, 192-tile wireframe, $\ell = 0, 1, 2, 3$ partial wave contributions, full partial wave result.

reinstated the pressure relief condition, placed the same plume at a center depth of 2 m, and examined the frequency dependence of its scattering response in the vertical plane. Figure (3) illustrates the result in the vicinity of the second resonance. The figure is a polar plot of $|F|$ as a function of receiver elevation throughout the vertical half-plane, both for back-scattering forward scattering. The source is at 2° grazing elevation. The resonance seen here agrees in location and width with Fig. (2) and has the expected p-wave angular symmetry, modified by the reflecting sea surface.

Figure (4) illustrates the dependence on the center depth d that we observed for near-grazing elevations. The source and receiver are held at grazing angles of 2.5°, the frequency is 125 Hz and, except as noted, other parameters remain the same. Assuming a power law, $|F(d)| \propto d^p$, the exponent can be assessed by plotting $p = |F(d)|'d/|F(d)|$ versus d. This is done in Fig. (4) with the depth derivative estimated by finite differences. First we ran a benchmark case with a plume diameter of only $D = 0.5$ m.

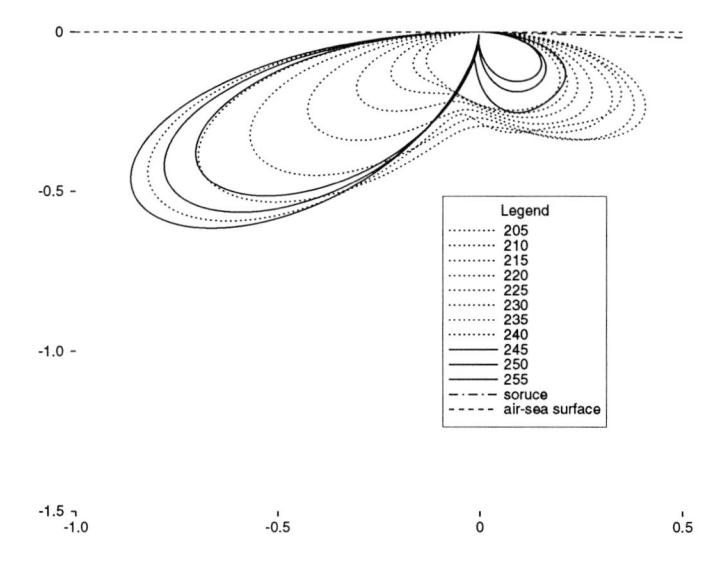

Fig. 3 – Scattering response in the vertical plane for the uniform spherical plume at a center depth of 2 m. Frequency runs from 205 Hz to 255 Hz in 5 Hz steps as indicated in the legend. Back-scattering is to the right.

Since $D \ll d, \lambda_\mathrm{E}$, this response has the $p = 2$ (quadratic) dependence characteristic of a point target in this geometry. Next, we reinstated the 2 m diameter and found the result closer to $p = 1$. Since this plume is resonant at 125 Hz, it was not obvious whether this linear depth dependence was produced by the resonance or simply by the larger size. We ran a third depth simulation to find out. Here we increased the interior sound speed to 1000 m/s, which shifted the lowest resonance well above the operating frequency. The result for this large non-resonant plume was roughly linear for $d \approx 2$ m but approached quadratic by $d \approx 4$ m. The linear dependence seems to be in part a result of large D/d. Resonance, however, enhances it, possibly by giving the plume a larger effective diameter. Of course, a finite diameter and a resonant interior do more than just alter the depth dependence of a plume's scattering response. They both raise the absolute level too. At the center of the depth interval, $d = 3$ m, the large non-resonant plume and the large resonant one scatter at levels 10 dB and 22 dB, respectively, greater that the "point-target" plume.

In addition to the results shown here, we have also done simulations involving plumes with non-symmetric ellipsoidal shapes and vertical sound speed gradients up to $300\,(\mathrm{m/s})/\mathrm{m}$. BIRPS is designed to accommodate multiple coherent arrivals; i.e., multiple $\hat{s}$ directions with separate amplitudes and phases. We have used it in this

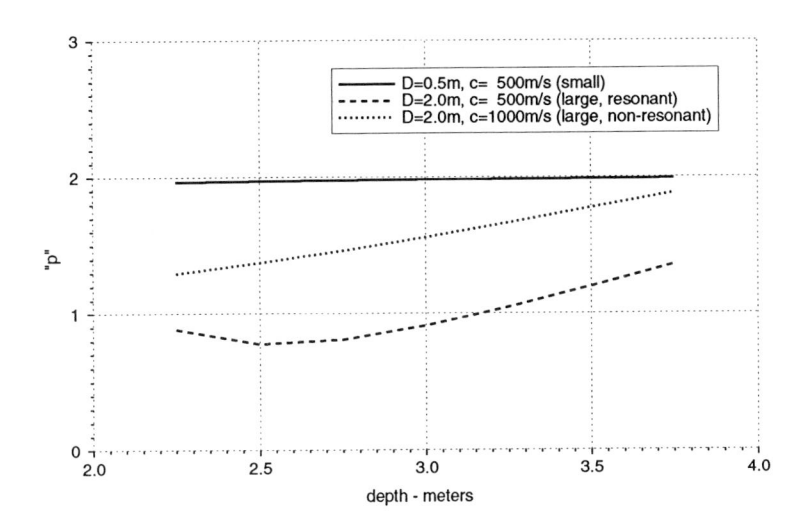

Fig. 4 – Power law dependence of the near-grazing scattering response

capacity, with the arrivals supplied by an elementary propagation model, to simulate scattering from plumes in a bottom-limited environment [2].

SUMMARY AND CONCLUSIONS

We have developed a numerical implementation of the boundary integral treatment of scattering from an acoustically penetrable bubble structure of very general type. The underlying theory includes the effects of collective bubble resonances and our implementation provides options for the presence of a pressure-release sea surface and for coherent multipath arrivals. The implementation, in the form of the BIRPS software, provides a numerical modeling component for the investigation of low frequency surface reverberation and has the potential for wider application to the study of resonant scattering from other complex bodies. We have simulated the complex scattering amplitudes of model "intermediate" plumes at frequencies up to a few hundred Hertz for a range of size, shape, depth, and interior sound speed profile. We find these plumes to have multiple weak-to-moderate resonances in the sub-kHz band. The non-resonant scattering response for near-grazing geometries is about 10 dB higher than point-target levels and tends to be linear in the depth near 2 m, changing to quadratic by 4 m. Resonance adds approximately another 10 dB to the level and extends the linear dependence to greater depths. Typically, the lowest-frequency resonance is a low-Q feature whose response has a weak frequency dependence with a vertical dipole-like dependence on elevation angle. As frequency is increased, the next resonance encountered has a higher Q and exhibits forward and backward scattering lobes. As the frequency is swept upward through such a resonance, the dominant component of the response shifts from backscattering to forward scattering at a rate dependent on the Q value. Features with a significant impact on the resonance spectrum of a plume, and on the scattering response in general, are the "primary parameters" (size, shape, mean interior sound speed) and the depth. Strong interior sound speed gradients enhance the Q of the resonant responses, while weak gradients have no appreciable effect.

We have focused this effort on the resonance effects of bubble structures and have neglected "tenuous" bubble clouds (the large Langmuir features) for that reason. This is only a matter of emphasis. BIRPS is certainly also applicable to tenuous clouds and may be useful as a means of validating the approximations made in other approaches. The principal limitations of BIRPS at the present stage of development are that (a) it treats the sea surface as a smooth flat plane, (b) it deals with plumes as static structures, and (c) it cannot accommodate extremely strong interior sound speed gradients. Exploratory work is underway to relax these limitations. If these efforts succeed, we should in future be able to simulate scattering from "strong" plumes and even individual "mega-bubbles" and to address issues of surface swell and roughness.

ACKNOWLEDGEMENTS

This work was supported by the Office of Naval Research, Code 1125OA and by the Office of Naval Technology, Code 234.

REFERENCES

1. Kittappa, R. and Kleinman, R. E. (1975) 'Acoustic scattering by penetrable homogeneous objects', J. Math. Phys. 16, 421–432.

2. Gragg, R. F. and Wurmser, D. (1992) 'Theory and numerical modeling of low-frequency acoustic scattering from bubble plumes near the sea surface', Naval Research Laboratory Report 9391, in publication.

DISCRETE BACKSCATTER CAN BE DOMINANT IN ROUGH BOTTOM REVERBERATION

Ira Dyer, Arthur B. Baggeroer, Henrik Schmidt, J. Robert Fricke,
Nazan Ozluer, and Dominique Giannoni
Department of Ocean Engineering,
Massachusetts Institute of Technology
Cambridge, Massachusetts, 02139 USA

ABSTRACT. Rough bottom acoustic backscatter observed in the 200-300 Hz frequency range has a discrete character. This is predicted by a simple facet model that has the form of Lambert's Law, but biased by the facet's slope angle, which then imparts a finite backscatter value as grazing angle approaches zero. Backscatter in this model is dominated by planes about one wavelength in size that, for certain orientations, scatter sufficiently in the back direction. Other orientations of larger facets scatter predominantly in the forward direction, while smaller facets of any orientation scatter too weakly to be important. The data, and the model, appear consistent with observations by others on rough bottom backscatter over a wide frequency range.

Introduction

In the Acoustic Reverberation Special Research Program (ARSRP), acoustic signals backscattered from the bottom were designed and processed to localize bottom features radially to within a few meters, with spatial resolution widths, again radially, as small as about 40 m. Upon study of data acquired in the Reconnaissance Cruise of August 1991 on the western flanks of the Mid Atlantic Ridge (26^0N, 47^0W), we find the backscatter to be dominated by discrete events. We give examples of the observed discrete process, and offer preliminary physical explanations and discussion.

1. Data

Beamformed match-filtered data at 230 Hz are shown in Figure 1 in the time range corresponding to direct insonification of the bottom with the source's main lobe. (More detail is in Section 2.) The data appear partly lineated, perhaps reflecting geological orientations imparted by spreading processes of the nearby Mid Atlantic Ridge. With selection of a range of time corresponding to the smallest grazing angles of the insonification, Figure 2, we see the lineations more clearly. Another view of this, entailing a smaller number of beams and a shorter range of time, Figure 3, shows the backscatter to be a discrete process.

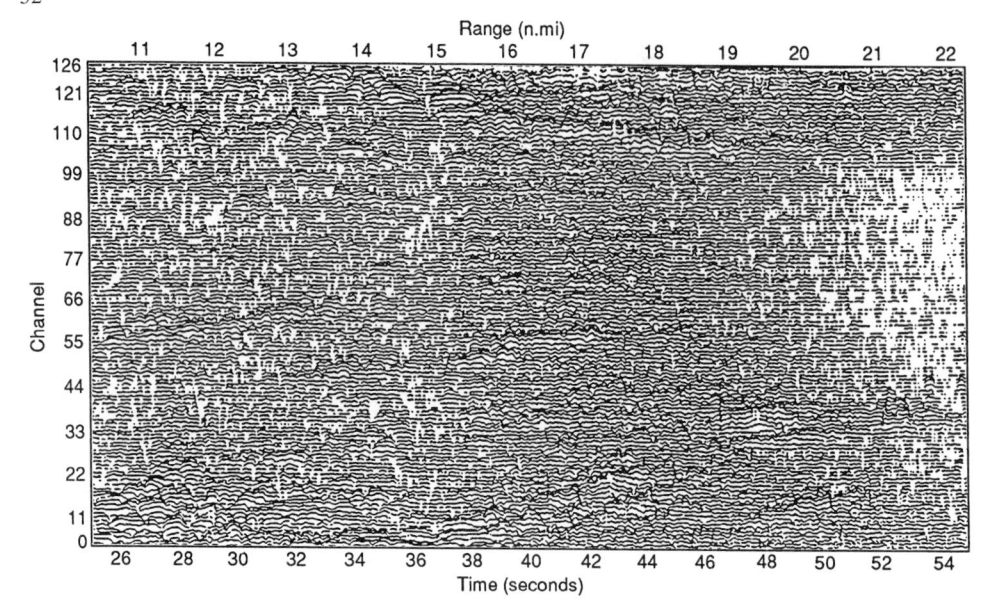

Figure 1. Match-filtered backscatter level in 126 beams of the horizontal line array, for 40s of round-trip time after initiation of a hyperbolic FM signal centered on 230 Hz. Beam 0 is steered forward (toward the tow ship) and 126 is aft, with 62 and 63 partially overlapped in the broadside direction. The fathometer return can be distinguished, and lineations can be discerned starting at about 24s. (Data from Run 12, J220, 2331 Z.)

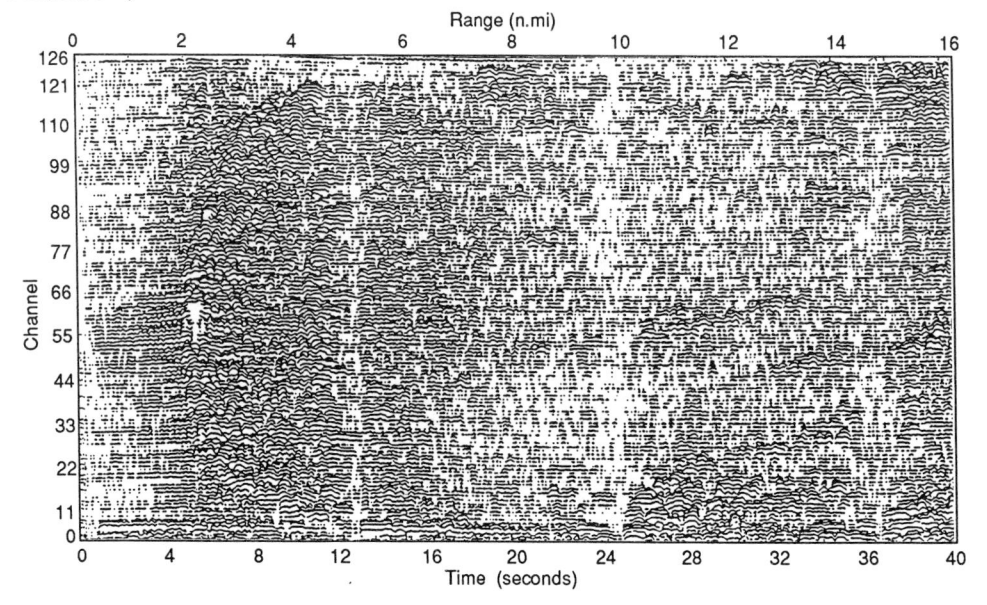

Figure 2. Same data as Figure 1, but selected to show lineations to about 50s, in the direct insonification zone at the smallest grazing angles.

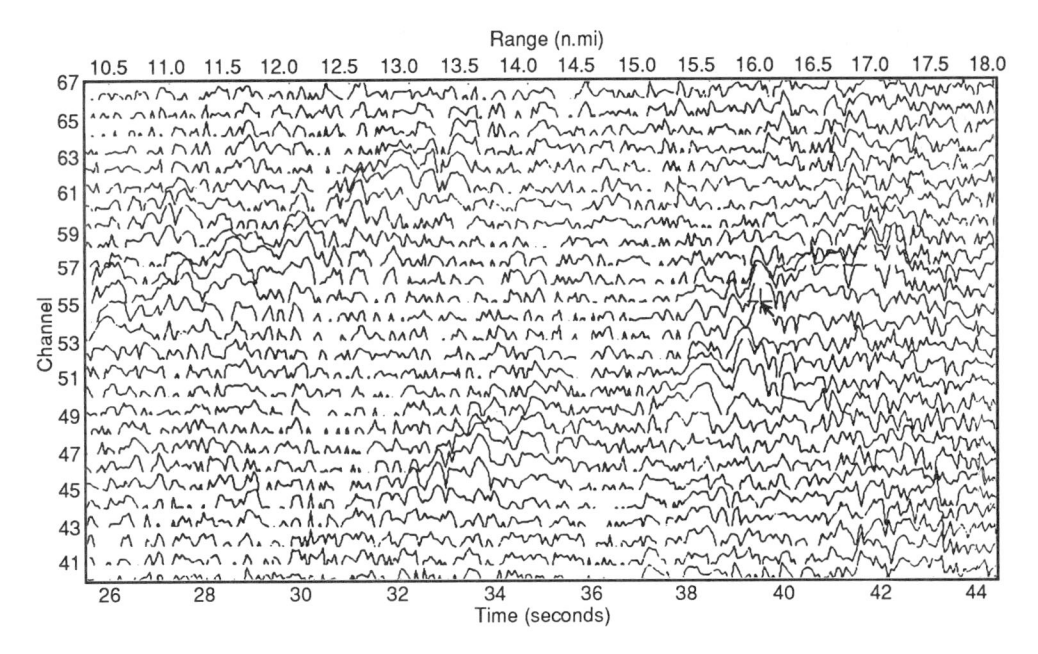

Figure 3. Backscatter level versus time, shown to emphasize the discrete backscatter returns. Beams shown are 40 to 67 for time from 25 to 45s. The levels shown are in a range of 28 dB above an arbitrary threshold.

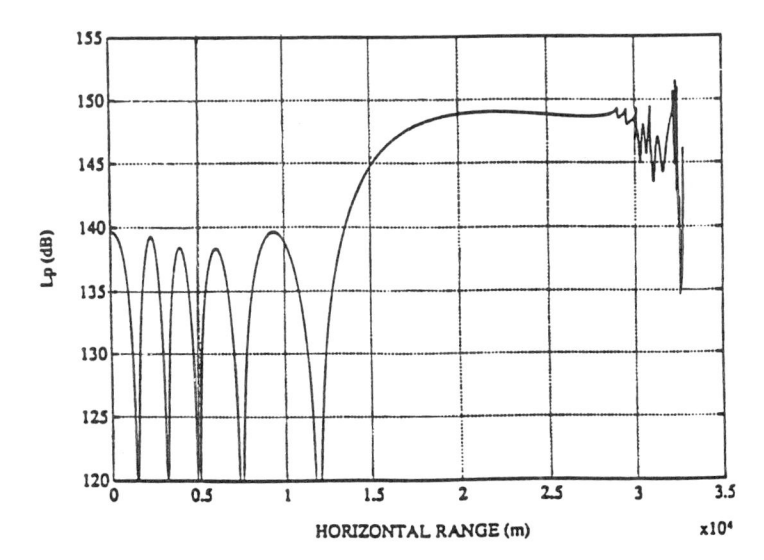

Figure 4. Incident sound field in dB re 1 µPa at a depth of 4000 m, versus range from the source. The on-axis source level is 232 dB re 1µPa and 1 m, and the field is calculated from ray theory with use of a 9-segment linear approximation to the measured sound speed profile. Caustics render the calculation inapplicable beyond 28 km. The source is steered down to 6º from the horizontal.

The data shown are representative of a much broader data set [1]. At times, sound can backscatter from as many as 10 subsequent interactions with the bottom, with range periodicity governed by the convergence zone radius. As far out as can be observed, the data appear to maintain their discrete character.

A systematic study of the amplitudes, number density, and other statistics of these discrete events is underway. Based on very early results, the amplitudes correspond to a target strength of 16 dB $\pm$ 5 dB re 1 m, a remarkably large value. The number density is about n = 10 per km^2. The latter is most readily obtained from the first or direct insonification zone, because there the azimuthal window Rϕ has a width of about 1 km (R ~ 30 km, ϕ ~ 1/30 rad), and the probability of overlapping two or more discrete events is small.

2. Bottom Insonification

While the state of both the sound source (a vertical line array, VLA) and receiver (a horizontal line array, HLA) was monitored, we must predict the sound field incident upon the bottom, an unsure process given that the bottom locally varies in depth, and its roughness might shield parts of it. The sound speed profile and mean depth, however, is well known, and leads to useful incident field predictions. In Figure 4 we show the field from coherent addition of the 10 individual sound sources in the VLA, for a 9-segment linear approximation to the measured profile. Based on ray bundle theory, the predicted incident field at 4000 m depth provides a basis upon which to normalize the backscatter field. At about 28 km the ray analysis shows that one or more eigenrays from source to bottom develops caustics, at which, and beyond, the ray analysis fails. Figure 5 is a SAFARI output for the same situation; the two agree well for horizontal ranges less than 28km, and the latter, being wave-based, provides results until the bundle lifts off the nominal bottom depth at 32 km. (Actually Figure 5 includes a basalt half-space at 4000 m, so that interference patterns from its reflections are also seen.)

We do not believe that caustics or other wave-field effects can explain the discrete nature of the observed backscatter. None of 126 beams in Figure 2 shows any difference in backscatter at the predicted inception of caustic behavior. Further, in another study, lineations such as in Figure 2 are shown to correspond with major topographic features of the bottom, some of them at depths locally as small as 3000 m, well outside the strong caustic formation region; the lineations cannot be ascribed to wave-field complexity. Thus, while we remain open to effects of such complexity, the essential discrete character of the backscatter is laid to the sound wave interacting with the rough bottom.

3. Physical Interpretation and Discussion

We hypothesize that backscatter is dominated by a "self selection" facet process. In this theory:

- the rough bottom is represented by a series of flat planes (facets) of various shapes, sizes, and orientations

- each facet is locally reacting (i.e. elastic waves excited in the

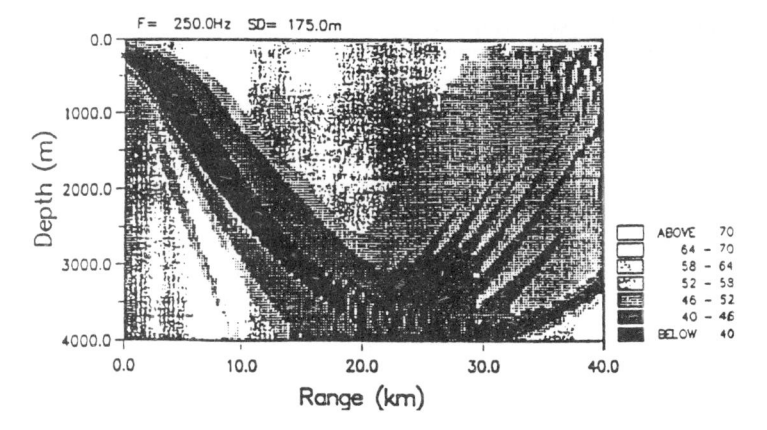

Figure 5. Incident and reflected sound field for a half-space of basalt at 4000 m, and a source steering angle of 6° down. Contours give the transmission loss in dB re 1 m, upon adding 10 log R, where R is horizontal range in m. The measured profile is used in this wave-based calculation.

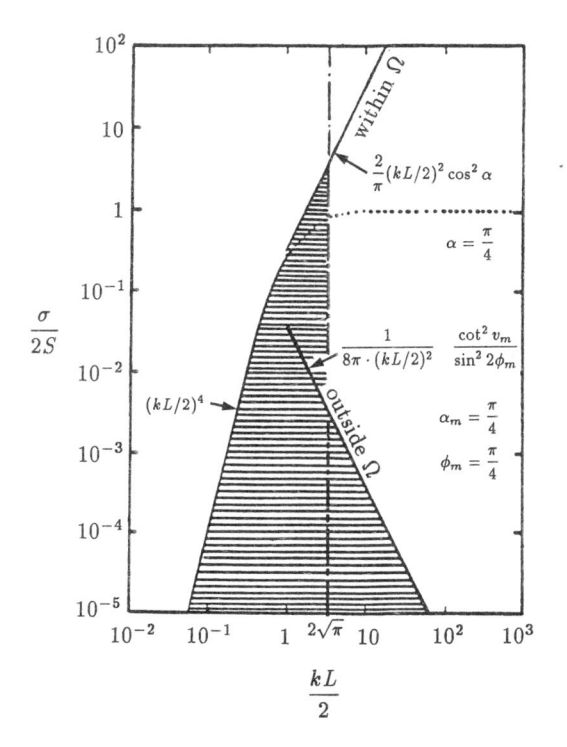

Figure 6. Diagram of the facet backscattering theory, where $S=L^2$ is the facet area and L its mean size, k the acoustic wavenumber, and α the angle between the facet normal and the incident wave direction. We show a peak defined by the facet's main scattering lobe, which is affected by transition from inclusion to exclusion, in the main lobe, of the backscatter direction.

microfractured basalt below the facet is ignored) and, for simplicity, each facet is perfectly reflecting

- the facets large compared to wavelength scatter mainly forward into a small solid angle Ω

- the facets small compared to wavelength scatter uniformly in a 2π solid angle, but do so weakly

- only those facets comparable to wavelength can scatter strongly in the back direction

- there is thus an apparent spatial filter that selects among all facets those whose size gives rise to important backscatter, i.e. size about one wavelength, and leads to a discrete backscatter pattern, narrowly distributed in scattering cross-section (target strength)

Figure 6 summarizes this theory, showing the forward scatter solid angle Ω, and the narrow distribution that results in scattering cross-section σ. While a rectangular facet shape is used in the theory presented, the narrow distribution is robust in facet shape, shape affecting only the high frequency skirt of the result labeled "outside Ω".

The transition from low to high frequency behavior is abrupt. Crudely, when the mean facet size L is about one wavelength λ, and its local normal is at an angle α no more than about $\pi/3$ from the incident direction, strong backscatter can occur. For smaller and larger facet size, backscatter decreases dramatically.

If we assume that for our data $\alpha = \pi/3$, we predict a scattering cross section σ that maximizes or self-selects at

$$\sigma \approx 5\lambda^2 \tag{1}$$

and we get a target strength T of

$$T = 10 \log \frac{\sigma}{2\pi} = 16 \text{ dB re 1 m} \tag{2}$$

a value equal to the mean of data we have studied thus far at 230 Hz.

Equation (1) can be put in Lambert's Law form, as indeed Figure 6 suggests to be a result of the model:

$$\sigma_0 = \sigma \frac{\sin^2(\gamma + \theta)}{\sin^2(\gamma + 5°)} \approx 20 \lambda^2 \sin^2(25° + \theta) \tag{3}$$

In this equation θ is the incident grazing angle with respect to the mean depth of the ocean (5° is estimated for our data from the profile), and γ is the facet slope angle.

The second part of the equation is specific for $\gamma = 25°$, and is used solely to emphasize the Lambert's Law nature of the result only for $\theta > \gamma$. Statistics on γ would be needed to average Equation 3 over a large footprint, but it is already clear that as $\theta \to 0$ the facet backscatter model is constant and larger than the pure Lambert's Law prediction.

We return to our data at $\theta \approx 5°$, and note that the integrated or total backscatter from many facets (the usual mode of low resolution measurements), would lead to a scattering strength of

$$SS = 10 \log \frac{n\sigma}{2\pi}, \tag{4}$$

$$SS = 10 \log (8 \times 10^{-6} \lambda^2), \text{ dB} \tag{5}$$

in which λ is in m and where the estimated number density $n = 10^{-5} m^{-2}$, and from which we get

$$SS = -34 \text{ dB} \tag{6}$$

This value at 230 Hz agrees reasonably with data for rough bottoms at other frequencies (Stanic et. al., [2], for coarse shell bottoms at frequencies from 20 to 180 kHz, Dyer et. al., [3], for a rough plate hinge in the Arctic at 9 Hz, and Jiang, [4], for a fracture zone near the Mid Arctic Ridge at 8 Hz.), each at θ approximately equal to five degrees. Can rough bottom backscatter be encompassed over such a wide frequency range by a result as simple as Equations (4) and (5)? It could only happen if the statistics for γ are reasonably alike for these disparate cases, and if $n \sim \lambda^{-2}$, a not unreasonable guess since the larger the facet, the smaller its probability of being there. The supposition begs clarification that we are presently pursuing.

4. References

1) Jerald W. Caruthers and Ira Dyer, 7 April 1992, Acoustic Reconnaissance Cruise, Monitoring Support Software Supplement, prepared for ONR.
2) S. Stanic, K. B. Briggs, P. Fleischer, W. B. Sawyer, and R. I. Ray, (1989), "High frequency acoustic backscattering from a coarse shell ocean bottom", J. Acoust. Soc. Am., 85, 125-136.
3) I. Dyer, A. B. Baggeroer, J. D. Zittel, and R. J. Williams, (1982), "Acoustic backscattering from the basin and margins of the Arctic Ocean," J. Geophys. Res., 87, 9477-9488.
4) Sheng-li Jiang, May 1988,"Backscattering of sound in the eastern Arctic Ocean," Ph.D. thesis submitted to M.I.T. (I. Dyer, supervisor).

Work supported by ONR, Dr. Marshall Orr, grant supervisor.

SEA-BOTTOM REVERBERATION: THE ROLE OF VOLUME INHOMOGENEITIES OF THE SEDIMENT.

M. GENSANE
Thomson-Sintra ASM
1 av. Aristide Briand
94117 Arcueil. France

ABSTRACT. The bottom backscattering strength due to the upper sediment volume is expressed by a rather simple expression, under assumptions valid in many practical cases. An analysis of some published sediment core measurements, allows us to obtain experimental orders of magnitude for the velocity and density fluctuations, and their spatial correlation function. The bottom strengths calculated with these experimental values confirm the importance of the role of sediment volume scattering in the general phenomenon.

1. INTRODUCTION

Bottom scattering is due to the two following types of inhomogeneities: 1) sea-sediment interface roughness, which affects the reflected energy; 2) sediment volume inhomogeneities (velocity and density fluctuations), which affect the refracted energy.
Most of the works dealing with the physical interpretation of bottom scattering concern the first phenomenon. But the second has drawn attention for a dozen of years ([2] to [5]). Particularly, if one considers the backscattering strength measurements as a function of the grazing angle, it has been shown ([3], [4]) that the shape of the obtained curve cannot be theoretically explained, except by introducing the sediment volume fluctuations.

Some questions remain. One is to verify if the volume fluctuation values necessary for the model to fit the experimental curves, correspond to reality. Another one comes from the great number of parameters: the same scattering strength may be obtained by many ways; so, how to choose among them? The main topic of our study is to attempt to answer these questions.

2. BACKSCATTERING THEORY

The volume inhomogeneities are supposed to be random fluctuations of the velocity and density (Δc and $\Delta \rho$) around their mean values c_0 and ρ_0 in the sediment. They are characterized by a spatial correlation function $C(R)$, and its three-dimensional Fourier transform $G(K)$. This correlation function leads to the notion of correlation lengths (vertical and horizontal). Following Chernov's theory ([1], [2]) we may express the bottom backscattering coefficient m_s, due to the volume fluctuations. We suppose a constant attenuation α in the sediment. With a harmonic incident wave:

$$m_s = m_v \sin\theta / 4\alpha, \quad \text{with} \quad m_v = 2\pi k^4 G(2k) \qquad (1)$$

Here 10 log m_s is the classical bottom backscattering strength, while 10 log m_v is the volume backscattering strength. k is the wave number of the incident signal. θ is the grazing angle.

Expression 1 is valid with the following hypotheses and approximations: 1) Simple scattering. 2) Plane compressional waves. 3) $\Delta c \ll c_0$ and $\Delta\rho \ll \rho_0$. 4) The dimensions of the scattering volume are much smaller than the distance to the emitter, and much greater than the correlation lengths of the volume fluctuations. 5) The correlation lengths are smaller than the "attenuation length" (defined, for example, as the propagation length where the signal is reduced by 3 dB).

Within the limits of this theory, it is possible to take into account the incident pulse duration, the refraction, and the large-scale interface roughness. The evanescent wave may also be considered [5]. Here, application to moderate grazing angles, refraction ratios very close to 1, and not too short pulses, let formula 1 be quite good.

We think it useful to remark that, in the derivation of formula 1 from the Chernov's differential equation, we have not seen any restrictive condition about the ratio of the wavelength to the fluctuations correlation length, except hypothesis n°5: the attenuation order of magnitude being 1 dB/wavelength, it means that the formula 1 holds for correlation lengths up to, say, 3 times the wavelength.

We shall suppose an exponential volumic correlation (in polar coordinates r, z):

$$C(R) = \mu^2 \exp\left(-\frac{|z|}{a} - \frac{r}{b}\right) \qquad (2)$$

a, b are the vertical and horizontal correlation lengths. The Fourier transform is:

$$G(K) = \mu^2 F_1(K_z) \cdot F_2(K_r)$$
$$F_1 = a / \pi(1 + K_z^2 a^2) \quad ; \quad F_2 = b^2 / 2\pi(1 + K_r^2 b^2)^{3/2} \qquad (3)$$

In the case of backscattering, $K_z = 2k\sin\theta$, $K_r = 2k\cos\theta$.
μ is the volume fluctuation amplitude. Designating σ_c and σ_ρ the standard deviations, and supposing that Δc and $\Delta\rho$ are perfectly correlated:

$$\mu = (\sigma_c/c_0) + (\sigma_\rho/\rho_0) \qquad (4)$$

It is interesting to note that the function $m_s(k)$ has a maximum. The equation $d(m_s)/dk = 0$ gives the solution

$$k^2 = \frac{1 + [1 + 24(b/a)^2\cot^2\theta]^{1/2}}{(4b\cos\theta)^2} \tag{5}$$

3. SEDIMENT VOLUME INHOMOGENEITIES

Most of the literature dealing with measurements of geoacoustic properties give only mean values (often functions of depth). Some of them ([6] to [11]) allow us to get numerical values of the vertical fluctuations. Even one [7] gives an idea of the horizontal fluctuation. We have digitized them, corrected them to make $c_0(z)$ and $\rho_0(z)$ constant, and estimated the characteristic parameters.

Concerning vertical correlation, table 1 summarizes: 1) the experimental conditions: mean properties of the upper sediment (when known); characteristics of the cores (number, height, vertical sample interval); 2) the estimated values: σ_c, σ_ρ, μ (from expression 4), vertical correlation length a, after fitting by an exponential correlation (formula 2), and cross-correlation between Δc and $\Delta \rho$, when it is possible. Figure 1 shows how the spectrum (formula 3, function F_1) fits to the spectrum of some of the measurements.

First, we observe that the refraction index is very close to 1, often even lower: this proves the correctness of some of the above approximations.

The correlation between Δc and $\Delta \rho$ is far from being complete. However, we cannot expect a core sampling to keep exactly the in-situ sediment

TABLE 1: VOLUME FLUCTUATIONS

ref. numb.	sediment mean ρ	n	cores numb	H (m)	ΔH (cm)	σc (m/s)	σρ (kg/m3)	μ	a (cm)	Δc/Δρ corr.	σρ/σc
[6]	1.5	.98	1	4	10	8.5	53	.04	<10	.25	6.2
[7]	clay, sand		18	2	2	14	-	.04	3 to 10	-	-
[8]	1.4-1.8	< 1	17	12	2&5	16	60	.05	5 to 40	.56	3.7
[9]	1.4-1.7	< 1	10	10	15	50	120	.11	<15, to 30	.80	2.4
[10]	clay		3	≥1.5	10	-	250	.23	15	-	-
[11]	-	1-1.12	14	<1.5	5	30	-	.08	7 to 15	-	-

ρ: density. n: refraction index c(sed.)/c(water). H: core height; ΔH: sample interval.
a: correlation length (exponential fitting). Δc/Δρ corr.: cross-correlation coefficient.
When either c or ρ is not measured, μ is computed assuming $\sigma_\rho/\sigma_c = 3$.

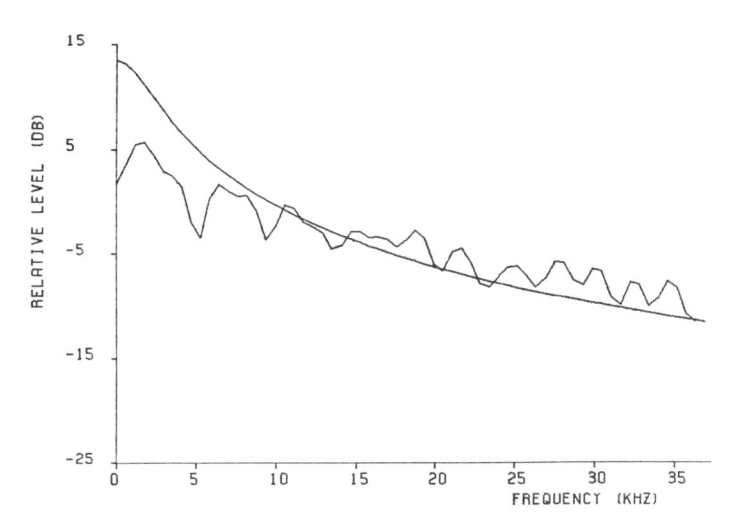

FIGURE 1. Example of vertical correlation spectrum. Mean of measurements from ref. [7]. Smooth curve: exponential correlation fitting, with a = 5.7 cm.

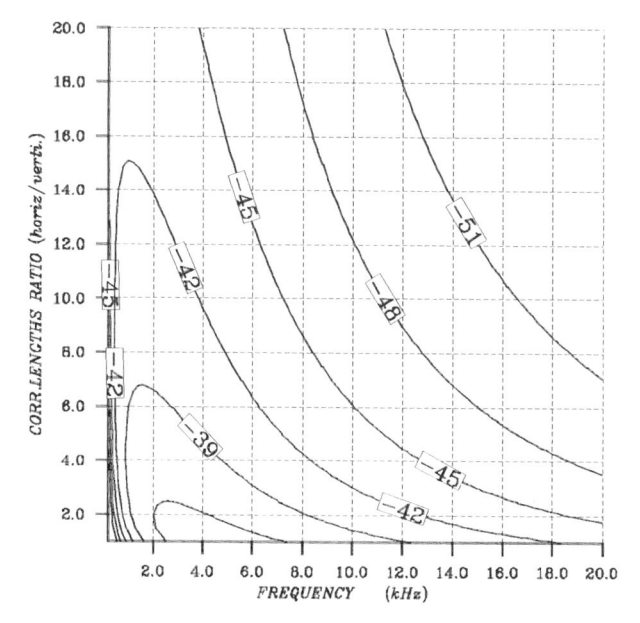

FIGURE 2. Bottom backscattering strength m_s (contour lines, in dB). Example for: $\theta = 30°$, α: 1 dB/wavelength, a = 5 cm, μ = 5%.

properties; so, from the measured correlation coefficients we may roughly suppose that this correlation exists.
The σ_p/σ_c ratio is about 3, which is close to the value of 2.53 found by Hines [5] and deduced from Hamilton's works.
The exponential shape fitting seems to be rather good. The correlation lengths are rather dispersed, from several centimeters to several decimeters. In fact, they may be partly related to the core height.

Horizontal correlation can be deduced from Anderson & Hampton [7], who have analyzed five series of cores. In each series, the cores were extracted at horizontal distances Δr = "30 to 60 cm". We have computed the cross-correlation coefficients of the vertical profiles taken two by two. We found that, for the consecutive pairs (Δr separation), half of them have a coefficient greater than 50%. In the case of the 5 pairs seeming to be separated by $2\Delta r$, only one has a coefficient above 50%. So in this example, one may conclude that the horizontal correlation length is about Δr, that is roughly 5 times the vertical one.

4. BOTTOM BACKSCATTERING STRENGTH: results and discussion

Volume backscattering is likely to be predominant at moderate grazing angles [4]. So let us consider the measured strengths at 30° grazing for example. Above 10 kHz, up to several hundreds of kHz, the bottom strengths are from -35 to -25 dB [5]. Below, they decrease with frequency; at 1 kHz, they range between -45 and -35 dB [12].

Let us apply formula 1 (together with 3, 4) to the experimental values of table 1. So we fix μ and a, and we may vary the frequency f and the anisotropy ratio r = b/a.
For a = 5 cm, μ = 0.05 (typical from [6], [7], [8]), we obtain the plot figure 2. It appears that the computation leads to backscattering strengths of the same order of magnitude as these measured near 1 kHz; but towards higher frequencies the computed strengths become too low. Moreover, for a given value of r, m_s diminishes when f increases,which does not correspond to the measurements.
For a = 15 cm, μ = 0.15 (typical from [9], [10], and still [8] partly), the plot would be similar, with greater strengths (7 to 10 dB), that leads the strengths computed at f $\geq$ 10 kHz to reach the -35 to -25 dB interval. For this, however, the anisotropy ratio must approach 1. Also, if we observe the profiles $\Delta c(z)$ or $\Delta \rho(z)$ in [7], [8], [9], we see that the fluctuation amplitude is lower in the upper sediment layer (which is the only influential layer above 10 kHz because of the attenuation). Stating that the volume scattering may really be dominant over the interface scattering, we deduce that we must introduce correlation lengths much smaller than those found in table 1, with fluctuation amplitudes around 10%. Notice that this is consistant with the values considered by Hines [5] in his model: he put a $\leq$ 1 cm, and μ $\simeq$ 8%. Similarly, at frequencies much lower than 1 kHz, from figure 2 it appears necessary to use greater correlation lengths (up to 1 m).

This leads us, after [2], to turn towards the definition of a multi--scale volume correlation (a becoming a random variable) with μ and r

varying with a. r may be stated as decreasing with a, and tending to 1. Towards the low frequencies we may use the measured correlation spectra to fit this multi-scale function. Towards the higher frequencies, to have an experimental reference, there does not appear to be a lot of measurements ! So, we could use the fact that, probably, the correlation length acting at a given frequency would be the one which makes m_s maximum (that is, a kind of resonance). For example, applying formula 5, at $\theta = 30°$, $r = 5$:
  for $f = 1$ kHz , $a = 9$ cm
  for $f = 10$ kHz , $a = 1$ cm
With $r = 1$, and comparing with the values chosen by Hines (and qualified by him "somewhat arbitrary"):
  for $f = 30$ kHz , $a = 0.7$ cm (Hines: $a = 1.$ cm)
  for $f = 500$ kHz , $a = 0.04$ cm (Hines: $a = 0.05$ cm)
Obviously, Hines' values may be considered as not arbitrary.

5. CONCLUSION

Measured surperficial sediment inhomogeneities show fluctuations of 4 to more than 10 %, with vertical correlation lengths from 3 to 50 cm (the horizontal one being appreciably greater). The spatial correlation is correctly fitted, in the first approximation, by an exponential function. With these experimental values, we have confirmed that, at moderate grazing angles and frequencies around 1 kHz (say, 0.5 to 5 kHz), the sediment volume is most likely the predominant cause of bottom scattering. For other frequencies, the use of a multi-scale correlation model should be necessary, and we gave some of the starting points to establish it on the basis of realistic values.

ACKNOWLEDGEMENTS: This work was supported by the Direction des Recherches et Etudes Techniques under contract 87/015.18.

REFERENCES:

  [1] Chernov, L.A. (1960), Wave propagation in a random medium, McGraw - Hill, New York.
  [2] Ivakin, A.N. and Lysanov, Yu.P. (1981), Sov.Phys.Ac. 27(1), 61-64.
  [3] Crowther, P.A. (1983), in N.G. Pace (ed.), Acoustics and the sea--bed, Bath University press, 147-156.
  [4] Jackson, D.R., Winebrenner, D.P. and Ishimaru, A. (1986), J.Acoust.Soc.Am. 79(5), 1410-1422.
  [5] Hines, P.C. (1990), J.Acoust.Soc.Am. 88(1), 324-334.
  [6] Winokur, R.S. and Bohn, J.C. (1968), J.Acoust.Soc.Am. 80(4), 1130-1138.
  [7] Anderson, A.L. and Hampton, L.D. (1974), in L.D. Hampton (ed.), Physics of sound in marine sediments, Plenum press, 357-372.
  [8] Akal, T. (1974), in same as [7], 447-480.
  [9] Tucholke, B.E. (1980), J.Acoust.Soc.Am. 68(5), 1376-1390.
  [10] Winn, K. and Becker, G. (1983), in same as [3], 107-114.
  [11] Thiele, R. (1983), in same as [3], 207-214.
  [12] Merklinger, H.M. (1968), J.Acoust.Soc.Am. 44(2), 508-513.

THE PERTURBATION CHARACTERIZATION OF OCEAN REVERBERATIONS

E. Y. T. KUO

Naval Undersea Warfare Center Detachment
New London, CT 06320
U.S.A.

ABSTRACT The usefulness of two classical perturbation methods used in past characterization of ocean reverberations will be demonstrated. Then, an extended approach using the same perturbation method will be used to characterize bubbly ocean surface reverberation.

1. INTRODUCTION

Section 2 briefly discusses the two classical surface perturbation scattering methods (Marsh [1]/Kuo [2] and Bass [3]) and compares predictions of ocean surface reverberation with experimental data. This is followed by the presentation of some representative results obtained by applying the Marsh/Kuo perturbation method to model acoustic scattering from (1) thick and thin sedimentary layered rough ocean bottoms and (2) Arctic under-ice cover.

Section 3 discusses a new application of the same perturbation method in characterizing bubbly ocean surface reverberation. Section 4 follows with the conclusions.

2. PREVIOUS CHARACTERIZATION OF OCEAN REVERBERATIONS

2.1 Perturbation Methods

Given an incident plane wave at a rough surface, the reflected wave field (p) consists of specular (coherent) and off-specular (incoherent) component wave fields. The specular component predicts the reflection coefficient, while the off-specular component predicts scattering strength and signal spreads. Two classical perturbation methods by which p can be obtained are briefly compared.

Marsh [1]/Kuo [2] assumed a plane-wave expansion integral-form solution. After substitution into a set of rough surface boundary conditions, the boundary condition and differential amplitude of plane waves in the integral form solution were perturbed to the second order in terms of appropriate statistical boundary properties such as random height and slope. Sequentially, in ascending perturbation order, these unknown differential amplitudes of different order were solved and substituted back into the assumed solution to obtain p. Through a statistical analysis, this solution gave a reflected-field spectrum that consisted of the specular and off-specular components.

Bass [3] assumed $p = p_0 + p_s$, where p_0 and p_s represented mean and stochastic (perturbation) fields. Following substitution of this solution form in rough surface boundary conditions, the results were perturbed in terms of p_s and roughness height. Brekhovskikh and Lysanov [4] performed this perturbation to the first order to obtain p_s for the scattering strength, and utilized the conservation of energy flux to obtain the reflection coefficient. Both Wenzel [5] and Kuperman [6] performed these perturbations to the second order and obtained p_0 for the reflection coefficient and p_s for the scattering strength. However, they utilized different methods in decoupling p_0 and p_s in their boundary equations.

2.2 Ocean Surface Reverberation

The previous results of applying the above methods to predict ocean surface reverberation were shown by Kuo [7] to be consistent, not only between the two methods (if different additional approximations beyond perturbation were taken into account), but also with the existing data at the time.

2.3 Ocean Bottom Reverberation

A dominant parameter that influences the modeling method is the sediment thickness. A thick sedimentary bottom can be adequately modeled by a shear-free layer, while a thin sedimentary bottom requires inclusion of shear effects. Brief descriptions with representative results for both ocean bottom types follow.

For a thick sedimentary bottom, the reflected field was obtained by Kuo [2] under the shear-free assumption. The off-specular component was used to predict backscattering strength, a bottom roughness spectrum, acoustic properties of the bottom, and arrival time spread (Kuo *et al.*[8]). The bottom loss predicted by the specular component was successfully compared (Kuo [9]) with the loss measured by Tattersall [10].

For a thin sedimentary bottom, the reflected field was obtained by Kuo [11] under the assumption of three layers that included shear effects in the sedimentary layer. Dominant bottom-loss mechanisms were found to occur at the basalt-sediment interface, which has a large acoustic impedance discontinuity. Besides the well-known scattering loss mechanism, another loss mechanism was caused by absorption of shear waves converted at the interface. The roughness of the interface was found to alter the conversion process because of variations in the local incident angles.

2.4 The Arctic Under-Ice Reverberation

In contrast to the previous assumption of a pressure-releasing under-ice boundary that ignored the physical existence of ice cover, a thin ice plate of a given density and complex longitudinal plate wave velocity was assumed for a low-frequency acoustic wave. It was a logical and practical first step before application of the complex layer model used in the three-layer ocean bottom modeling. After perturbation of appropriate boundary conditions, which included a plate equation, Kuo [12] obtained specular and off-specular components of the reflected field.

The specular component resulted in reflection coefficients for various grazing angles that were strongly affected by the boundary roughness, ice absorption, and ice thickness. It was found that roughness scattering loss increased with the grazing angle; however, at small grazing angles the effects of ice absorption and thickness prevailed. The latter phenomenon made it possible to compare these predictions with experimental data obtained at small grazing angles.

The off-specular component resulted from a point source, and point receiver geometry gave signal spreading in arrival time, arriving angle, and Doppler frequency shift. Thicker ice tended to spread scattered energy in a wider angular region and the Doppler frequency-shift spectrum spread with arrival time. The maximum scattered signal level arrived from an off-specular direction, as found from other types of rough boundaries.

3. THE CHARACTERIZATION OF BUBBLY OCEAN SURFACE REVERBERATION

Mechanisms for the ocean reverberations addressed to this point are limited to boundary reverberations. Recent observations [13], [14] of high-sea-state ocean surface reverberations indicate that micro-bubble effects on acoustic waves should be included. In principle, micro-bubble effects can be modeled in two ways according to the assumption made. When a distinct submerged rough interface exists between bubbly and bubble-free waters, the effect can be modeled similarly to Kuo's layered bottom [11]. When mixing does not take place, the ocean surface layer is populated by bubble plumes that appear randomly in space and time. Initial tasks for modeling the latter situation have been performed to the extent of obtaining the random reflected field solution in terms of measurable physical parameters.

For low frequency acoustic waves, the volumetrically scattered field can be represented by Born's approximation in many situations; see, for example, Batchelor [15]. Recent investigators McDonald [13] and Henyey [14], by using the same approximation, reported their findings on the volumetrically scattered field from scattering volumes attached to a flat ocean surface. Because a dominant part of the micro-bubble population originates from surface wave actions, surface roughness and the corresponding surface roughness reverberation effects cannot be ignored in the analysis. The proposed model here includes both surface roughness and volumetric scattering effects.

Consider an incident plane wave entering the bubbly surface layer from below. The objective is to estimate the total field of reflected waves coming down from the bubbly layer. The incident wave will be volumetrically scattered. Then the up-going volumetrically scattered waves will be surface scattered, and volumetrically scattered again. This process of alternating single volumetric and surface scattering continues. As described by Kuo [2], the single surface scattering can be symbolically characterized by a multiplication factor of $(1 + dA + dAdA^*)$, where dA is a Fourier Stieltjes component of the scattering surface roughness. The single volumetric scattering can be shown to be symbolically characterized by a multiplication factor $d\Gamma$, which is a Fourier Stieltjes component of $(\delta c/c_0)$, the fractional acoustic velocity fluctuation. Because both dA and $d\Gamma$ are considered small perturbation parameters, the scattered field generated by the alternating scattering process will eventually become negligible — beyond second order in dA and $d\Gamma$.

According to the volumetric perturbation described by Batchelor [15], the incident wave will penetrate the bubbly layer approximately unchanged despite the ongoing volumetric scattering described above. On arrival at the ocean surface, the incident wave will be surface-scattered, then volumetrically scattered. This process continues as described previously, but the scattered fields become negligible when the second order in dA and $d\Gamma$ is reached.

The total field results from a combination of all the down-going waves generated by every surface and volumetric scattering process. After re-arrangement according to perturbation orders, the reflected field velocity potential (ϕ_r) can be symbolically expressed as

$$\phi_r \sim 1 + d\Gamma + dA + d\Gamma dA^* + d\Gamma d\Gamma^* + dAdA^*,$$

in which the $d\Gamma$ term can be identified as the basis for the analysis of McDonald [14] and Henyey [15]. Because the properties of the medium and the surface roughness vary randomly as a result of the turbulent wind action, $d\Gamma$, dA, and ϕ_r take different values from one realization to another; therefore, only the statistical properties of $d\Gamma$, dA, and ϕ_r

are relevant. Statistical analysis is proposed to predict reflection loss, scattering strength, and signal spreading.

Because direct measurement of either $(\delta c/c_0)$ or $d\Gamma$ is difficult, it is desirable to express them in terms of measurable parameters. For low frequency acoustic waves, the number of bubbles per unit volume, N, was effectively recommended by Carey et al. [16] as the dominant parameter. A relationship

$$(\delta c/c_0) \sim N$$

can be found from Clay/Medwin [17]. N by itself is not an easily measurable parameter, but it can be related to other measurable parameters.

Because the occurrence frequency of wave-breaking and bubble clouds is the same, according to Thorpe/Humphries [18], it will be possible to relate N to its surface value N_0. Thorpe [19] and Farmer/Vagle [20] proposed a semi-empirical relationship

$$N = N_0 e^{-\beta z},$$

where β is a depth-wise decay constant and z is the depth. N_0, in turn, can be related to W, which is that fraction of the sea surface covered by whitecaps. Wu [21] effectively proposed a relationship:

$$N_0 \sim W.$$

Combining the above expressions produces one desired relationship,

$$(\delta c/c_0) \sim e^{-\beta z} W,$$

in which W is measurable by means of photography.

Thorpe [19], [22] also found N to be proportional to M_v, the scattering cross-section/volume. From the above discussions it is plausible to express M_v by $e^{-\beta z} M_{v,2D}$, which explicitly shows the horizontal statistical nature; namely,

$$N \sim M_v \sim e^{-\beta z} M_{v,2D}.$$

Substitution results in the second desired relationship,

$$(\delta c/c_0) \sim e^{-\beta z} M_{v,2D},$$

in which $M_{v,2D}$ can be obtained from measurable M_v by using high frequency underwater acoustics.

As flow separation from the roughness element was found by Wu [23] and Banner/Melville [24] to trigger wave breaking, another measurable surface parameter relating to N_0 or N may be the wave-breaking roughness η_k. Roughness elements are short waves riding on long waves and are believed to play a major role in energy transfer from winds to surface waves. By combining information contained in references [25] through [27], N_0 is found to be proportional to $\eta_k^{1.5 \sim 1.2}$, which may be approximated by η_k for simplicity; namely,

$$N_0 \sim \eta_k^{1.5 \sim 1.2} \sim \eta_k.$$

Substitution results in the third desired relationship,

$$(\delta c/c_0) \sim e^{-\beta z} \eta_k.$$

In summary, $(\delta c/c_0)$ can be expressed in terms of a measurable random parameter, X, as in

$$(\delta c/c_0) \sim e^{-\beta z} X(\underline{x}_{2D}),$$

where

$$X = W, \eta_k, M_{v,2D}$$

and

$$\underline{x}_{2D} = (x,y) = \text{horizontal coordinates}.$$

This, in turn, gives

$$d\Gamma \sim d\Gamma_x = \left(\tfrac{1}{2\pi}\right)^2 \int e^{i\underline{k}_{2D} \cdot \underline{x}_{2D}} X(\underline{x}_{2D}) d\underline{x}_{2D} d\underline{k}_{2D}$$

and

$$\phi_r \sim 1 + d\Gamma_x + dA + d\Gamma_x dA^* + d\Gamma_x d\Gamma_x^* + dA dA^*,$$

which forms the basis of the proposed tasks. The proposed tasks are to statistically characterize the bubbly ocean surface reverberation (reflection loss, scattering strength, and signal spreadings) in terms of spectra describing the scattering roughness, whitecap coverage, wave breaking roughness, scattering cross-section, and their correlations.

4. CONCLUSIONS

Despite its inherent approximation, the perturbation method has been shown to successfully predict measured rough boundary reverberations characterized by reflection loss, scattering strength, and signal spreading in arrival time, angle, and frequency shift. The extended use of the perturbation method is shown to enable the characterization of the bubbly ocean surface reverberation in terms of measurable parameters and their correlations.

5. REFERENCES

[1] Marsh, H.W. (1961) 'Exact solution of wave scattering by irregular surfaces,' *J. Acoust. Soc. Amer.*, 33, 330-333.

[2] Kuo, E.Y.T. (1964) 'Wave scattering and transmission at irregular surfaces,' *J. Acoust. Soc. Amer.*, 36, 2135-2142.

[3] Bass, F.G. (1960) 'Boundary conditions for the average electromagnetic field on a surface with random irregularities and with impedance fluctuations,' Izv. Vuzov, Radio Fizika, 3, 72-78.

[4] Brekhovskikh, L. and Lysanov, Y. (1982) *Fundamentals of Ocean Acoustics*, Springer-Verlag, Berlin.

[5] Wenzel, A.R. (1974) 'Smoothed boundary conditions for randomly rough surfaces,' *J. Math. Phys.*, 15, 317-323.
[6] Kuperman, W.A. (1975) 'Coherent component of specular reflection and transmission at a randomly rough two-fluid interface,' *J. Acoust. Soc. Amer.*, 58, 365-370.
[7] Kuo, E.Y.T. (1988) 'Sea surface scattering and propagation loss: review, update, and new predictions,' *IEEE J. Oceanic Eng.*, 13, 229-234.
[8] Kuo, E.Y.T. and Jensen, E.P. (1986) 'A relationship between off specular scattering and time spread phenomena,' NUWC Technical Memorandum 861175, Naval Undersea Warfare Center Detachment, New London, CT.
[9] Kuo, E.Y.T. (1986) 'Further insights on scattering from single transmitting rough boundary,' NUWC Technical Memorandum 871118, Naval Undersea Warfare Center Detachment, New London, CT.
[10] Tattersall, J.M. (1986) personal communication, Naval Undersea Warfare Center Detachment, New London, CT.
[11] Kuo, E.Y.T. (1992) 'Acoustic wave scattering from two solid boundaries at the ocean bottom: reflection loss,' *IEEE J. Oceanic Eng.*, 17, 159-170.
[12] Kuo, E.Y.T. (1990) 'Low frequency acoustic wave-scattering phenomena under ice cover,' *IEEE J. Oceanic Eng.*, 15, 361-372.
[13] McDonald, B.E. (1991) 'Echoes from vertically stratified subresonant bubble clouds: a model for surface reverberation,' *J. Acoust. Soc. Am.*, 89, 617-619.
[14] Henyey, F.S. (1991) 'Acoustic scattering from ocean microbubble plumes in the 100 Hz to 2 kHz region,' *J. Acoust. Soc. Am.*, 90, 399-405.
[15] Batchelor, G.K. (1956) 'Wave scattering due to turbulence,' Symposium on Naval Hydrodynamics, Washington, D.C.
[16] Carey, W.M., Fitzgerald, J.W., and Browning, D.G. (1990) 'Low frequency noise from breaking waves,' NUWC TD 8783, Naval Undersea Warfare Center Detachment, New London, CT.
[17] Clay, C.S. and Medwin, H. (1977) *Acoustical Oceanography: Principles and Applications*, John Wiley & Sons, New York.
[18] Thorpe, S.A. and Humphries, P.N. (1980) 'Bubbles and breaking waves,' *Nature*, 283, 463-465.
[19] Thorpe, S.A. (1982) 'On the clouds of bubbles formed by breaking wind-waves in deep water, and their role in air-sea gas transfer,' *Phil. Trans. R. Soc. Lond.*, A304, 155-210.
[20] Farmer, D.M. and Vagle, S. (1989) 'Waveguide propagation of ambient sound in the ocean-surface bubble layer,' *J. Acoust. Soc. Amer.*, 86, 1897-1908.
[21] Wu, J. (1992) 'Individual characteristics of whitecaps and volumetric description of bubbles,' *IEEE J. Oceanic Engineering*, 17, 150-158.
[22] Thorpe, S.A. (1984) 'On the determination of Kv in the near surface ocean from acoustic measurements of bubbles,' *J. Physical Oceanography*, 14, 855-863.
[23] Wu, J. (1969) 'A criterion for determining airflow separation from wind waves,' *Tellus*, 21, 707-714.
[24] Banner, M.L. and Melville, W.K. (1976) 'On the separation of airflow over water waves,' *J. Fluid Mech.*, 77, 825-842.
[25] Wu, J. (1980) 'Wind-stress coefficients over sea surface near neutral conditions — a revisit,' *J. Physical Oceanography*, 10, 727-740.
[26] Charnock, H. (1955) 'Wind stress on a water surface,' *Quart. J. Roy. Meteor. Soc.*, 81, 639-640.
[27] Schlichting, H. (1968) *Boundary-Layer Theory*, McGraw-Hill, New York.

THE REDUCTION OF SURFACE BACKSCATTER AND AMBIENT NOISE BY MONOMOLECULAR FILMS ON THE OCEAN SURFACE

Jim Rohr, Robert Kolesar and Frank Ryan
Naval Command, Control and Ocean Surveillance Center
RDT&E Division, San Diego, California 92152-5000

ABSTRACT. The opposing wavelength dependence of acoustic and radar backscatter from slick-forming films is discussed in light of ambient noise measurements collected beneath similar films. Generally, a significant reduction ($\approx$6 dB) in ambient noise ($\geq$1 kHz) is observed beneath the film--apparently as a result of a decrease in frequency of air entraining events occurring within the film covered area. It is hypothesized that the film's indirect effect on bubble generation, rather than its direct attenuation of ocean-surface fine structure, is responsible for the *increasing* acoustic backscatter reductions observed with *increasing* ensonifying wavelength.

1. INTRODUCTION

Seafarer's accounts of the calming effect of surface-active materials, such as olive oil (thus the adage: "pouring oil on troubled waters"), extend throughout recorded history [1]. Modern day laboratory and at-sea measurements have confirmed the damping of capillary and short gravity waves within a slicked area [2,3]. There is also evidence that possibly through nonlinear wave-wave and wind-wave interactions, larger waves may be affected by the film as well [2]. Scott [1] has argued that the films, through damping the small-scale surface roughness which ride on the larger waves, may reduce the likelihood that the larger waves will break.

Through their influence on the surface wave field, surface slicks are well known to affect the scattering of visible and microwave electromagnetic waves. Huhnerfuss et al. [3] have compared X and L band radar backscatter from a slick-forming film of oleyl alcohol (an unsaturated fatty alcohol which resembles natural slicks in its physico-chemical behavior). A greater reduction in backscatter was noted [3] for X band radar indicating that shorter wavelengths are more intensely damped by the film.

Complementary acoustic backscatter data at comparable wavelengths from an underwater source have also been obtained [4]. It was inexplicably found, contrary to the radar results, that as the ensonifying wavelength diminished so did the reduction of acoustic backscatter from the slick (also composed of oleyl alcohol). Subsequent to the acoustic backscatter experiment there have been a series of ambient noise measurements taken beneath identical and similar slick-forming materials, which suggest that bubble production within the slick can be dramatically reduced. The aim of this paper is to present, in light of the ambient noise measurements, a possible explanation for the opposing wavelength dependence between radar and acoustic backscatter, at intermediate grazing angles, from monomolecular films.

2. SURFACE BACKSCATTER REDUCTIONS FROM A MONOMOLECULAR FILM

A mechanically steerable unit, instrumented with transducer arrays providing 30, 20 and 10 kHz (λ~5, 7.5 and 15 cm) sources of sound, was lowered from a ship to depths of 33.5 and 67 m. The center of the main beam lobe intersected the ocean surface at a horizontal distance of about 75 m from the vessel. Oleyl alcohol was dispensed from the downwind side of the moored ship to form a large slick to its lee. The transducer arrays, for several minute durations, were alternately aimed down wind into the slick, and then at right angles to the wind so that the main lobe would intersect the surface outside the slick. Prior to the deployment of the slick, down and cross wind comparisons of surface reverberation obtained at 30 kHz were made. The differences were of marginal statistical significance. The sea state was estimated to be about 2 throughout the test.

During the experiment a series of tone bursts of 3 ms duration were transmitted, and the resulting reverberation recorded in an intermediate-band FM format. The source frequencies were periodically cycled after each frequency had pulsed for about 30 seconds. Subsequent to the experiment, the recorded reverberation was enveloped, detected and digitized. The digitized reverberation envelopes were aligned temporally with respect to the times at which their respective tone bursts were transmitted and those within each group were then averaged in power for comparison.

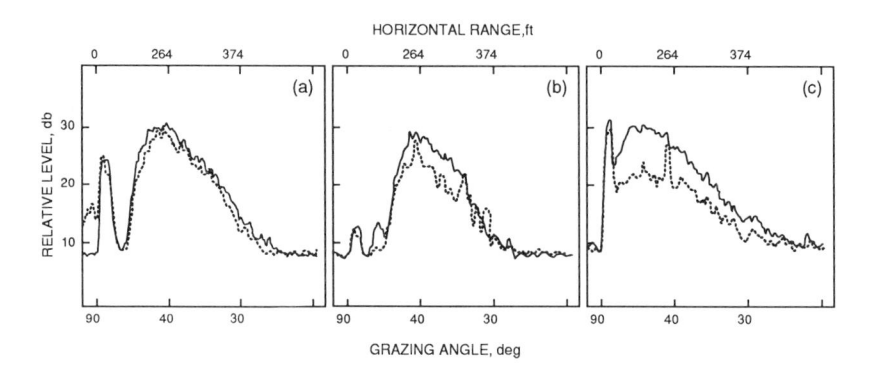

Figure 1. Reverberation envelope with (---) and without (—) surface film, (a) 30 kHz, (b) 20 kHz, (c) 10 kHz.

Figures 1a-c are representative of the intermediate grazing angle data recorded (source at 67 m depth) for 30, 20 and 10 kHz respectively. Each curve on these plots represents a power average of about 115 individual reverberation envelopes. In each case the dotted curve represents reverberation from under the slick, while the solid line represents reverberation outside it. The vertical axis scale on these plots is in decibels with some arbitrary reference, which is fixed for all the data recorded at a given frequency. The lower horizontal scale shows the angle of incidence at the sea surface, and the upper horizontal scale shows the horizontal range to the scattering surface.

Throughout the ensonifying frequencies employed, and at intermediate grazing angles to the ocean surface, there is always less backscatter from the direction of the slick. Maximum backscatter reductions of 2 dB at 30 kHz, 4 to 5 dB at 20 kHz, and about 8 dB at 10 kHz were found. In light of the radar backscatter experiments [2,3] from identical films and for similar grazing angles, we had expected the film's effect to increase with increasing frequency (decreasing wavelength). At shallow grazing angles (≤20°, source at 33.5 m) negligible differences in backscatter between slicked and normal ocean surfaces were observed.

3. AMBIENT NOISE REDUCTIONS BENEATH A MONOMOLECULAR FILM

It was thought that the reduced surface agitation within a film-covered area would have some quieting effect on the ambient noise beneath it. The basic format for the ambient noise experiments was to deploy pairs of sonobuoys, each containing an omnidirectional hydrophone at 9 m depth, about 1 km apart, in the deep ocean far from shipping and shore noise. Pre- and post-slick measurements were simultaneously obtained by a sensor which would have a slick spread above it, and another (reference sensor) located far from its influence.

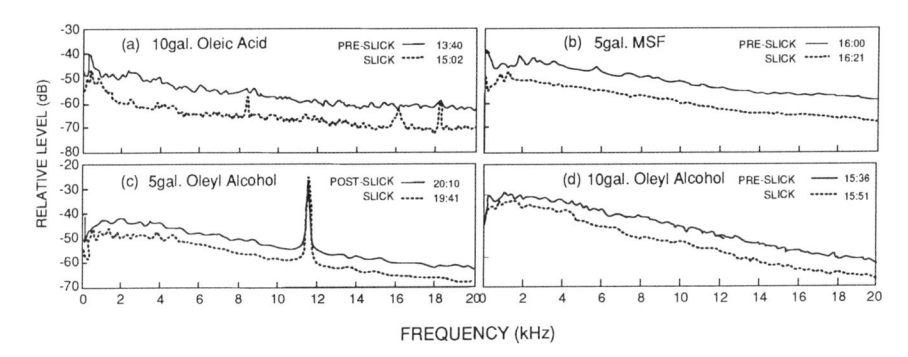

Figure 2. Ambient noise measurements with (---) and without (—) surface film, wind speeds: (a) 3m/s; (b) 4m/s; (c) 6m/s; (d) 7m/s.

Figure 2 is representative [5] of the general character of ambient noise reductions found beneath monomolecular films during sea states 1 through 3 (wind speeds 3 - 7 m/s). The reference ambient noise measurements taken simultaneously, changed little during these same times (not shown, see [5]). Ambient noise reductions beneath the films generally began between 1 and 2 kHz, exhibited little frequency dependence, and achieved levels of around 6 dB. In a series of experiments using the same film but employing hydrophones having a greater frequency range, the reduction (albeit smaller) in ambient noise beneath the films was found [5] to extend to at least to 70 kHz. The spike in Fig. 2c between 11 and 12 kHz is a result of an on-board echo sounder. While the slick could not, as expected, affect the level of this underwater sound source, it dramatically increased its signal-to-noise ratio.

When wind speeds were greater than 5 m/s, scattered whitecaps were generally common outside the slicks and only occasionally observed within them. At wind speeds less than about 2 m/s none of the films tested were found to have any effect on the underlying ambient noise field. This wind speed is near the onset of microbreaking [6]. The acoustic signatures of these small-scale wave breaking events are distinguished by individual oscillating bubbles whose composite spectra are reminiscent of ocean ambient noise [6]. In light of these findings and the near constant ambient noise reductions found beneath monomolecular films, it has been proposed [5] that the number of bubble-entraining wave breaking events can be greatly reduced within the slick.

4. DISCUSSION

Both radar and acoustic reverberation from the sea surface have been often attributed to Bragg diffraction [7]. The Bragg wavelengths, λ_B, responsible for the backscattered radar signals reported by Huhnerfuss et al. [3] are about 12 cm (L band) and 2 cm (X band). Simultaneous wavestaff measurements within the oleyl alcohol slick show respective attenuations of about 30 and 60%, in reasonable agreement with the radar measurements [3]. The presence of a thin ($\approx$2nm) film of oleyl alcohol is not thought to have significant effects on either the dielectric [8] or sound propagation [9] properties of the air-seawater interface.

Discrepancies between predicted and observed acoustic backscattering strengths from the ocean surface (no slicks involved) have previously [7] been attributed to reverberation due to a near-surface bubble layer. We believe the present data supports this view in a rather unique way. It is assumed for the frequencies and sea state considered here, that the effect of bubbles on surface reverberation is principally through resonant scatter from individual bubbles. The bubble diameters which resonate at 30, 20 and 10 kHz are approximately 217, 325 and 650 µm respectively. A conservative estimate of the decrease of air entraining events occurring within a slick, resulting in the ambient noise reductions shown in Fig. 2, is about 50%. This will have a significant impact on the underlying bubble distribution with which the ensonifying wavelengths interact. Conversely, the effect of decreased wave breaking on radar backscatter, at the intermediate grazing angles of interest, is thought to be small [10].

A gross characterization of the impact of the slick on acoustic backscatter can be inferred from the following observations:

(a) the number of bubbles throughout the range of interest (roughly 200 to 700 µm) greatly increases with decreasing diameter [11];

(b) the rate of dissolution of ocean bubbles throughout this diameter range is approximately constant [12];

(c) and finally, the terminal rise velocities of 217, 325 and 650 µm bubbles increases significantly with size (about 2.1, 3.4 and 7.5 cm/sec respectively) [13].

These observations suggest that beneath the slick the relative decrease in the number of larger bubbles is significantly greater than that of smaller bubbles. This disparity will be further exaggerated by any air entraining process occurring within the slick, which results in the increased production of smaller bubbles [5].

A greater reduction in the number of larger diameter bubbles beneath the slick could account for the reduced acoustic backscatter at the lower frequencies observed. Extrapolating these results to higher frequencies suggests that the disparity between radar and acoustic sea surface backscatter should become even greater. This is corroborated by recent measurements [9] taken with a 248 kHz ($\lambda_B \approx \frac{1}{2}$ cm) side-scan sonar, where maximum reductions in mean backscatter from an ocean oil slick were only 1 to 2 dB (occurring between 45° and 65° from the horizon). In contrast, reductions of high frequency ($\lambda_B \approx 2$ cm) radar backscatter by oil films have been reported [8] to be more than an order of magnitude higher. For much lower acoustic frequencies a very different effect is observed. At 3 kHz ($\lambda_B \approx 30$ cm) backscatter from a naturally glossy sea surface ($U_{47} \approx 1$ m/s, grazing angle of 30°) has been reported [14] to be about 6 dB less, than when covered with small ripples ($U_{47} \approx 2.5$ m/s). It should be noted that the anomalous, acoustic-backscatter spikes reported [14] at wind speeds of 2.5 m/s, in the absence of whitecapping, could be the signature of individual microbreaking events.

Ironically at shallow grazing angles, where the subsurface scattering layer of bubbles has been believed to be most important [15], there was no significant effect by the film for any of the ensonifying wavelengths employed. Perhaps the greater number of bubbles encountered at shallow grazing angles, even with a reduced density beneath the film, still provide some saturation limit. Further experiments over a wide range of sea states (without a wind direction bias), and with concurrent bubble, surface wave, and acoustic and radar reverberation measurements are necessary before a sharp delineation between sea surface and bubble scattering processes can be made. (The importance of simultaneous wave height measurements is underlined by laboratory work with oleyl alcohol which suggests [16] that an increase in attenuation of wind waves in the short gravity range of interest is possible.) What *can* be said for certain is that it is impossible for both acoustic and radar reverberation to be governed by Bragg scattering, and yet be affected so differently by surface-active materials.

5. CONCLUSION

Reductions in acoustic backscatter from a monomolecular film have exhibited a wavelength dependence opposite to that found by wavestaff and radar data. Large reductions of ambient noise have also been found beneath slick forming films, and attributed to a dramatic decrease in the frequency of bubble entraining events occurring within the slick. It is argued that there is a greater reduction in the number of larger bubbles beneath the slick; and hypothesized that this is responsible for the greater reductions in acoustic backscatter found at lower frequencies (which correspond to the resonances of the larger diameter bubbles).

6. ACKNOWLEDGEMENTS

Financial support was provided by the Independent Research Program at NCCOSC, the U. S. Office of Naval Technology and the Defense Advanced Research Projects Agency, Space and Naval Warfare Systems Command.

7. REFERENCES

[1] Scott, J. C. (1987) 'Surface films in oceanography', J. Naval Sci. 13, 70-84.
[2] Huhnerfuss, H., Alpers, W., Jones, W., Lange, P. and Richter, K. (1981) 'The damping of ocean surface waves by a monomolecular film measured by wave staffs and microwave radars', J. Geophys. Res. 86 (C1), 429-438.
[3] Huhnerfuss, H., Alpers, W., Cross, A., Garrett, W., Keller, W., Lange, P., Plant, W., Schlude, F. and Schuler, D. (1983) 'The modification of X and L band radar signals by monomolecular films', J. Geophys. Res. 88 (C14), 9817-9822.
[4] Glass, R., Castile, B. and Rohr, J. (1988) 'The reduction of reverberation and ambient noise beneath ocean surface films', NOSC Tech. Doc. 1371.
[5] Rohr, J. and Detsch, R. (1992) 'A low sea-state study of the quieting effect of monomolecular films on the underlying ambient-noise field', accepted to the J. Acoust. Soc. Am.
[6] Updegraff, G. and Anderson, V. (1991) 'Bubble noise and wavelet spills recorded one meter below the ocean surface', J. Acoust. Soc. Am. 89, 2264-2279.
[7] McDaniel, S. and Gorman, A. (1982) 'Acoustic and radar sea surface backscatter', J. Geophys. Res. 87 (C8), 4127-4136.
[8] Johnson, J. and Croswell, W. (1982) 'Characteristics of 13.9 GHz radar scattering from oil films on the sea surface', Radio Science, 17 (3), 611-617.
[9] Belloul, M. and Thorpe, S. (1992) 'Acoustic observations of oil slicks at sea', J. Geophys. Res. 97 (C4), 5215-5220.
[10] Phillips, O. (1980) The Dynamics of the Upper Ocean, Cambridge University Press, Cambridge.
[11] Wu, J. (1988) 'Bubbles in the near-surface ocean: a general description', J. Geophy. Res. 93 (C1), 587-590.
[12] Harris, I. and Detsch, R. 'Small air bubbles in reagent grade water and seawater - Part II: Dissolution of 20 to 500 micron diameter bubbles at atmospheric pressure,' accepted to J. Geophys. Res.
[13] Detsch, R. (1991) 'Small air bubbles in reagent grade water and seawater part 1. Rise velocities of 20- to 1000 micron diameter bubbles', J. Geophys. Res. 96 (C5), 8901-8906.
[14] Nützel, B., Herwig, H., Monti, J. and Koenigs, P. (1987) 'The influence of surface roughness and bubbles on sea surface acoustic backscatter', NUSC Tech. Rep. 7955.
[15] Urick, R. (1967) Principles of Underwater Sound for Engineers, McGraw-Hill, N.Y.
[16] Huhnerfuss, H., Walter, W., Lange, P. and Alpers, W. (1987) 'Attenuation of wind waves by monomolecular sea slicks and the Marangoni effect', J. Geophys. Res. 92 (C4), 3961-3963.

Section 2

High Frequency Measurements and Mechanisms

HIGH FREQUENCY ACOUSTIC BOTTOM BACKSCATTER MECHANISMS AT SHALLOW GRAZING ANGLES

Nicholas P. Chotiros
Applied Research Laboratories, The University of Texas at Austin
Austin, Texas 78713-8029, U. S. A.

ABSTRACT The study of high frequency acoustic bottom backscatter at shallow grazing angles is only now emerging from infancy. With significant increases over the past ten years, the quantity of in situ data is just beginning to reach levels sufficient for meaningful analyses. Backscattering strength tends to follow Lambert's rule at shallow grazing angles. Two main groups emerge from the database, suggesting two independent scattering processes: one driven by grains and the other by bubbles.

1. INTRODUCTION

The mechanisms of high frequency acoustic backscatter at the ocean sediment interface remain elusive, particularly at shallow grazing angles. The term "high frequency" is arbitrarily defined as the range from 1 kHz to 1 MHz. Grazing angles less than about 20° are deemed to be "shallow" for the purposes of this discussion. This definition reflects the reality of high frequency sonar operations over the ocean bottom, where the overwhelming majority of the search area is interrogated at grazing angles from 20° to virtually zero. The grazing angle of 10° falls in the middle of this range and is a good representative angle for comparative studies.

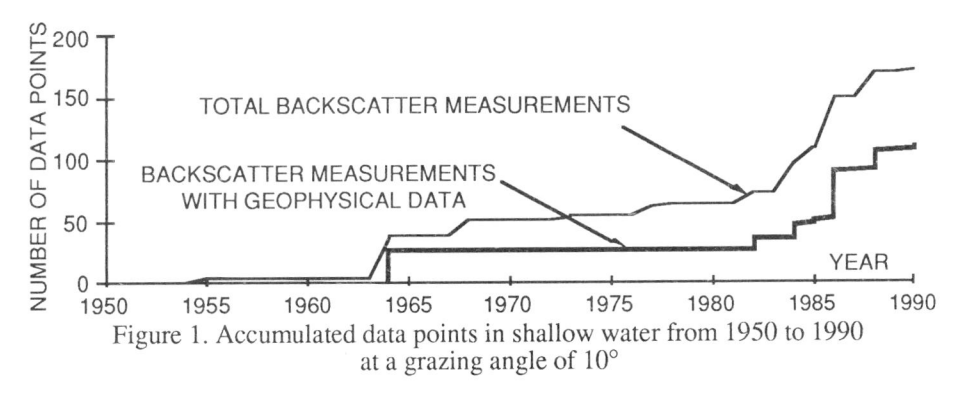

Figure 1. Accumulated data points in shallow water from 1950 to 1990 at a grazing angle of 10°

Since 1982, the quantity of data has more than doubled. The increase is illustrated in Fig. 1 which shows the accumulated number of data points at a grazing angle of 10°; each point is a backscatter measurement at one site and one frequency. The subset that have had geoacoustic data analysis support has tripled in the same period. These statistics indicate that there may at last be enough data to allow us to begin to discern meaningful empirical relationships between backscattering strength and geoacoustic parameters. Rather than being a "mature field", the subject is actually just emerging from infancy.

2. BACKSCATTERING FROM SAND AND GRAVEL

The extant database suggests that, at shallow grazing angles, there are two scattering mechanisms: one associated with sand grains and another associated with trapped gas bubbles. A few experimental observations suggest that stratification may further modify the acoustic backscattering process.

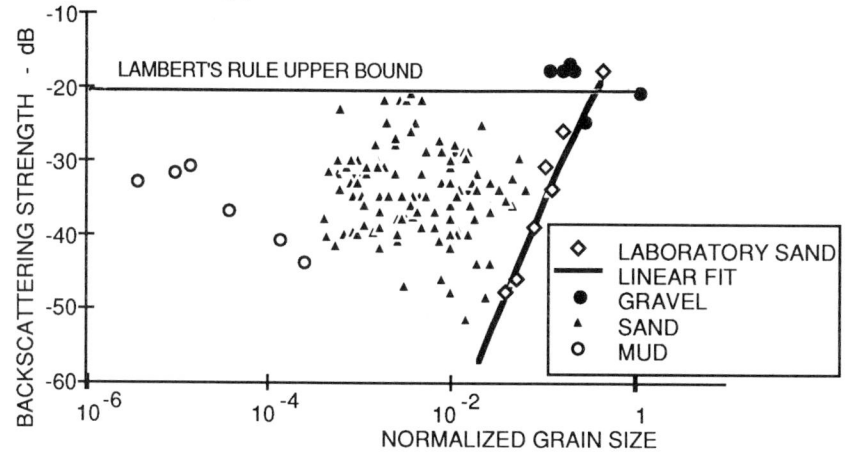

Figure 2. Bottom backscattering strength as a function of normalized grainsize at a grazing angle of 10°

2.1 Grains

In the early 1960's, Nolle[1] measured backscattering strength of sand in the laboratory, using cleaned, graded, compacted, degassed, water saturated sand, whose water-sediment interface had been scraped flat. Further analysis of his results shows that backscattering strength increased at the rate of 30 dB per decade of normalized grainsize, i.e. mean grain diameter divided by wavelength in water. Although in situ data showed that the backscattering strengths were always greater than those of laboratory sands of the same grainsize, it was thought that there may still be a similar relationship between backscattering strength and grainsize of ocean sediments. The available data up to 1982, and the measurements by McKinney and Anderson[2] in particular, tended to support this view.

The total in situ database, however, no longer appears to support any definite trends. Nolle's data and the extant database at a grazing angle of 10° are shown in Fig. 2 as a function of normalized grainsize. In addition to McKinney and Anderson, the sources include Jackson[3,4,5], Stanic[6,7], Muir et al.[8], Chotiros[9,10] and Boehme[11,12]. It is seen that the in situ backscattering strength values are greater than Nolle's laboratory results, and, except for gravel, less than a theoretical upper bound based on Lambert's rule. The data points are identified as mud, sand or gravel, based on grainsize. Sediments with mean grain diameters smaller than 8 μm (7ø) are labeled "mud", and larger than 1 mm (0ø) are called "gravel"; those in between are called "sand" and actually include a wide range of mixtures of sand, silt and clay.

2.2 Lambert's Rule

Nolle's data showed a monotonic increase in backscattering strength, approximately proportional to sine grazing angle squared, up to a grazing angle of about 30°. This is consistent with Lambert's rule which postulates that the reradiated energy has a directivity that varies as sine grazing angle; the second sine grazing angle term arises from the projection of the incident intensity on to the scattering surface. In view of the precautions that were taken to remove all extraneous scatterers, it would seem that there must be an association between sine-squared dependence and the mechanism of acoustic scattering by sand grains. The physical process that gives it this grazing angle dependence is not yet understood. The upper bound shown in Fig. 2 is a theoretical upper limit based on conservation of energy and the observed sine-squared dependence.

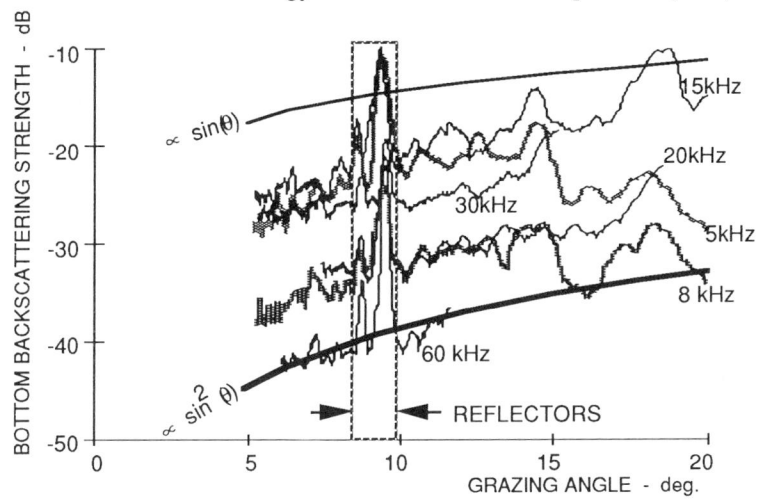

Figure 3. Example of bottom backscattering strength vs. grazing angle from a sandy sediment at frequencies from 5 to 60 kHz

In situ ocean bottom backscatter data from several sources including Mackenzie[13], Urick[14], and later from reports by Boehme and Chotiros[11] also show a tendency to follow Lambert's rule. An example of in situ backscattering strength, between grazing angles from 5° to 20°, measured over a stretch of sandy sediment off Panama City, Florida, in 1986, from Chotiros[9], at various frequencies from 5 to 60 kHz, is shown in Fig. 3. The peaks in the region between 8° and 10° should be ignored; they are caused by discrete reflectors placed on the sand. It is seen that most of the data below 15° shows a sine-squared dependence. At larger angles there are significant deviations from this trend. A possible mechanism that may account for such deviations is discussed in the next section.

2.3 Stratification

It is observed that in situ data do not always follow the sine-square dependence as consistently as the laboratory data. In particular, the data at 5 and 8 kHz in Fig. 3 appear to deviate from the sine-squared dependence above about 15°, beyond which the backscattering strength appears to decrease in an oscillatory manner. This behavior might be explainable in terms of sediment stratification, as proposed by McDaniel[15]. This is a very plausible explanation in view of the frequency dependence and oscillatory tendency, both of which are suggestive of a resonant ducting process.

2.4 Surface roughness

By the late 1970's, since very little correlation had been found between grainsize and backscattering strength from in situ data, it was deduced that there must be other geoacoustic parameters that have a more direct influence on backscattering. The obvious difference between laboratory sands and the ocean sediments is interface roughness. The composite roughness model was adapted by Jackson[16] to address this issue. Attempts to find a relationship between bottom roughness and in situ backscattering strength measurements have produced mixed results. Nolle[1] had experimented with sand with rough surfaces in the laboratory and found that roughness increased backscattering strength only at grazing angles greater than about 45°. The shortcomings of the roughness scattering model have not escaped notice, and the latest model by Jackson[17] employs an empirical fit that is only loosely based on roughness scattering theory for shallow grazing angles.

2.5 Gas bubbles

Although there are no obvious trends in the in situ data as shown in Fig. 2, there may yet be localized trends. This may be brought out by connecting the data points taken from the same site. For sand, it was found that the data fell into two groups. For clarity, these two groups are shown separated in Fig. 4. In group 1, the backscattering strength is characterized by a monotonic increase with grainsize, while in group 2, there appears to be a broad peak in the region of a normalized grainsize of 10^{-2}.

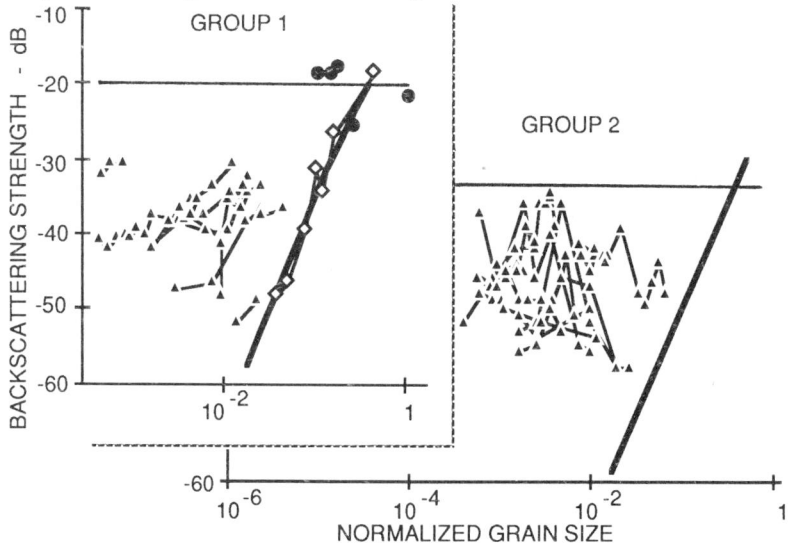

Figure 4. Two groups of backscattering data at a grazing angle of 10°

The existence of the two groups suggests that there may be two corresponding processes. Group 1, which includes Nolle's laboratory data, suggests a scattering process that is driven by the grains. The data from the gravel sites probably also belong in this group up to a normalized grainsize of one.

Group 2, with a broad peak in the region of 10^{-2} suggests a completely different scattering process. The resonance diameter of gas bubbles in the ocean, as noted by Medwin[18], lies in this region. The existence of gas in shallow water is well documented. Speculating that there were gas bubbles trapped between the sediment grains, their sizes would be limited to the size of the pore spaces, which, for sand, would be of the same order as the grains themselves. Initial estimates indicate that very small volume fractions, less than 10^{-5}, are needed to produce the observed levels of backscattering strength.

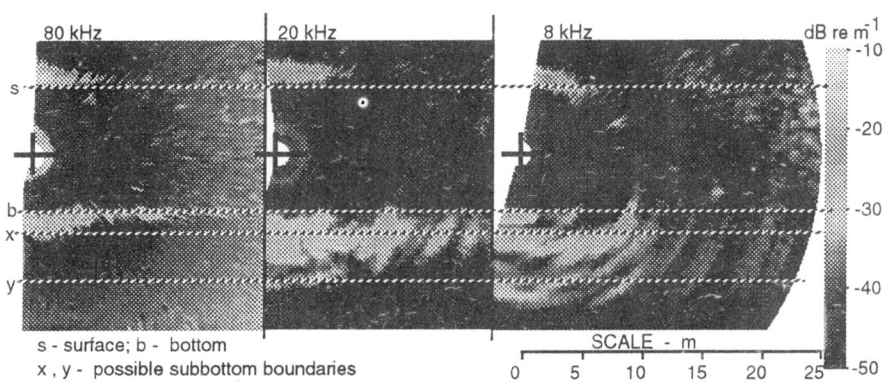

Figure 5. Volume backscattering strength of a mud site at three frequencies

3. BACKSCATTERING FROM MUD

The term "mud" is used to characterize a high porosity, flocculent structure of fine particles, and with a sound speed and density very close to that of water. Such a medium is acoustically practically transparent. It also has a low intrinsic attenuation and volume scattering strength. Yet, the in situ backscattering strength can be comparable to that of sand, as shown in Fig. 2. The scattering is presumed to be caused by inclusions such as larger particles and gas bubbles. This is illustrated in Fig. 5, which shows acoustic images constructed by sweeping a narrow beam transceiver in a vertical plane at a mud test site, at three different frequencies, 8, 20 and 80 kHz. The three images show varying degrees of penetration and scattering from within the volume of the mud. At 80 kHz and shallow grazing angles, the backscatter appears to come from the water-mud interface. The surface backscatter is also visible. At 20 kHz, it is clear that the scattering is from the top 2 m of the mud volume. At 8 kHz, the dominant scatterers are distributed over a larger range of depths; the specular returns suggest the likely existence of two boundaries.

4. CONCLUSIONS

Backscattering from sand at shallow grazing angles appears to have two intrinsic mechanisms: one associated with grains and another associated with trapped gas bubbles. Surface roughness by itself is unlikely to be a significant contributor. Fine-scale layering may modify the scattered field. Backscattering from mud is dominated by volume scattering due to inclusions of solid particles and gas bubbles.

5. ACKNOWLEDGEMENTS

The author thanks Drs. Robert Farwell, Steve Stanic, Peter Thorne, Hollis Boehme, Garland Barnard and Tom Muir for many useful discussions and encouragement. This work was supported by the Naval Research Laboratories/Stennis Space Center, U. S. A.

REFERENCES

1 Nolle, A. W. et al. (1963) 'Acoustical properties of water-filled sands', J. Acoust. Soc. Am., 35(9), 1394-1408
2 McKinney, C. M., Anderson, C. D. (1964) 'Measurement of backscattering of sound from the ocean bottom', J. Acoust. Soc. Am., 36(1), 158-163
3 Jackson, D. R. et al. (1986) 'High-frequency bottom backscatter measurements in shallow water', J. Acoust. Soc. Am. 80(4), 1188-1199
4 Jackson, D. R. (1986) 'High Frequency bottom backscattering strength at the Quinault range', Applied Physics Laboratory University of Washington, APL-UW-8-86
5 Jackson, D. R. (1986) 'High frequency bottom scattering in the Arafura sea', Applied Physics Laboratory University of Washington, APL-UW-5-86
6 Stanic, S. et al. (1988) 'Shallow water high-frequency bottom scattering off Panama City, Florida', J. Acoust. Soc. Am. 83, 2134-2144
7 Stanic, S. et al. (1989) 'High-frequency acoustic backscattering from a coarse shell bottom', J. Acoust. Soc. Am. 85(1), 125-136
8 Muir, T., Thompson L. A., Shooter J. A., DeMary T. E., 'Backscattering of sound in shallow water', submitted to J. Acoust. Soc. Am. Note: Since the data reported here were peak values of 4 independent samples, they were reduced by 3.4 dB.
9 Chotiros, N. P. (1990) 'High frequency bottom backscattering Panama City Experiment', Applied Research Laboratories University of Texas at Austin, ARL-TR-90-22
10 Chotiros, N. P. (1988) 'Analysis of bottom backscatter data from the Kings Bay Experiment', Applied Research Laboratories University of Texas at Austin, ARL-TR-88-6
11 Boehme, H., Chotiros, N. P. (1988) 'Acoustic backscattering at low grazing angles from the ocean bottom', J. Acoust. Soc. Am. 84(3), 1018-1029
12 Boehme, H. et al. (1985) 'Acoustic backscattering at low grazing angles from the ocean bottom. Pt.I. Bottom backscattering strength', J. Acoust. Soc. Am. 77(3), 962-974
13 Mackenzie, K. V. (1961) 'Bottom reverberation for 530 and 1030-cps sound in deep water', J. Acoust. Soc. Am., 33(11), 1498-1504
14 Urick, R. J. (1954) 'The backscattering strength of sound from a harbor bottom', J. Acoust. Soc. Am. 26(2), 231-235
15 McDaniel, S. T. (1992) 'Effect of surficial sediment layering on high-frequency seafloor reverberation', J. Acoust. Soc. Am. 91(3), 1353-1356
16 Jackson, D. R., Winebrenner, D. P., Ishimaru, A. (1986) 'Application of composite roughness model to high-frequency bottom backscattering', J. Acoust. Soc. Am., 79(5), 1410-1422
17 Jackson, D. R. (1987) 'Third Report on TTCP Bottom Scattering Measurements Model Development', Applied Physics Laboratory University of Washington, APL-UW-8708
18 Medwin, H. (1977) 'Acoustical determinations of bubble-size spectra', J. Acoust. Soc. Am. 62(4), 1041-1044

MEASUREMENTS OF HIGH FREQUENCY REVERBERATION IN SHALLOW WATER

J. G. Kelly, R. N. Carpenter, M. Buffman, and E. R. Levine
Naval Undersea Warfare Center
Newport, Rhode Island, 02841, USA

ABTRACT. In September 1991, the first of a series of measurements was made with high frequency (≥ 10 kHz) sonar in shallow waters (35-55 m) of Block Island Sound. Approximately 250 sonar echoes were collected with sonar heads mounted on a remotely controlled, submersed system towed by a surface ship over various bottom types. For ten types of transmit waveform, the output from each transducer in the sonars' (8x8) arrays was individually recorded digitally. Concurrent oceanographic and geologic measurements included vertical and horizontal profiles of conductivity, temperature, and depth (CTD); current; wave height spectra of the sea-surface; side scan sonar mapping and stereo photography of the bottom; ambient noise; bottom cores; underwater video. Air-filled steel spheres were moored at the site as a distributed target for acoustic imaging experiments. Moreover, data have been analyzed to characterize boundary reverberation for two cases representative of the variability encountered during the experiment.

1. INTRODUCTION

Recent interest in diverse shallow water areas has led to concentrated study of their unique acoustical and oceanographic properties. Accordingly, an experimental system was developed to investigate high frequency acoustics for application to sonar operations in shallow water. Furthermore, various shallow water sites have been selected for experiments with high frequency sonars mounted on a submersed platform towed by a surface ship. The frequency range and water depth of interest are 10-100 kHz and ≤100 m, respectively. Supporting the measurement of sonar performance, comprehensive environmental measurements were also taken. The first experiment was conducted in September 1991 at a shallow water test site approximately 3.5 km south of the eastern edge of Fisher's Island in Block Island Sound. The site is a 2 km x 4 km rectangle centered at 41° 15' N 71° 56' 24" W on the eastern U. S. continental shelf with bottom depths of 35-55 m. The experimental setup is illustrated in figure 1. Approximately 250 sonar echoes were collected over three days using several high frequency sonar heads on the platform, which was towed by the R/V Onrust, an 18-meter research vessel. Measurements were taken over various bottom materials: cobbles, gravel, shell, and sand. Ten transmit waveform types were used including cw pulses and fm coded pulses. The output from each transducer in the (8x8) sonar arrays was individually recorded digitally on 5 inch optical discs. Recording space - time samples in this way (as opposed to recording only preformed beam outputs, for example) is necessary for high resolution, three-dimensional imaging of targets and boundaries in shallow water. High speed data acquisition and signal processing technology is applied to preserve the integrity of the acoustic data for coherent spatial processing. This paper summarizes the experiment including a description of the instrumentation, measurements, signal processing, and examples of preliminary findings regarding acoustic imaging and reverberation in shallow water.

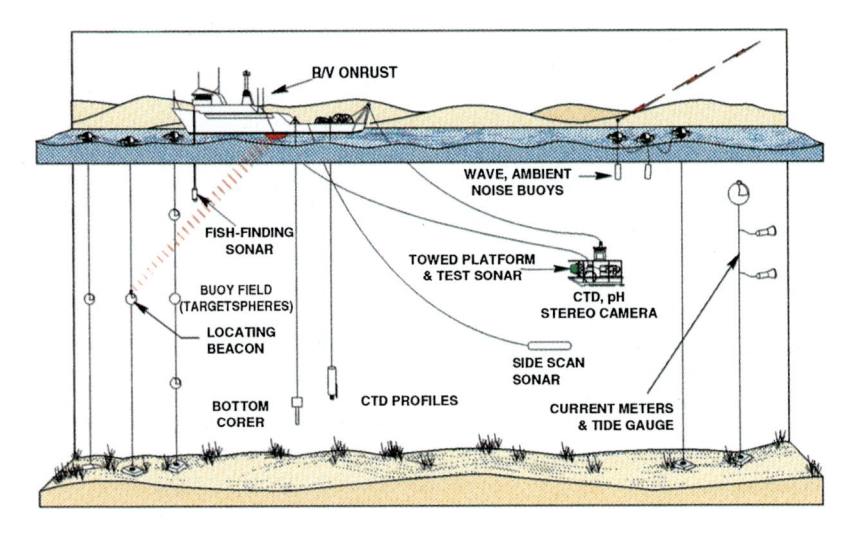

Fig. 1. Experimental setup

2. MEASUREMENTS AND INSTRUMENTATION

The measurement instrumentation consists of shipboard data acquisition and analysis subsystems, integrated navigation components, oceanographic sensors, and a modified sonar head mounted on the towed platform (figures 2 and 3). The complete system can be deployed from a small ship like a trawler and transported in standard air shipping containers.

The acoustic analog data are transmitted from the instrumented sonar head over cable to the host ship for signal conditioning, processing, and recording. The sonar heads are remotely controlled from topside for both pan and tilt with custom gimbals (±90° horizontal; ±30° vertical) for several sizes of sonar heads ($\leq$ .5 m dia.). The platform (the TSS-1000 from Deep Sea Systems, Inc.) operates in sea state 3, to an operational depth of 120 m, and at speeds to 1.5 m/s. Data from GPS and acoustic transponders (mounted on ship, target, and towed platform) were integrated by computer for precise navigation and tracking. The sonar heads were modified to provide individual transducer output voltage, preamplified and differentially driven with a variable gain (10-40 dB) remotely selectable aboard ship. All 64 analog signals are transmitted to the shipboard electronics over a 500 feet long (1 inch dia.) data cable, with two additional cables for power and control. Each of the 64 channels contains a differential amplifier/filter; an analog signal conditioner (72 dB programmable gain and an anti-aliasing filter with bandwidth programmable from 1-150 kHz); a digital data acquisition processor with a sampling rate to 1 mHz (12 bit/sample), a programmable digital filter, and random access buffer memory (0.5 mbyte). Among all channels, gain and phase match to within 0.1 dB and 1°, respectively. Ten types of waveforms, designed for imaging and scattering function estimation, were transmitted: cw, linear stepped fm, and various hopped-frequency codes; duration and bandwidth range from 2-560 ms and 4-2000 Hz, respectively. Horizontal and vertical transmit beam patterns were selectable from beamwidths of 15°- 60°.

Fig. 2. Towed platform on R/V Onrust. 　　Fig. 3. Towed platform with sonar head.

Environmental data include vertical and horizontal samples of sound speed, directional surface waves, near-surface ambient noise, current speed and direction, tidal height, and bottom geoacoustic characteristics (figure 1). Vertical sound speed profiles were measured within 1 hr of sonar measurements, deployments alternating with acoustic pings (figure 4). The geological data were obtained immediately after the acoustic tests. Sound speed variability was determined using two Seabird self-recording CTD instruments: a Seacat, deployed from the R/V Onrust, for the vertical profiles and a Sealogger, on the TSS-1000, for the continuous horizontal profiles. A comparison of synoptic sound speed measurements made within several meters at the towed platform depth shows that the two instruments agree to within ± 0.07 m/s. During the experiment, the principal cause of vertical variability in sound speed was the onset of the late summer - fall water column transition brought on by storms (e. g., the evolution of cooler, fresher water, especially in the near - surface layer). This seasonal transition is evidenced in vertical sound speed profiles from 24 and 28 September 1991, designated as ssp(a) and ssp(b), respectively (figure 5); these are used in the comparisons of modeled and measured reverberation shown subsequently. Within the site, horizontal variability in the sound speed field at the source depth was estimated to be approximately 0.5 m/s. Bottom morphology was determined with a 100 kHz EG&G side scan sonar and a Photosea M2000 stereo camera. Sediment layers in the upper 1.5 m of the bottom were determined from six gravity cores, which, after preliminary analysis, indicate sand, shells, gravel, and cobbles.

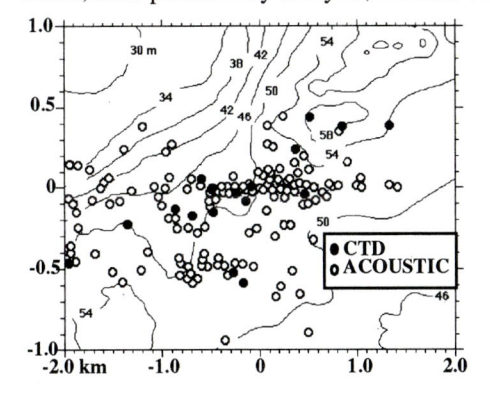

Fig. 4. Sonar and CTD measurements. 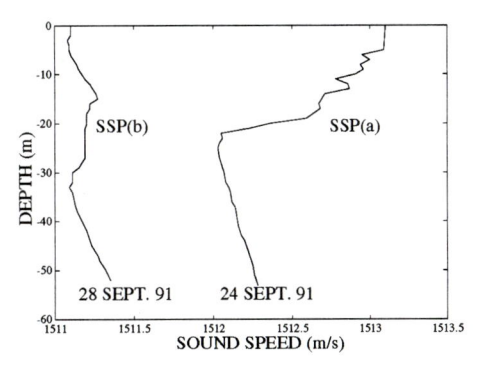　　Fig. 5. Vertical profiles of sound speed.

Measurements of directional wind waves were made with an Endeco 956 wave rider buoy tethered to a surface mooring (figure 1); data were telemetered to shore 22 km from the site. Synoptic wind and sea state observations were also made from the R/V Onrust; significant wave height was 0.2-2.6 m.

3. ACOUSTIC IMAGING

For temporal signal processing in the laboratory, the digital data are band-pass filtered, and analytic signals are generated and matched filtered for each of the 64 channels. Then spatial processing (time delay and sum beamforming) is performed to generate acoustic images of the water column. Multiple beams are formed simultaneously in azimuth and elevation at increments of 1/6 of the null-to-null beamwidth of approximately 29° (for the four-wavelength aperture of the arrays used) - an interpolation finer than conventional Rayleigh spacing that produces more detailed acoustic images.[1] Figure 6 shows the setup of the five spheres (air-filled, 30 inch diameter, .5 inch steel) in the experiment and the orientation of the sonar's transmit beam pattern, which has side lobes illuminating both the surface and bottom. (The spheres were moored at the origin in figure 4.) The main beam illuminates only the three spheres in the middle of the water column. Figure 7 is an example of an acoustic image of the spheres using a 2 ms cw pulse. Receive beams are formed at 5° increments between ±90° in azimuth and elevation (37^2 beams). The intensity of a beam's output is computed approximately every 0.5 m in range between 6-100 m . Furthermore, range, azimuth, and elevation data are converted to rectangular coordinates with color coding of the acoustic intensity. Without additional processing, only the closest target is visible, the others obscured by the strong bottom and surface reverberation returns caused by the transmit sidelobes. To achieve the image in figure 7, further processing is needed: data corresponding to surface scattering and to bottom scattering have been projected onto their respective planes. The remaining volume data are displayed between the two surfaces. The acoustic intensities for the three regions of the image are separately normalized (and color coded). The three targets are clearly visible.

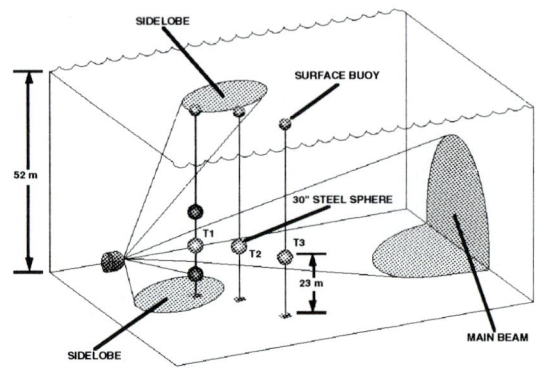

Fig. 6. Arrangement of acoustic targets.

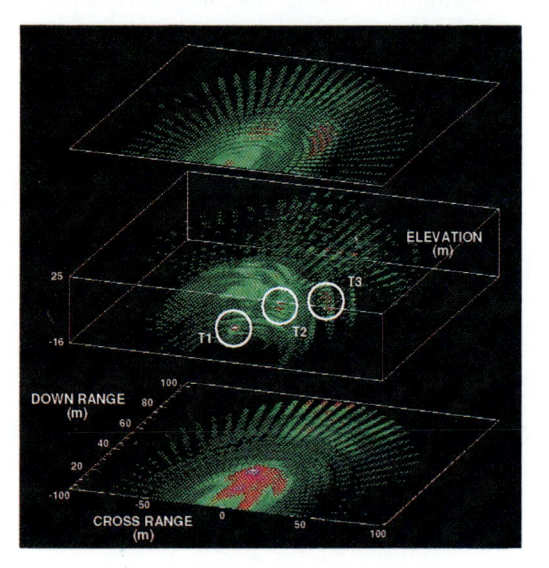

Fig. 7. Acoustic image of spheres.

4. REVERBERATION

Figure 8 shows two temporal records of measured reverberation level (instantaneous power) for a cw pulse of duration 85 ms: one from a single (arbitrary) transducer, another the (incoherent) average of the individual levels from all transducers (52 in this case) in the array. The spatial average clearly reduces the variance of the estimate of the reverberation level, demonstrating the advantage of multichannel recording. For the same waveform, figure 9 shows a comparison between measured levels (spatial averaging with further temporal averaging for 42 ms) and results from the Generic Sonar Model (GSM).[2] For backscattering and forward scattering from both the sea-surface and the bottom, the GSM was augmented by various models [3-6] as compiled and extended by the Applied Research Laboratory, University of Washington. Shown are two cases representative of the variability observed in the experiment: (a) a downward refracting sound speed profile ssp(a) with wind speed of 1.4 m/s (significant wave height of 0.0 m); (b) an upward refracting profile ssp(b) with wind speed of 4.1 m/s (0.5 m). The in-band ambient noise (including the ship's contribution), measured by the sonar itself, was about 80 dB // µPa and insignificant. In both cases, water depth is 50 m; source depth is 25 m; from analysis of core material, the surface sediment on the bottom is sand and nonuniformly distributed shells. In the model, only medium sand is assumed, the shells' contribution undetermined. In case (b), measurement and model - particularly the slopes - agree well, but for a constant bias of about 3 dB, owing to an unexplained, systematic error either in the model or in the calibration of the instrumentation. In case (b), the results are consistent with predominance (after about 0.3 s) of surface effects, particularly, backscattering and forward reflection losses due to bubble layers; however, in case(a), where agreement between model and measurement is poor, bottom scattering is probably predominant. This conjecture is consistent with the representation in figure 10 of reverberation intensity $I_r(\theta, t)$ versus vertical receive angle θ and time t (attained by forming vertical beams over $\pm 90°$ at $3°$ increments and displaying, in logarithmic gray level,

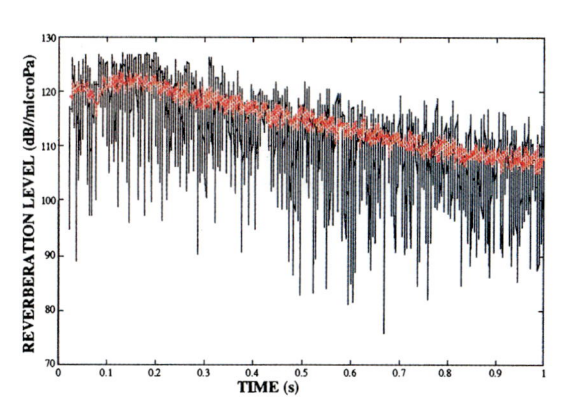

Fig. 8. Reverberation level vs time: spatial average over 52 channels compared with single channel.

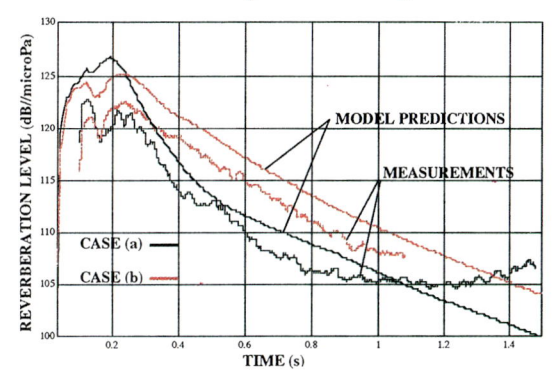

Fig. 9. Reverberation level vs. time: comparison of measurement with generic sonar model.[2]

{$i_r(\theta, t) = I_r(\theta, t)/\max_\theta I_r(\theta, t): i_r(\cdot) \geq .9$, for all θ, t}). The distinct bands evident in the measured levels correlate well with the significant eigenrays, corresponding to sea-surface and bottom sources, calculated from the GSM.[2] This reveals, in case(a), the aforementioned predominance of the bottom-induced effects; thus, the poor agreement seen in figure 9 may be explained, in part, by inadequate models of bottom backscattering and forward scattering. In the model, bottom homogeneity is assumed. If this is, indeed, a weak approximation to the actual distribution of surface sediment, then sonar performance models must be accordingly refined to include spatial variation of bottom backscattering and reflection loss.

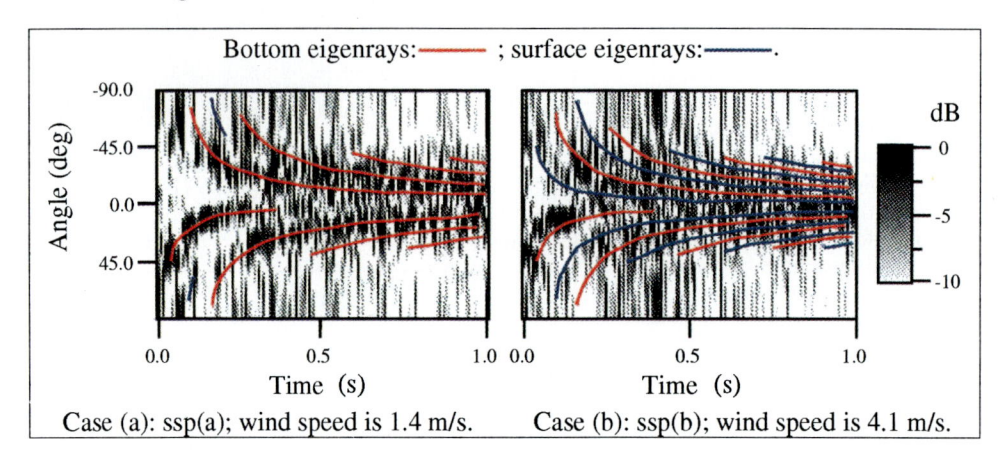

Fig. 10. Measured reverberation (logarithmic gray level within 10 dB of instantaneous maximum) vs. vertical angle and time for two cases. Eigenrays calculated from GSM.[2]

REFERENCES

1. Kelly, J. G., R. N. Carpenter, and J. A. Tague (1992): Object classification and acoustic imaging with active sonar. J. Acoust. Soc. Am. 91 (4), Pt. 1, 2073-2081.
2. Weinberg, H. (1985): Generic sonar model. NUSC TD 5971D.
3. Jackson, D.R. and P. D. Mourad (1989): Models for bottom loss and backscattering. APL-UW TR 8919.
4. McDaniel, S. T. (1986): Diffractive corrections to the high-frequency Kirchhoff approximation. J. Acoust. Soc. Am. 79, 952-957.
5. Thorsos, E. I. (1984): Surface forward scattering and reflection. APL-UW 7-83.
6. McDaniel, S. T. and A. D. Gorman (1982): Acoustic and radar sea-surface backscatter. J. Geophys. Res. 87, 4127-4136.

ACKNOWLEDGEMENT

The authors thank Dr. D. N. Connors, M. J. Griffin, and M. R. Medeiros for their contributions. This project was supported by Dr. W. I. Roderick, Associate Technical Director, Research and Technology, NUWC.

FREQUENCY RESPONSE MEASUREMENTS ON BACKSCATTERING FROM A SHALLOW SEA FLOOR, USING A PARAMETRIC SOURCE

T. G. Muir, L. A. Thompson, J. A. Shooter, T. E. DeMary, and R. J. Wyber*
Applied Research Laboratories, The University of Texas at Austin, Austin, TX 78712, U.S.A.

* on leave from Royal Australian Naval Research Laboratory, Darlinghurst, New South Wales, 2010, Australia

ABSTRACT. Measurements are reported on the backscattering of sound from a sand bottom overlying the continental shelf near San Diego, California. Data were acquired in the 1-10 kHz band with a directive parametric source having a sidelobe-free beam, 4° in half-power width, at a grazing angle of 10°. The bottom was a flat layer of coarse grained sand and clay that contained many shell fragments, overlying a hard sub-bottom layer, some 5 m deep. The mean bottom backscattering coefficient at 1 kHz was -39 dB re 1 m^2 and increased with frequency to the power of 0.5. Statistical data in the form of probability distributions show a Rayleigh distribution for ensembles of peak acoustic amplitudes at each frequency studied.

1. INTRODUCTION

A significant impediment to the use of active sonar in shallow water is the high level of reverberation that results from the scattering of sound from hard sand bottoms. Previous studies of bottom backscattering from sand bottoms in coastal waters have been reported in several noted papers.[1-10] The present work utilizes a directive parametric research tool[11] to eliminate surface and sub-bottom reflections, focusing on bottom backscattering.

2. ANALYTICAL MODEL

Historically, the scattering strength of a surface has been described in terms of a coefficient characterizing the particular surface plus a term quantifying the areal extent of surface insonification. The appropriate sonar equations appear in Ref. 1. The geometry of the present experiment is depicted in Fig. 1. The incident pressure at the insonified patch is the measured free field pressure at the same range, which is in the parametric interaction region of the nonlinear source. The backscattered pressure is measured at a receiver collocated with the source, and spherical spreading from the insonified patch to the receiver is assumed.

3. TEST SITE

The test site was a shallow water area overlying the continental shelf off San Diego. The specific location was 2 nmi due West from the inlet of Mission Bay, where the water was 40 m deep. A comprehensive report, Ref. 12, describes the oceanography of this area. Surficial sediments in this area consist of a medium to coarse sand, mantled with organic

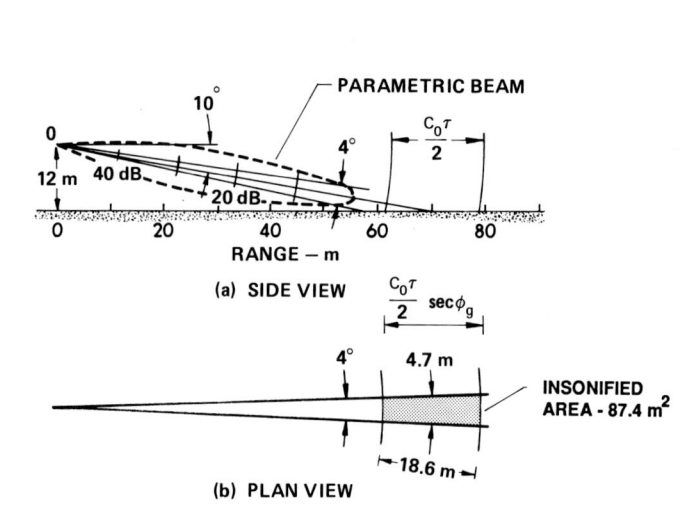

Figure 1. Backscattering geometry

detritus and containing many shell fragments. The mean grain size is very near the value of 0.57 mm (= 0.81 on the phi scale) reported from prior work[13] done only a few hundred meters from the present site. Other prior measurements[13] on this sediment indicate a density of 2.06 gm/cc, a sound velocity of 1817 ms, a porosity of 38%, and a compressional wave attenuation very near 0.9 dB per wavelength, all at a sediment depth of 30 cm. The bottom is quite flat, with occasional rounded sand ripples some 2 to 5 cm in height and separated by distances of 40-80 cm. The main sand layer is approximately 5 m thick, overlying bedrock composed of sandstone, weathered conglomerates, and shales. Cores taken nearby show the presence of thin layers of shell fragments as well as interbedded layers of sand-silt-clay and fine sand.

4. MEASUREMENT TECHNIQUE

The parametric source and associated equipment were deployed from the Ocean Research Buoy (ORB) operated by the Marine Physical Laboratory of the Scripps Institute of Oceanography, University of California at San Diego. The acoustic source was oriented to achieve a grazing angle of 10° and was lowered through a center well in ORB by means of a steel cable. Reception of echoes was achieved with an array of ten omnidirectional hydrophones placed around the outer edge of the parametric source. The effective half-power beamwidth of the apparatus was that of the parametric beams, which was constant at about 4° at the range of the measurements at hand.

The source was energized under the direction of a microcomputer to provide a stepped frequency sequence of cw pulses, each of 25 ms duration. This enables near simultaneous acquisition of frequency response data. The discrete parametric frequencies in the sequence were 1.1, 1.5, 1.9, 2.3, 2.7, 3.1, 3.9, and 4.7 kHz. Data at a primary frequency of 10 kHz were also acquired simultaneously. After reception by the broadband hydrophone array, the data were filtered through a bank of passive filters, each having a bandpass of 50 Hz, each tuned to a discrete frequency. The filtered data was recorded with a multi-channel tape recorder and analyzed with digital techniques. The processing of digitized data involved the analysis of peak echo returns in 500 sample ensembles. Data entering each ensemble was gated with a time window appropriate for the travel time to and from the resolution cell on the scattering surface. The width of the window was large enough to accommodate travel time variance due to vertical motion of the supporting platform in ocean swell.

5. BOTTOM BACKSCATTERING DATA

Results for the frequency response of the peak bottom backscattering coefficients are depicted in Fig. 2. Each data point represents the average of each ensemble. It is observed that the mean backscattering strength, S_s, of the present data varies from -39 to -34 dB re 1 m^2 and increases with frequency as the power of 0.5 (i.e., $S_s \sim 10 \log f^{0.5}$). It is also seen that the data fall close to the 0.5 power dependence, with little scatter.

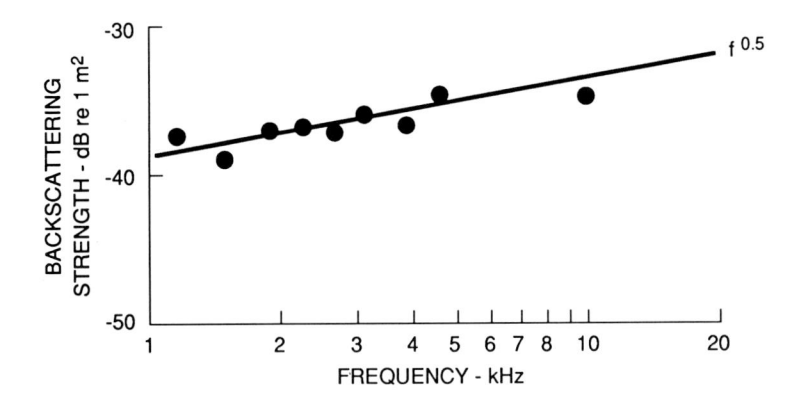

Figure 2. Comparison of bottom backscattering results at 10° grazing angle

The present results are compared to many of those previously reported for sand in Fig. 3. Most of the data points shown were acquired at or near 10° grazing angle. The data of ref. 5 were acquired at 70° grazing angle and were scaled in the plot to 10° grazing by use of Lambert's law ($S_s \sim 10 \log \sin^2 \phi_g$, where ϕ_g is the grazing angle).

Although considerable data points are shown in Fig. 3, and although there is considerable spread, even within most of the same data sets, the data comparison shows that some trends may be beginning to emerge. Most of the data in the 1 to 10 kHz band seems best fitted to a 0.5 power law frequency dependence, and this tends to continue well into the hundreds of kHz range, for the high backscattering strength data sets. However, a grouping of lower backscattering strength data sets in the 10 to 300 kHz range seems better fitted to a 1.0 power law frequency dependence ($S_s \sim 10 \log f^{1.0}$).

It is expected that backscattering in different frequency regimes derives from different scattering mechanisms. One could argue that less frequency dependence is obtained at low frequencies where surface roughness scales are smaller than acoustic wavelengths, and sub-surface inhomogeneities are the dominant scatterers, with greater penetration and insonification of more inhomogeneities as the wavelength increases. The high frequency dependence might be stronger, due to surface roughness scaling with wavelength. Unfortunately, the spread of data points in most of the data sets is too large to permit a quantitative assessment of trends in frequency dependence. This points out the importance of adequate consideration of statistical methods in backscattering measurements. It is essential to acquire statistically independent samples in ensembles large enough to provide good estimates of statistical parameters (means, higher moments, etc.), so that trends at a given experimental site can be adequately determined.

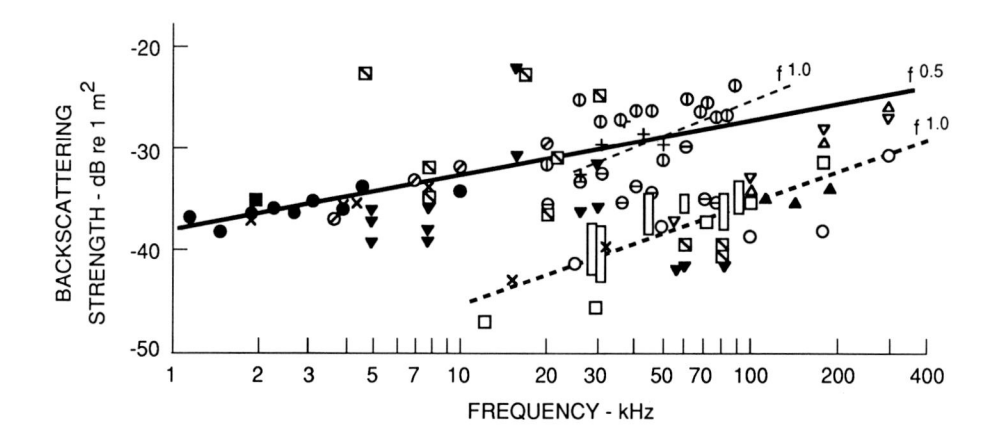

Figure 3. Comparison of bottom backscattering from sand at 10° grazing angle.
●- present data, ○□▽△- Ref. 1, []- Ref. 2-4, ✗- Ref. 5, ⊖ ⊕ +- Ref. 6, ⊘- Ref. 7, ■- Ref. 8, ▲- Ref. 9, ▼- Ref. 10, ⊠- Ref. 11

For the present data set, the bottom backscattering statistics for peak echo amplitudes were tested for goodness of fit to Rayleigh and Maxwell distributions and were found to follow a Rayleigh distribution. Some results are shown in Fig. 4, in terms of the cumulative probability distribution curves. The probability axis in these curves has a nonlinear scale so that a truly Rayleigh distribution plots to a straight line as shown by the dashed segments. Each curve represents a different acoustic frequency, with frequency increasing to the right, as is shown on the slanted scale.

Although all of the curves follow the Rayleigh distribution form to a good approximation, some discrepancies are noted. First, none of the measured curves approach zero at low backscattering amplitudes. One could argue that zero echo pressure is impossible to measure, especially since noise is always present. However, the signal to noise in these measurements was quite high (typically ~25 dB), which indicates that the low amplitude departures above the Rayleigh line in the cumulative probabilities (say 0 to 0.3 or 0.5) are in fact due to the prevalence of a few small scatterers in each sample. No other persistent, non-random departures from the Rayleigh distribution appear in the backscattering data.

Figure 4. Cumulative probability distribution for peak backscattering returns at several frequencies.

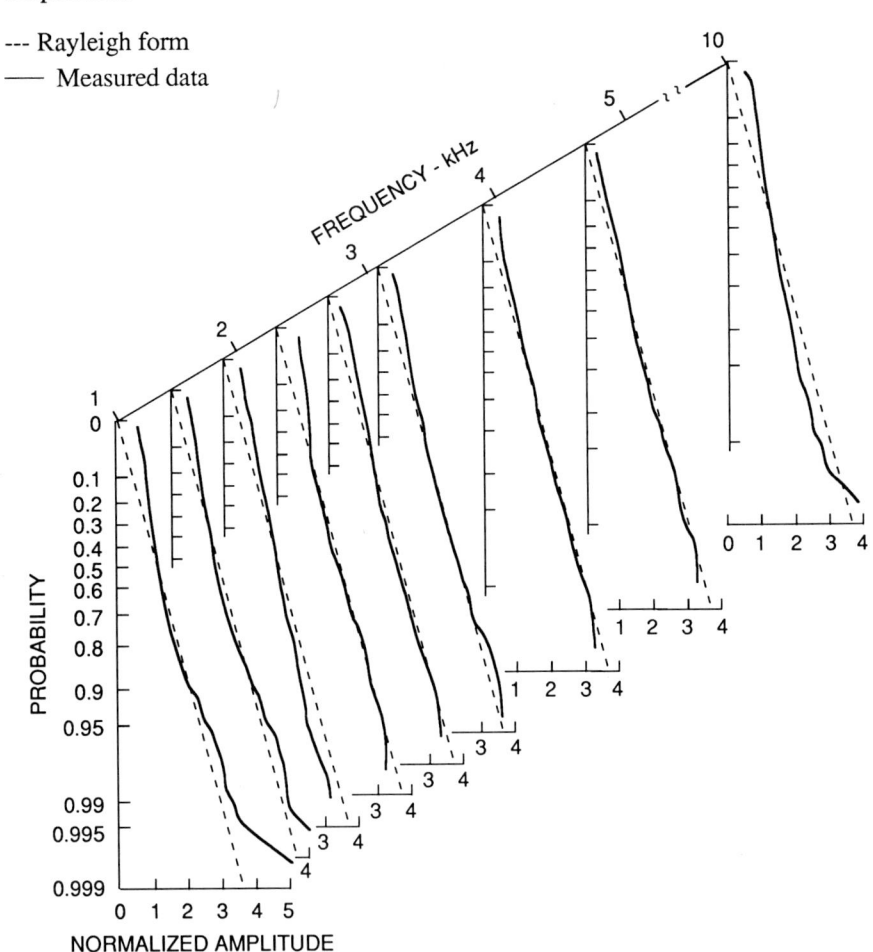

6. SUMMARY AND CONCLUSIONS

The utilization of a directive, parametric research tool enabled measurements of bottom backscattering on a sand bottom in shallow water, at frequencies from 10 kHz down to 1 kHz, at a grazing angle of 10°. The sidelobe-free beam enabled the elimination of aliasing due to sidelobe contamination of ocean surface and sub-bottom echoes. In the measurements and analysis, consideration was given to the stochastic nature of backscattering and the importance of adequate sampling and statistical data analysis.

Data on the mean value of the peak backscattering coefficient for a sand sediment in the 1 to 10 kHz band was found to vary from -39 to -34 dB re 1 m^2, increasing as the frequency to the power of 0.5. Here, the scale of bottom roughness was small in comparison to acoustic wavelength, so sub-bottom heterogeneities arise as a probable scattering mechanism. Some prior backscattering data at higher frequencies shows a linear increase with frequency. In the high frequency regime, bottom surface roughness is a viable mechanism. The present data showed that the statistical distribution of peak echo amplitudes follows a Rayleigh form.

7. ACKNOWLEDGEMENTS

This work was sponsored by the U.S. Navy Office of Naval Research. The authors acknowledge helpful comments from Dr. Nick Chotiros.

8. REFERENCES

1. McKinney, C. M., and Anderson C. D. (1964). "Measurements of backscattering of sound from the ocean bottom," J. Acoust. Soc. Am. **36**, 158-163.
2. Boehme, H., et.al. (1985) ."Acoustic backscattering at low grazing angles from the ocean bottom. Part I. Bottom backscattering strength," J. Acoust. Soc. Am. **77**(3), 962-974.
3. Chotiros, N. P., et.al. (1985). "Acoustic backscattering at low grazing angles from the ocean bottom. Part II. Statistical characteristics of bottom backscatter at a shallow water site," J. Acoust. Soc. Am. **77**(3), 975-982.
4. Boehme, H. and Chotiros, N. P. (1988). "Acoustic backscattering at low grazing angles from the ocean bottom," J. Acoust. Soc. Am. **84**(3), 1018-1029.
5. Bunchuk, A. V., and Zhitkovskii, Yu. Yu. (1980). "Sound scattering by the ocean bottom in shallow water regions," Sov. Phys. Acoust. **26**(5), 363-370.
6. Jackson, D. R., et.al. (1986). "High-frequency bottom backscatter measurements in shallow water," J. Acoust. Soc. Am. **80**(4), 1188-1199.
7. Monti, J. M. and Nutzel, B. (1984). "Acoustic scattering from the sea floor," U.S. Naval Undersea Systems Center Technical Document #7293.
8. Smailes, I. C. (1978). "Bottom reverberation measurements at low grazing angles in the NE Atlantic and Mediterranean Sea," J. Acoust. Soc. Am. **64**, 1482-.
9. Stanic, S., et.al. (1988). "Shallow water high-frequency bottom backscattering off Panama City, Florida," J. Acoust. Soc. Am. **83**, 2134-2144.
10. Chotiros, N. P. and Boehme, H. (1988). "Analysis of bottom backscattering data from the Kings Bay experiment," Applied Research Laboratories, the University of Texas at Austin, Technical Report No. 88-6 (ARL-TR-88-6).
11. Muir, T. G., et.al. (1980). "A low frequency parametric research tool for ocean acoustics," in W. A. Kuperman and F. B. Jensen, eds., Bottom Interacting Ocean Acoustics, Plenum Press, New York.
12. La Fond, E. C. (1965). "The U.S. Navy Electronic Laboratory's Oceanographic Tower," Navy Electronic Laboratory Report #1324.
13. Hamilton, E. L., (1972). "Compressional wave attenuation in marine sediments," Geophysics **37**, 620-646.

MEASUREMENTS OF ACOUSTIC BACKSCATTERING OF THE NEAR-SURFACE LAYER

BERND NÜTZEL AND HEINZ HERWIG

Federal Armed Forces Institute
for Underwater Sound and Geophysical Research
Klausdorfer Weg 2-24, 2300 Kiel 1, Germany

ABSTRACT A selcted set of results from monostatic acoustic near-surface scattering experiments conducted in the North Sea is presented. Scattering strength values were measured as a function of frequency (3 kHz - 80 kHz) and grazing angle (10° to 90°) for a large variety of environmental conditions.
The investigation of principle scattering processes points out three wind speed regimes for which the sea surface or near-surface bubbles are the dominant scattering mechanisms. The corresponding wind speeds for surface scattering and saturation onset are presented as a function of frequency.

1. INTRODUCTION

Numerous theoretical and experimental studies have been conducted to investigate the acoustic reverberation originating at or near the sea surface. The results indicate that surface roughness and air bubbles - caused by breaking waves - must be considered to explain the environmental dependence of acoustic scattering on sonar engineering parameters such as frequency, grazing angle, and system design characteristics. In our investigation special emphasis was laid on experiments at sea to determine the regimes in which bubbles or the sea surface is the dominant scattering mechanism. A selected set of results from monostatic reverberation experiments conducted in the North Sea during the last years is presented.

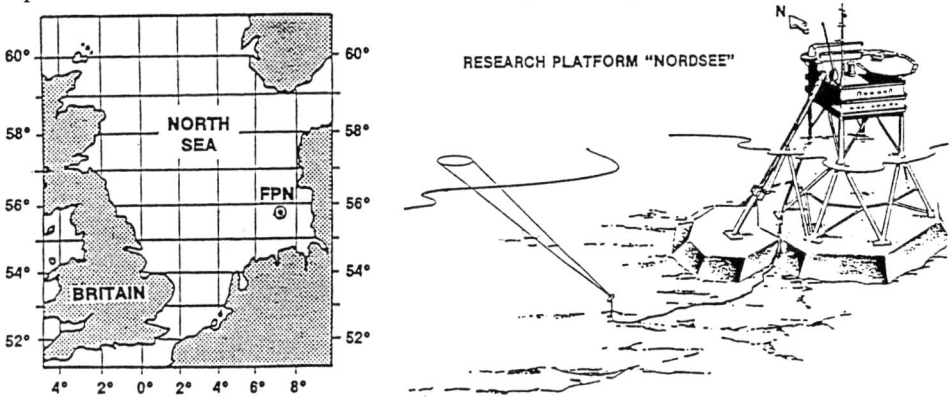

Figure 1. Measurement location

Figure 2. Experimental setup

2. EXPERIMENTAL SETUP

The experiments were conducted in the shallow water of the North Sea. The experimental site, shown in figure 1, is the location of the German research platform NORDSEE, which is approximately 40 nautical miles west of the German and

Danish coasts. The water depth in this position is approximately 30 m. The data were obtained utilizing a high-resolution (narrow beamwidth) pulsed parametric sonar, which was driven either in conventional or parametric mode [1]. This system was installed atop a 7.5 m tower and implanted in a distance of 160 m from the platform (see figure 2). The parametric array could be remotely trained in elevation and azimuth. The wind speed was measured at a height of 47 m above the sea surface and the wave height was recorded from Waverider buoy data.

3. RESULTS

From the data set obtained during the recent years appropriate parameter combinations of frequency, grazing angle, and environmental condition are selected to demonstrate under which conditions near-surface bubbles or the sea surface is the dominant scattering mechanism.

In figure 3 the difference of backscattered signals measured at normal incidence with and without bubbles within the near-surface layer is illustrated. For these different conditions the envelopes of two time series averaged over 750 pings are shown. The arrival time difference of the individual echoes caused by differing wave heights was removed using a thresholding technique. The left curve shows the average return from the surface when no bubbles are present. This can be seen by the nearly constant steep slope of the leading edge of the echo. The right curve represents the average return from a high sea state condition when breaking waves occur. The onset of scattered energy indicates the presence of bubbles down to 3.5 m below the surface.

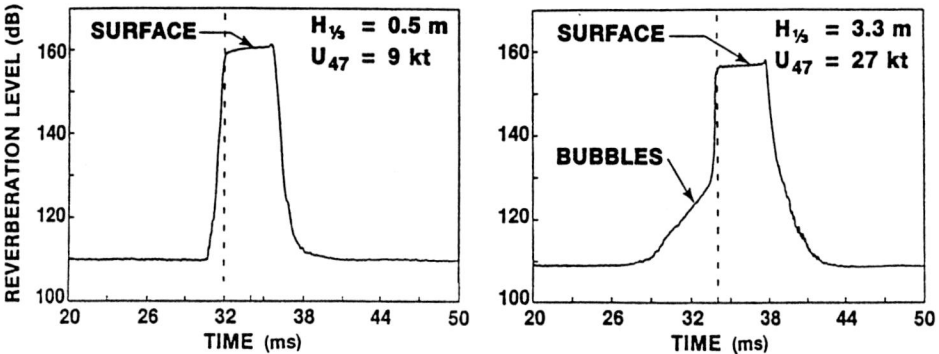

Figure 3. Ensemble averaged acoustic reverberation at normal incidence for two environmental conditions, Frequency 18 kHz, Ensemble 750

Figures 4a to 4c illustrate the increasing bubble scattering with wind speed in a 3-dimensional representation for a transmitted frequency of 40 kHz. At a wind speed of 17 knots only few bubble clouds occur. These can be detected up to a water depth of about 2 m. At 20 knots (s. fig. 4b) the number of bubble clouds increases and they are measurable down to a depth of 5 m. The relatively constant echoes from the sea surface show that the scattering from the surface stays obviously unaffected by these bubble clouds. As can be seen in figure 4c at a wind speed of 28 knots there are always bubbles present, penetrating down to more than 10 m and masking the surface acoustically for periods of some minutes.

To demonstrate the influence of bubbles on backscattering at lower frequencies, the backscatter history is depicted in figure 5. The backscattering strength is computed for each individual ping at the time when the maximum of the average envelope occurs and plotted versus time. For comparison the solid line represents the level of 110 dB. The data were taken at 10 kHz at a grazing angle of 30°. As can be seen, there are almost no fluctuations at the lowest wind speed of 7 knots. At 24 knots high backscatter periods are present. At a wind speed of 43 knots the high backscatter periods, caused by bubble clouds do not change much in level but they stay longer and drop down only for short times.

The temporal fluctuations are described by the coefficient of variation (CoV) which is shown in figure 6 for frequencies of 3, 10, and 18 kHz as a function of wind speed. The three curves all show the same trend. At low wind speed when the backscattering is only caused by the sea surface, the fluctuations are very low. They increase rapidly with increasing wind speed due to intermitting scattering mechanisms: low backscattering caused by the sea surface interrupted by high backscatter levels caused by bubble plumes. The periods of these high backscatter levels become longer until only bubbles contribute to the backscattering. Thus, the CoV approaches the theoretical value of 0.52. As can be seen this wind speed is 18 knots for 18 kHz data and 35 knots for 10 kHz data. At 3 kHz the CoV still decreases up to a wind speed of 45 knots.

The absolute backscattering values versus wind speed for two frequencies and a grazing angle of 45° is shown in Figures 7a and 7b. The backscattering strength is computed from the equation given at the top of the figure [2]. Each data point represents the average from 3000 echoes within a time of 20 minutes. As can be seen in figure 7a at 50 kHz the backscattering strength increases with wind speed up to 10 knots. In the wind speed regime from 10 knots to 30 knots there is only weak dependence, if any. Above 30 knots the backscattering strength decreases. Figure 7b illustrates

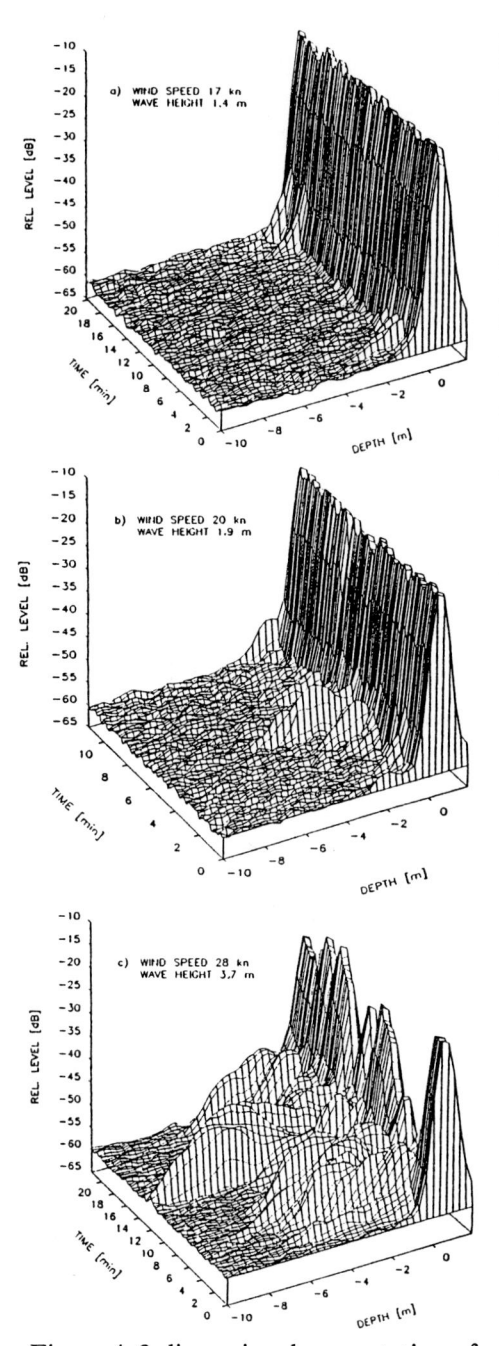

Figure 4. 3-dimensional presentation of normal incidence reverberation, Frequency 40 kHz

the backscattering for a frequency of 26.5 kHz. The data show the same trend for comparable wind speed regimes.

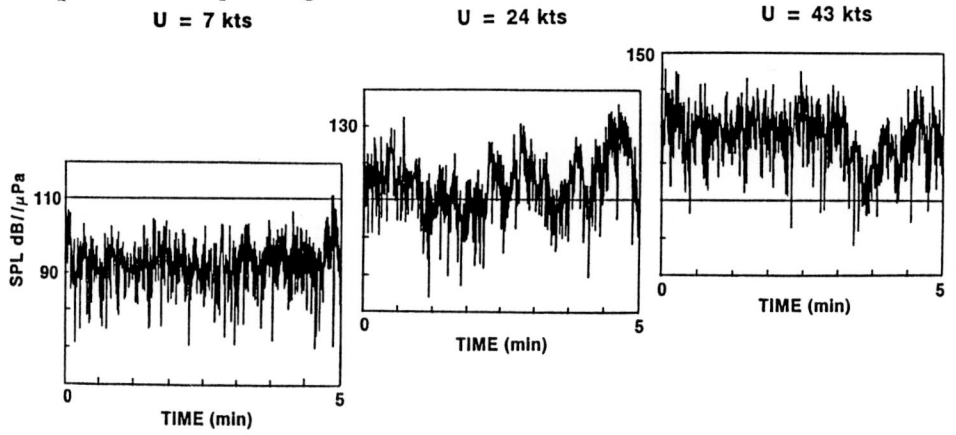

Figure 5. Temporal variability of backscatter at 10 kHz for a 30° grazing angle at three wind speeds

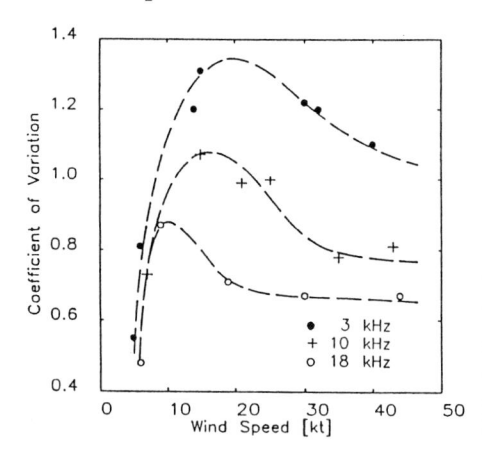

Figure 6. Coefficient of variation as a function of wind speed

4. DISCUSSION

The backscattering dependence on wind speed for grazing angles below 50° shown in figure 7a points out the following principle scattering mechanisms: At very low wind speeds the sea surface is calm and almost all of the energy is reflected in the forward direction. Thus, the backscattering strength is very low. At a wind speed u_r ripples appear on the sea surface and cause the backscattering strength to increase dramatically as illustrated in figure 8. This effect arises for all frequencies at a wind speed $u_r=5$ knots. With increasing wind speed the Rayleigh roughness parameter increases which raises the amount of backscattered energy. At a wind speed u_b bubble clouds or plumes are generated which penetrate into the water column. In the wind speed regime from u_b to u_a the backcsattering strength no longer increases, i.e. saturation occurs. It shall be mentioned here that the boundaries of this regime are frequency dependent. In this regime bubbles and the sea surface contribute to the scattering. At even higher wind speeds above u_a the backscattering strength decreases. We assumed in the data reduction process that the transmission loss is only range dependent. Simultaneous measurements from the same location show a sudden very strong increase in the bubble induced attenuation for the high wind speed regime, which would translate to a lower backscattering strength [3].

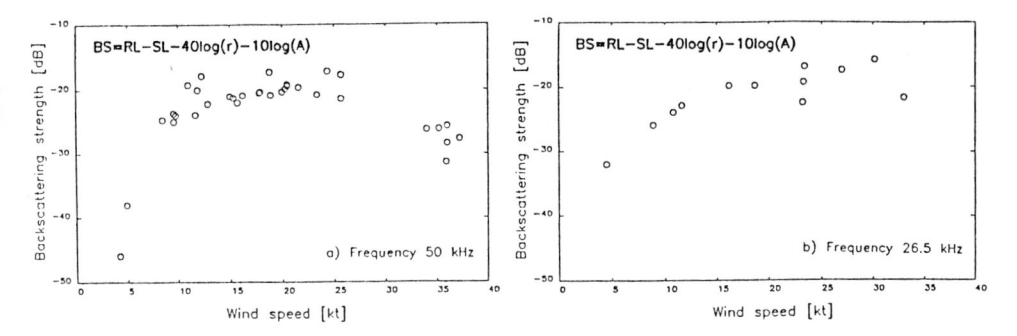

Figure 7. Backscattering strength as a function of wind speed for two frequencies and a grazing angle of 45°

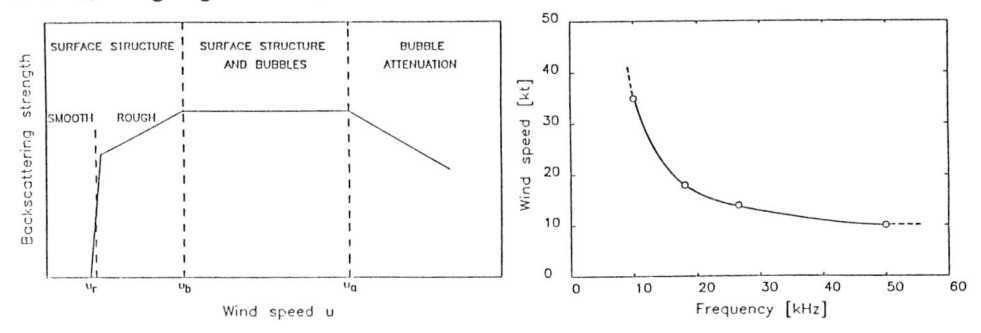

Figure 8. Principal mechanisms affecting backscattering strength as a function of wind speed

Figure 9. Frequency dependence of saturation onset at u_b

The frequency dependence of the saturation onset at u_b is shown in figure 9 for a grazing angle of 45°. As can be seen, the saturation onset obviously approaches a wind speed of 10 knots for increasing frequencies. At lower frequencies saturation occurs at very high wind speeds which corresponds to a decreasing influence of bubbles in this frequency regime. The curve is based on the analysis of a quarter million of individual pings. It shall be mentioned here that measurements at a wind speed of 60 knots and a frequency of 5 kHz did not show saturation.

5. CONCLUSIONS

It appears that the platform NORDSEE is an ideal research tool for scattering experiments at open sea conditions even in the high wind speed regime.
At grazing angles below 50° there are three wind speed regimes where different mechanisms dominate the shape of the scattering strength curve: sea surface roughness at low wind speeds, sea surface and bubbles in a moderate wind speed regime, bubbles at high wind speeds. The boundaries of these regimes are frequency dependent.
At grazing angles above 70° this shape of the scattering strength curve in the low wind speed regime is expected to be different: the backscattering strength decreases with increasing wind speed. Even at normal incidence bubble attenuation screens

the surface acoustically at high wind speeds and high frequencies. Thus, the contribution of bubbles to the backscattering process converts to a lower backscattering strength when computed in the conventional way.
It was shown that bubbles have a strong influence on scattering strength even at low frequencies.

REFERENCES

[1] Nützel,B.
Herwig,H.
Monti,J.M.
Koenigs,P.D.
The Influence of Surface Roughness and Bubbles on Sea Surface Acoustic Backscattering, FWG-Bericht 1987-3, Forschungsanstalt der Bundeswehr für Wasserschall- und Geophysik,Kiel,1987 also NUSC TR 7955, Naval Underwater Systems Center, New London, CT, 1987

[2] Urick,R.J.
Principles of Underwater Sound, McGraw-Hill Inc., 1983, 3. ed., p.247

[3] Herwig,H.
Nützel,B.
The Influence of Bubbles on Acoustic Propagation and Scattering, Underwater Acoustic Data Processing, Kluwer Academic Publishers, 1989, Y.T.Chan, pp.105-111

Section 3

Reverberation Modelling

NUMERICAL MODELING OF THREE-DIMENSIONAL REVERBERATION FROM BOTTOM FACETS

HENRIK SCHMIDT
Massachusetts Institute of Technology
Cambridge, MA 02139
USA

ABSTRACT. The three-dimensional acoustic reverberation from bottom facets in a stratified, seismo-acoustic ocean environment is modeled using a hybrid approach where the environment is divided into two regions, one being a stratified model of the environment outside the scatterer, and the other being the scattering facet itself. A boundary element formulation is used for the facet boundary separating the two domains, leading to an expression for the reverberant field in terms of superposition of fields produced by a distribution of sources on the boundary. The basic component of the model is the SAFARI code for propagation in stratified media, and the hybrid model therefore directly provides the spectral decomposition of the reverberant field in terms of modal components, head waves as well as seismic interface waves. The model is used to determine the importance of facet geometry and composition, as well as bottom stratification. The three-dimensional modal conversion produced by line-type facets such as ridges is addressed using a spectral composition approach, basically decomposing this class of 3-dimensional reverberation problems into a superposition of two-dimensional problems. It is demonstrated that the elastic properties of the facet and the bottom stratification are extremely important factors in shaping the spatial distribution of the reverberant field, and further it is demonstrated that the scattering problem is strongly three-dimensional, making the classical target strength concept totally inadequate for bistatic active scenarios in particular.

1. Introduction

Acoustic reverberation from the ocean boundaries is the most severe limiting factor in relation to the use of active sonar systems for target detection, localization and classification. The ocean acoustic reverberation is controlled by three major factors. The first is the propagation from the source to the reverberant feature, the second is the scattering process and the third is the propagation to the receiver. Traditionally the scattering process has been represented by a single parameter in the sonar equation, the *target strength*, which together with the transmission losses along the propagation path from the source to the scatterer and from the scatterer to the receiver has defined the reverberation level. The target strength concept has been adequate for classical sonar systems based on the amplitudes of the returned signals and traditional array processing for bearing estimation. Here, reverberation modeling has been performed using the target strength concept of the sonar equation. Thus, the propagation from the source to the scatterer is represented by a transmission loss as predicted by one of the standard propagation models. The scatterer is then represented by a point source with a strength equal to the sum of the incident field in dB and the *target strength*. The reverberant field at the receiver is then determined using a standard propagation model.

However, in the mid-frequency regime of 1-1000 Hz, the target strength of the bottom scattering features is very similar to the target strength of a submarine, a factor which severely limits the use of classical active systems in large parts of the ocean environment.

The adaptation to active systems of the modern, high resolution array processing schemes developed for passive systems has the potential of vastly improving the system performance. For example, *matched field processing* (MFP) explores the multipath structure of the field to enable localization of a source in all spatial coordinates as well as environmental parameters. MFP can be directly applied to the active scenario for discrimination between a target within the water column and reverberation facets at the surface or the bottom, provided the latter can be properly described by a point source representation. However, this is in general not the case for the simple reason that the dominant scattering features have a size comparable to a few wavelengths and therefore produces a field with a non-uniform radiation pattern. The scattering problem is inherently three-dimensional, and the bottom reverberation is strongly affected by the elastic properties of the sea bed, and these are features that MFP could explore to enable discrimination between reverberation returns and target returns, even in cases where the target is close to the ocean boundaries. To explore the full potential of MFP it is therefore necessary to properly model the spatial distribution of the scattered field, as well as the ocean waveguide propagation.

Unfortunately, the classical propagation models are not capable of handling the scattering problem properly. Models such as the FFP [1,2], Normal Mode [3] and the classical PE [4] are applicable only to forward, 2-dimensional propagation in range-independent or weakly range-dependent environments. There has therefore in recent years been increased effort in expanding the acoustic modeling capability to incorporate such effects. The Coupled mode approach of Evans *et al.* [5] allows for modeling the backscattering from distinct changes in the environment in a 2-dimensional ocean model. Collins *et al.* [6] has introduced a single scattering approximation to the acoustic PE to obtain the same capability more efficiently. None of these models incorporate the elastic effects in the bottom. Collins *et al.* [7] have developed an elastic PE, but the single scattering approximation necessary to expand the applicability to reverberation problems is non-trivial. The only models capable of solving the full 3-D, elastic scattering problems are the discrete methods such as Finite Difference Methods [8] or Finite Element Methods [9]. However, these methods are extremely computationally extensive, even in 2-D, and must therefore be applied in hybrid schemes where the propagation to and from the scatterer is treated by one of the classical models, with only a small region surrounding the scattered feature being modeled with the discrete model.

We will here apply a different hybrid approach, recently developed by Gerstoft and Schmidt [10], to model the 3-dimensional seismo-acoustic reverberation from

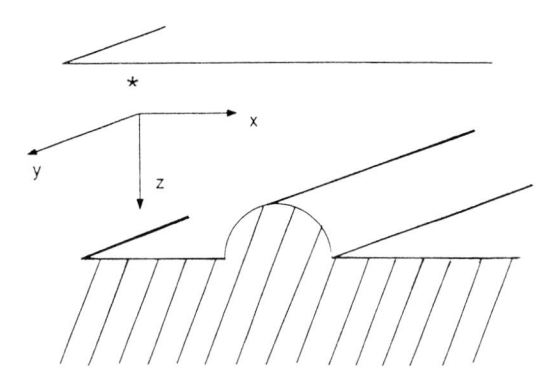

Figure 1. Environmental model for reverberation of a point source field in a horizontally stratified ocean environment with an infinitely long bottom facet

discrete inhomogeneities - or facets - in an elastic sea bed. The environment is divided into an exterior and an interior region, separated by the boundary of the facet. Both regions are assumed to be range-independent stratified media. A boundary integral representation is then used for the reverberant field, equivalent to a superposition of the field produced in the waveguide by a distribution of sources around the facet, allowing for the field to be computed by a standard propagation model such as SAFARI [12]. This method therefore also directly provides the spectral decomposition of the reverberant field in terms of normal modes, head waves and seismic interface waves.

2. Hybrid Reverberation Model

The propagation and scattering in the environment shown in Fig.1 is controlled by the 3-dimensional acoustic Helmholtz equation

$$\left[\frac{\partial^2}{\partial x^2} + \frac{\partial^2}{\partial y^2} + \frac{\partial^2}{\partial z^2} + k^2\right]\psi(x,y,z;\omega) = 0, \tag{1}$$

and its elastic equivalents for the compressional potential and two shear potentials. Since the environment is range independent in the y–direction, we can apply the spatial Fourier transform in this coordinate to Eq. (1). The total field then has the integral representation

$$\psi(x,y,z;\omega) = \int_{-\infty}^{\infty} \psi(x,z;k_y,\omega), \tag{2}$$

where the integration kernel is the solution to the 2-dimensional Helmholtz equation

$$\left[\frac{\partial^2}{\partial x^2} + \frac{\partial^2}{\partial z^2} + (k^2 - k_y^2)\right]\psi(x,y,z;\omega) = 0. \tag{3}$$

Eq. (3) is identical to the standard Helmholtz equation for a 2-dimensional problem, but with the medium wavenumber term k^2 replaced by $(k^2 - k_y^2)$. We can therefore determine the total field as a superposition of 2-D solutions to problems with the same environmental geometry, but varying medium properties. For acoustic problems the 2-D problems can be solved using any method including backscattering such as the 2-way PE [6] or Coupled Modes [5]. However, for elastic bottoms the potential for the horizontally polarized shear waves and the associated additional boundary conditions must be incorporated to properly compute the out of plane components for $k_y \neq 0$. The hybrid boundary integral - wavenumber integration approach recently developed for the 2-D problem by Gerstoft and Schmidt [10] can straightforwardly be extended to incorporate these additional components, and this is the modeling approach taken here.

The boundary integral method (BIM) is based on Green's theorem which states that within a volume V bounded by the surface S, the acoustic field is given by the integral representation,

$$\psi(\mathbf{r}) = \int_S \left[G_\omega(\mathbf{r},\mathbf{r}_S) \frac{\partial \psi(\mathbf{r}_S)}{\partial n_S} - \psi(\mathbf{r}_S) \frac{\partial G_\omega(\mathbf{r},\mathbf{r}_S)}{\partial n_S} \right] dS$$
$$- \int_V f(\mathbf{r}_V) G_\omega(\mathbf{r},\mathbf{r}_V) dV \ , \tag{4}$$

where $G_\omega(\mathbf{r},\mathbf{r}')$ is any *Green's function* satisfying the Helmholtz equation within the volume V, and n_S is the outward pointing normal at the point $\mathbf{r}_S$ on the boundary S. The volume integral accounts for the volume sources within the volume V.

The most common numerical implementation of BIM is the Boundary Element Method (BEM), so named due to its resemblance with the Finite Element Method (FEM). Here, a discrete form of Eq. (4) is obtained by dividing the boundary S into panels with a limited number of degrees of freedom, leading to a matrix equation for the field $\psi(\mathbf{r}_S)$ and its normal derivative on the boundary. Once solved, the field within the volume can be found by evaluating Eq. (4) for any receiver position $\mathbf{r}$.

The efficiency of the boundary element method relies on a simple expression for the Green's functions, and BEM has therefore mainly been used for scattering by objects and structures in an infinite medium where the Green's function is given in closed form. In the ocean environment, the Green's function is a very complicated function of space and frequency, requiring numerical solution in itself. The importance of BEM in underwater acoustics has therefore been extremely limited. However, the increased availability of computational power and efficient Green's function generators has yielded the possibility of applying BEM to scattering by rough ocean boundaries [11] and from objects within the ocean waveguide and the ocean bottom [10].

The environment shown in Fig.1 is an example of an ocean which is globally range independent, except for a local feature or *facet*. Here, we can directly separate the environment into an external and internal stratified environment, separated by a boundary surrounding the facet. We can then use Green's theorem for both regions, with the boundary conditions expressed through the boundary element method (BEM) described above, and the Green's functions determined very efficiently by a WI model such as SAFARI [12]. Such a model, capable of providing frequency and time solutions to problems involving seismo-acoustic scattering and reverberation

□ F = 250 Hz, SD = 175 m

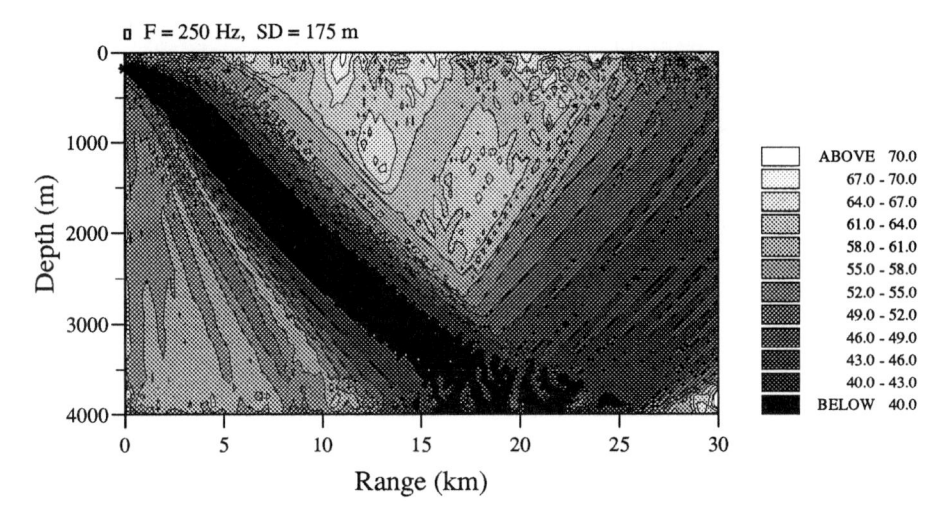

□	ABOVE 70.0
	67.0 - 70.0
	64.0 - 67.0
	61.0 - 64.0
	58.0 - 61.0
	55.0 - 58.0
	52.0 - 55.0
	49.0 - 52.0
	46.0 - 49.0
	43.0 - 46.0
	40.0 - 43.0
■	BELOW 40.0

Figure 2. Transmission loss contours of incident and reflected field in the ARSRP scenario. The source array is steered down at 6° to produce a field incident at low grazing angle on the bottom

was recently developed by Gerstoft and Schmidt [10]. This model has been extended to include the additional elastic boundary conditions for the decomposed Helmholtz equation, Eq. (3), and is applied to analyze the basic physics of the 3-dimensional acoustic reverberation from ridge facets on and within the sea bed.

3. Bottom Facet Reverberation

The hybrid boundary element - wavenumber integration model has been used to simulate reverberation from hypothetical ridge structures on and within the sea bed in the mid-Atlantic area where the ONR-ARSRP reconnaissance experiment was carried out in the summer of 1991. The water depth is 4000 m, and the bottom was characterized by a very thin sediment layer overlying a basalt subbottom. A narrow-band acoustic field of 250 Hz was generated by a vertical array of 10 sources, phased to generate a beam incident on the bottom at very low grazing angles. Fig.2 show contours of the transmission loss for the incident and reflected field under the assumption of horizontal stratification, and with the cylindrical spreading removed.

An infinite ridge of height 6 m is assumed to be present in the insonified area, and two cases are considered as shown in Fig.3. The first case has no sediment layer, but a semi-cylindrical ridge of 6 m height is present at a range of 20 km, at the center of the incident beam. In the second case the same basalt is covered by a 10 m sediment layer.

Fig.4 shows contours of the transmission loss of the scattered field for the two bottom models. Fig.4(a) shows the result for the environment in Fig.3(a), i.e. without sediment layer, and similarly Fig.4(b) shows the result for the sediment-covered facet in Fig.4(b). The receiving horizontal array in the ARSRP experiment was

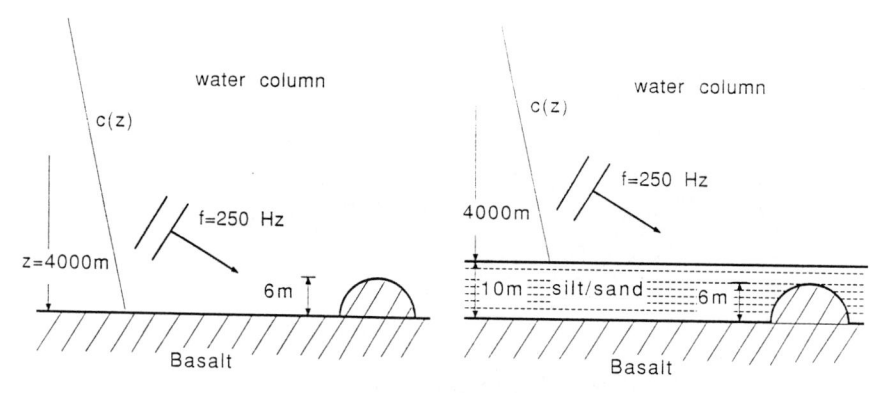

Figure 3. Bottom facet models used for modeling deep ocean reverberation for a field incident at low grazing angles. (a) Ridge of height 6 m on a basalt halfspace. (b) Same basalt ridge covered by 10 m of silt/sand sediment.

towed at a depth of 170 m, and it is clear from Fig.4 that the sediment lowers the reverberation detected at the receivers by more than 10 dB. This is due to the fact that the incident field is strongly evanescent in the sediment layer.

When accounting for the geometrical spreading the result in Fig.4(a) translates into a target strength of 16 dB for the 6 m ridge, which is consistent with the experimental results as reported by Dyer et al. [13]. Similarly the target strength for the sediment-covered ridge is approximately 5 dB. The results in Fig.4 also clearly illustrates the limitation of the classical target strength concept. Thus, the radiation pattern of the scattered field is far from being omnidirectional. Although the strong reverberation at the steeper angles arrives at the receiving array after one or more bottom bounces, these multipath arrivals could be included in a matched field processing scheme for a more complete characterization of the scatterer, which may then in turn enable discrimination between a reverberation return and a real target. The spatial variability of the reverberant field in Fig.4 also makes it clear that the radiation pattern is of particular importance for bistatic scenarios.

The results shown here are computed for a plane perpendicular to the ridge, but initial results for the full 3-D scenario have indicated a strong 3-dimensional nature of the reverberant field. For example the individual modes excited by the scatterer will propagate at different horizontal angles. For horizontal angles of incidence close to the axial direction of the ridge the scattering into waterborne modes becomes less significant, but the various sub-sonic seismic waves will still be strongly excited by the scatterer. These features are obviously of less importance for monostatic scenarios where the reverberant field will be dominated by scattering from features perpendicular to the direction to the source. However, for bistatic scenarios this inherent 3-D nature of the scattering problem becomes extremely important.

4. Conclusion

A Fourier synthesis approach has been developed for modeling the 3-dimensional

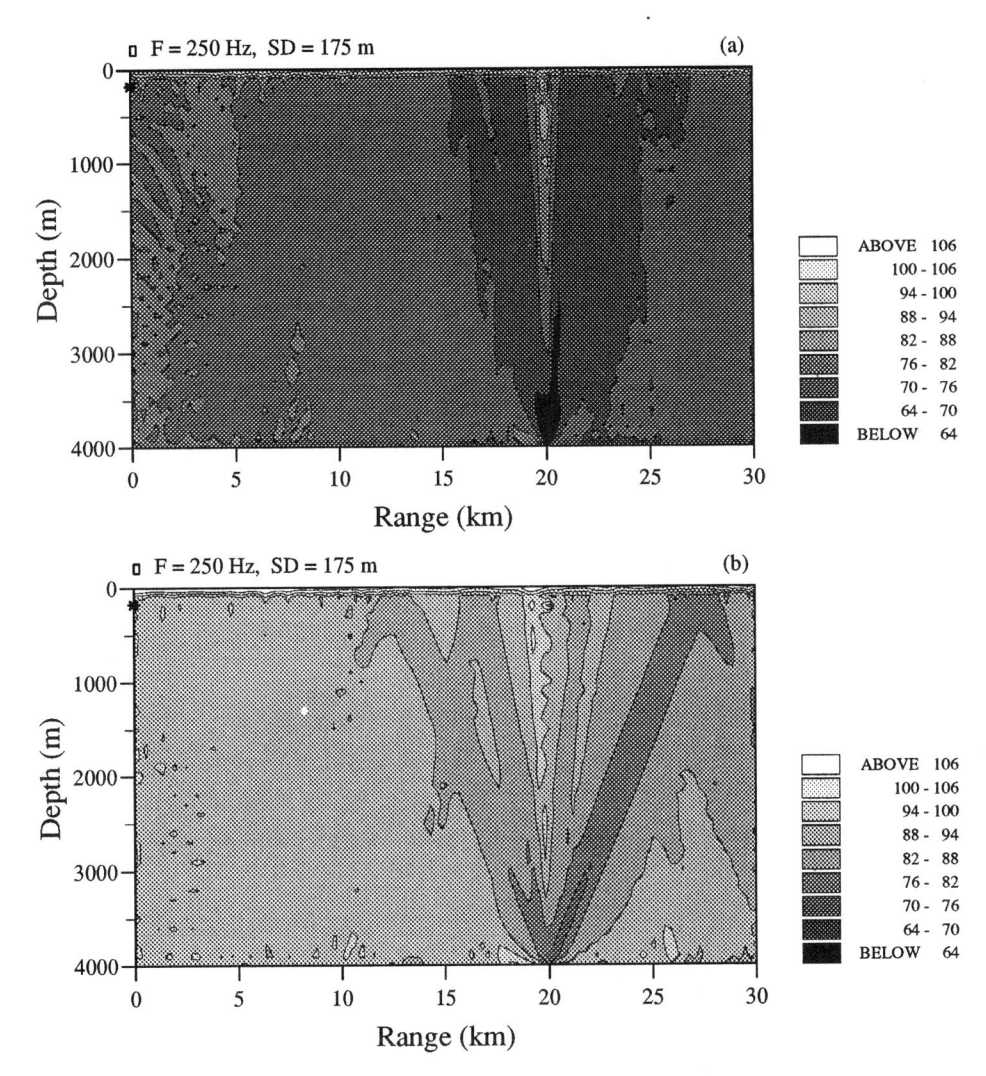

Fig.4. Transmission loss contours for reverberant field in ARSRP environment. (a) Basalt ridge of height 6m on Basalt halfspace. (b) Same ridge covered by 10 m silt/sand sediment.

reverberation from infinitely long bottom facets. This approach decomposes the full problem into a number of 2-dimensional problems the solution to which are superimposed to yield the total reverberant field. A hybrid boundary element - wavenumber integration model is used to solve the individual 2-D problems. The model has been applied to simulate monostatic reverberation in a mid-Atlantic environment, and it has been demonstrated that elongated bottom features of a height comparable to the acoustic wavelength will have a target strength for low incident grazing angles which compares well with those experimentally determined. Initial results for the full bistatic scenarios have demonstrated the strong 3-D nature of the bottom reverberation problem, suggesting that the traditional 2-dimensional modeling approaches are of very limited importance in such scenarios.

References

[1] F. DiNapoli and R. Deavenport, "Theoretical and numerical green's function solution in a plane multilayered medium." J. Acoust. Soc. Am., 73:92–105, 1980.

[2] H. Schmidt and F.B. Jensen, "A full wave solution for propagation in multilayered viscoelastic media with application to Gaussian beam reflection at fluid-solid interfaces." J. Acoust. Soc. Am., 77:813–825, 1985.

[3] A.D. Pierce, "Augmented adiabatic mode theory for upslope propagation from a point source in variable-depth shallow water overlying a fluid bottom." J. Acoust. Soc. Am., 74:1837–1847, 1983.

[4] F.D. Tappert, "The parabolic approximation method." In J.B. Keller and J.S. Papadakis, editors, *Wave Propagation in Underwater Acoustics*. Springer-Verlag, New York, 1977.

[5] R.B. Evans, "A coubled mode solution for propagation in a waveguide with stepwise depth variations of a penetrable bottom." J. Acoust. Soc. Am., 74:188–195, 1983.

[6] M.D. Collins and R.B. Evans, "A two way parabolic equation for acoustic backscattering in the ocean," J. Acoust. Soc. Am. **91**, 1357–1368 1992.

[7] M.D. Collins and W.A. Kuperman, "A stable higher-order elastic parabolic equation with application to Scholte wave propagation." In J.M. Hovem, M.D. Richardson, and R.D. Stoll, editors, *Shear Waves in Sediments*. Kluwer, Doordrecht, NL, 1991.

[8] R.A. Stephen, "A review of finite difference methods for seismo-acoustics problems at the seafloor," Rev. of Geophys., **V 26**, 445–458, 1988.

[9] O.C. Zienkiewicz, *The Finite Element Method,* McGraw-hill, London,UK, 1977.

[10] P. Gerstoft and H. Schmidt, "A boundary element approach to ocean seismoacoustic facet reverberation." J. Acoust. Soc. Am., 89:1629–1642, 1991.

[11] T.W. Dawson and J.A. Fawcett, "A boundary integral equation method for acoustic scattering in a waveguide with nonplanar surfaces." J. Acoust. Soc. Am., 87:1110–1125, 1990.

[12] H. Schmidt, "SAFARI: Seismo-acoustic fast field algorithm for range independent environments. user's guide." SR 113, SACLANT ASW Research Centre, La Spezia, Italy, 1987.

[13] I. Dyer, A.B. Baggeroer, "Analysis and interpretation of bottom backscattering from the Atlantic Natural Laboratory." In this volume, 1992.

NUMERICAL SIMULATIONS OF LOWER-FREQUENCY ACOUSTIC PROPAGATION AND BACKSCATTER FROM SOLITARY INTERNAL WAVES IN A SHALLOW WATER ENVIRONMENT

Stanley A. Chin-Bing and David B. King
Naval Research Laboratory
Stennis Space Center, MS 39529, U.S.A.

Joseph E. Murphy
University of New Orleans
New Orleans, LA 70148, U.S.A.

ABSTRACT.

The effects of a solitary internal wave on the low-frequency ocean acoustic field in a shallow-water waveguide have been examined using numerical simulations. The waveguide contained the inhomogeneous water column, range-dependent bathymetry, and inhomogeneous ocean bottom-subbottom. Environmental parameters (density, compressional and shear wave speeds and attenuations) in the bottom-subbottom varied in both range and depth. Small variations in range and depth along the ocean bottom-subbottom interfaces were used to simulate seafloor roughness. The coupled acoustic fields in this total waveguide simulation were calculated using full-wave two-way range-dependent finite-element ocean acoustic computer models (FOAM, FFRAME, SAFE).

1. INTRODUCTION

The existence of large amplitude internal solitons in shallow water have been documented in many coastal zones.[1] Recently resonant effects of these solitary internal waves on acoustic propagation have been used to explain unusually high attenuation in measurements made in the Yellow Sea.[2] In general, however, the study of the acoustic effects due to solitary internal waves in shallow water has been ignored. In this paper we attempt to quantify these acoustic effects by presenting the results of numerical simulations of acoustic propagation through and backscatter from solitary internal waves in a shallow-water environment. Since reverberation from internal wave fields at high frequency is quite likely, we concentrate this study on the lower frequencies. Simulation of such small effects requires acoustic computer models that are highly accurate and include the total solution to the acoustic wave equation. In addition, in shallow water the acoustic field is "molded" by the waveguide and this "effective waveguide" extends from the ocean-air surface down through the ocean subbottom and ocean seafloor basement. Thus, the ocean variability and the seafloor mechanisms are coupled. We have calculated these coupled acoustic fields in this total waveguide environment using the benchmark-accurate, full-wave, two-way, range-dependent, finite-element ocean acoustic computer models (FOAM, FFRAME, SAFE and PE-SAFE [3-6]). The 2-D waveguide included the inhomogeneous water column, range-dependent bathymetry, and inhomogeneous ocean bottom-subbottom. Environmental parameters (density, compressional and shear wave speeds and attenuations) in the bottom-subbottom varied in range and depth. Small environmental-bathymetric variations in range and depth along the ocean bottom interfaces were used to simulate seafloor roughness. Results of simulations on selected examples will serve to quantify effects and assist in classifying the acoustic importance of solitary internal waves in shallow water.

2. NUMERICAL MODELS

2.1 FINITE-ELEMENT SEISMO-ACOUSTIC MODELS

There were two primary types of cw acoustic models used in this study, each of which is based on the finite-element (FE) method. One type of FE model (FOAM [3,4] and FFRAME [5]) solves the scalar Helmholtz wave equation for a fluid-type ocean bottom. The other type of FE model (SAFE [6,7,8]) solves the elastic wave equation for a shear-supporting ocean bottom. A schematic diagram of the basic finite-element model geometry is shown in Fig. 1. The water column and ocean bottom are represented by a grid of FE elements. The environmental properties (density, compressional and shear wave speeds, compressional and shear wave attenuations, and density) can be different within each element. Since these FE models are axisymmetric, this means that the ocean and ocean-bottom environments can vary in both range and depth. The SAFE model shown in Fig. 1 employs rectangular FE elements that are adjusted in a stair-step fashion to approximate the ocean features (e.g., ocean-bottom bathymetry, sea-floor roughness). SAFE solves for the complex displacements from which complex pressure is calculated. In contrast, the fluid-bottom FE models, FOAM and FFRAME, use triangular elements that can be adjusted to lie exactly on the boundaries of the ocean features. In addition, FOAM and FFRAME directly solve for the complex pressure field. In this study, FFRAME was used to model the internal wave packet (solitons) since FFRAME could adjust its node position to represent the boundaries of the solitons. When this field from the soliton (as calculated by FFRAME) was propagated over an ocean-bottom region that supported roughness and shear, the SAFE model was used.

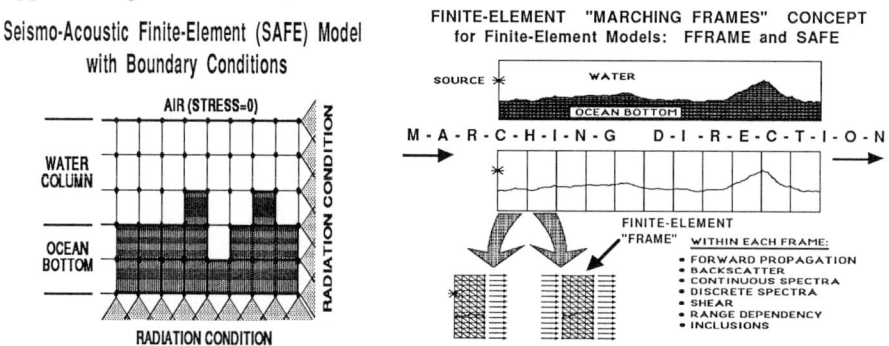

Fig. 1. SAFE model with boundary conditions.   Fig. 2. Marching Frames concept for FE.

To achieve longer range propagation, a "marching frames" approach (see Fig. 2) has been developed for both FOAM and SAFE -- the marching version of FOAM is known as the FFRAME model while the SAFE model has the marching frames approach incorporated in it. For either of these marching models, the FE model solves the problem in a selected range-dependent ocean region. The vertical field at the end of this region ("frame") is then used as the starting field for the next region. Thus, within each frame, the full-wave, 2-way, coupled, range-dependent complex pressure field is calculated. However, only the 1-way (outgoing) field is passed from frame-to-frame. In addition, a parabolic equation (PE) model [9] may be used to solve any particular frame and its field coupled to the FE model, and vice-versa. Finally, a mixture of scalar fluid models (e.g., FFRAME) and elastic models (e.g., SAFE) may be combined in this marching approach to provide long-range simulation over ocean bottom regions which contain shear in some regions and not in others. Of course the SAFE model could be used in its marching mode to solve both the shear and non-shear regions, but this is costly in computer time and memory since the SAFE model would be solving an elastic wave equation that in some regions has zero shear and shear attenuation. Nevertheless, the SAFE model can do this accurately as is shown in Fig. 3a. Figure 3b shows the accuracy of the SAFE model in propagating the water-borne field over a range-dependent shear-supporting bottom. Note the effect of the shear bottom as seen by comparing results with that predicted by the non-shear model. Figure 3c illustrates the ability of the SAFE model to include Scholte waves as well as to include shear properties in the bottom. Comparisons (Fig. 3) were made between the SAFE model and COUPLE [10], FEPES [11], and SAFARI [12], as indicated.

VERIFICATION THAT THE SAFE MODEL CAN SIMULATE SCENARIOS THAT DO NOT INCLUDE OCEAN-BOTTOM SHEAR WAVES

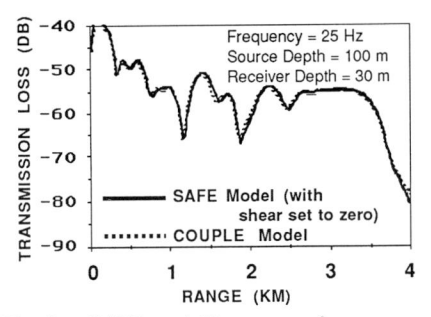

Fig. 3a. SAFE model is accurate for range-dependent fluid bottoms.

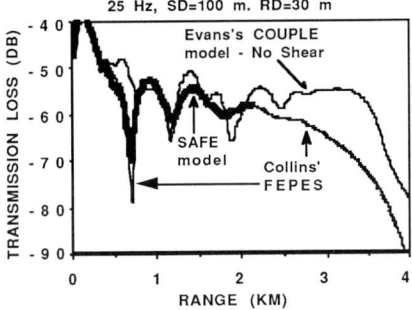

Fig. 3b. SAFE model is accurate for range-dependent shear bottoms.

VERIFICATION THAT THE SAFE MODEL CAN SIMULATE SCENARIOS THAT DO INCLUDE OCEAN-BOTTOM SHEAR INTERFACE WAVES

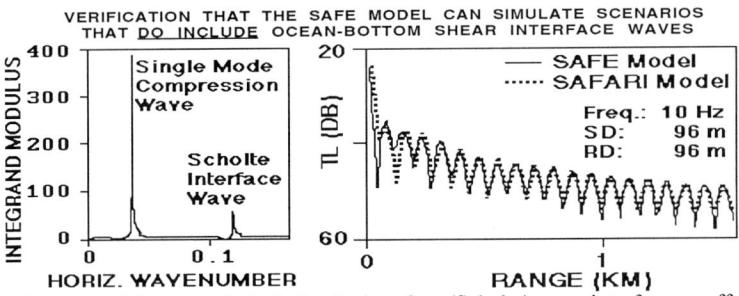

Fig. 3c. SAFE model accurately includes the interface (Scholte) wave interference effects.

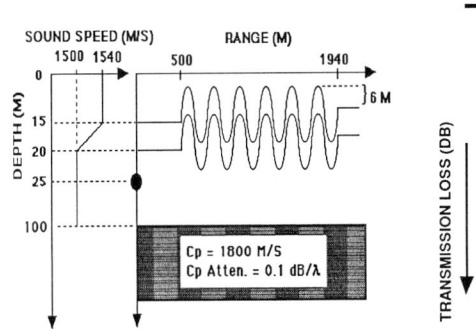

Fig. 4. Model of an internal waves packet.

Fig. 5. Effects of internal waves on TL.

2.2 INTERNAL WAVES MODEL.

The internal wave was represented by a form, $z(d) = d - A(d) \sin(2\pi r / \lambda)$, similar to that given in [1] and [2], where d refers to the depth of the horizontal interfaces along which the internal wave travels, z(d) is the actual depth of the internal wave boundary relative to the sea surface, r is the range at which the internal wave occurs, and A(d) and λ are the amplitude and wavelength, respectively, of the internal wave. The upper and lower interfaces of the internal wave were placed at d = 15 m and d = 20 m, with A(d) = 6 m and λ = 235 m for each interface. The internal wave was located at 500 m $\leq$ r $\leq$ 1940 m with a total length of 6λ. A schematic diagram of the internal wave and the corresponding sound speed profile is given in Fig.4. This equation, and the structure shown in Fig. 4 and used in Ref. 2, was originally posed by Lee [13].

3. NUMERICAL EXAMPLES

The FFRAME model was used to model the scenario depicted in Fig. 4. A sweep through the lower frequency range indicated that a small but visible change in the transmission loss (TL) occurred at 75 Hz, as is shown in Fig. 5. To access the effect of the waveguide environment on the detection of the soliton, a rough shear supporting ocean bottom was introduced, starting at the range of 2.2 km and ending at the range of 2.7 km. The interface between the ocean and the ocean bottom was modeled as a small, randomly rough surface. This was accomplished by adjusting the node positions of the SAFE model that lie on the ocean-bottom interface to correspond to depths calculated from a fractal equation model. The fractal equation model used to create the randomly rough surface is a subroutine in the SAFE model and node positions are done automatically. For this study the amplitudes of the roughness were $\leq$ 1.0 m about the mean depth (100 m) of the interface. A schematic of this region and the environmental parameters used are given in Fig. 6.

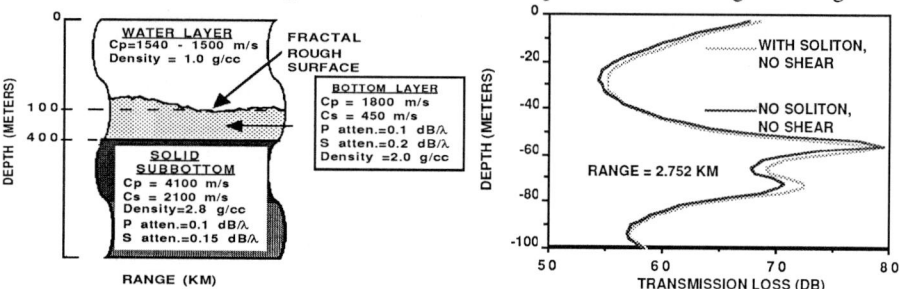

Fig. 6. Segment showing the rough shear region.   Fig. 7. Internal waves over no-shear region.

3.1 EFFECTS ON PROPAGATION

In the results and discussions that follow, the location of interest is constrained to the water-column. Thus, the effects of the soliton, the rough shear-supporting ocean bottom, and solid subbottom are all viewed as they would effect sensors placed in the water column.

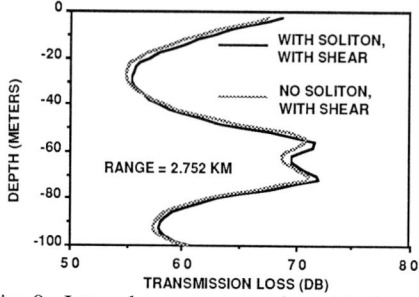

Fig. 8. Internal waves over region with shear. 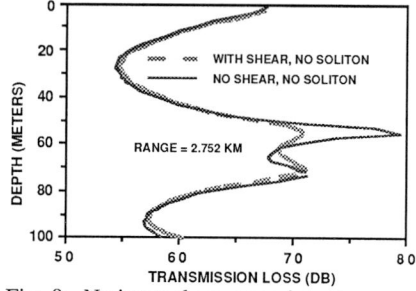   Fig. 9. No internal waves and no-shear.

3.1.1 *Due to the Soliton.*

The acoustic field for the region that included the soliton (Fig. 4) was calculated by FFRAME. This field was then used as the starting field for the SAFE model starting at a range of 2.2 km. Figure 7 compares the effects due primarily to the soliton. Shear speed and shear attenuations have been set to zero in the SAFE model. The observation range of 2.752 m is just after the acoustic field passes over the rough ocean bottom region. The effects shown in Fig. 7 are attributable to the soliton and, to a lesser degree, to the combined effects of the small roughness. Clearly the soliton only minimally effects the propagation. (Recall that these effects are due to the entire shallow water waveguide -- from the air-sea interface, through the sea-floor, and deep into the solid subbottom.)

3.1.2 *Due to the Ocean Bottom.*

With non-zero shear parameters (given in Fig. 6), a comparison of the results (shown in Fig. 8, between the presence-absence of the soliton) can be made as the acoustic field traversed a slightly rough, shear-supporting bottom region. Differences in TL resulting from the soliton are comparable to that shown in Fig. 7. However, the differences caused by the rough shear region are far more noticeable, as replotted in Fig. 9 for the two cases without the soliton. These results are plausible since the small bottom roughness does not influence the water-borne acoustic field when bottom-shear is absent; but, when shear is allowed, the roughness is large enough to change the angles at which the acoustic wave strikes and enters the shear-supporting bottom, thus causing an increase in shear-mode conversions. The null near the mid-depth of the water column (Fig. 9) can be attributed to the phase changes introduced by the rough shear-supporting bottom region.

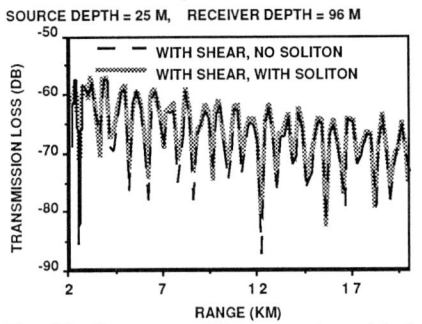

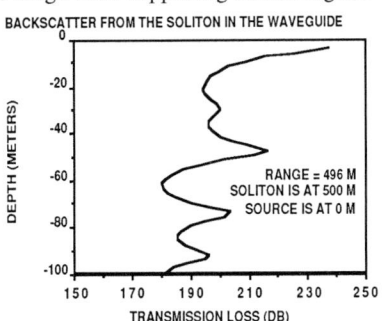

Fig. 10. Downrange TL over region with shear.   Fig. 11. Backscatter from internal waves.

The effects of the soliton on long-range propagation is shown in Fig. 10. The receiver was at a depth of 96 m (4 m above the bottom). The source was too far from the bottom to excite the Scholte wave in the shear-supporting region, thus an interference pattern (similar to Fig. 3c) would not occur. A receiver placement near the bottom was use to examine the possibility of using near-bottom sensors to monitor internal wave activity in shallow water. Near-bottom placement has the advantage of simplicity in mooring, stability with regard to drifting, and long-term endurance since that location is less likely to interfere with shipping. However, the results given in Fig. 10 indicate that the effects of the waveguide environment, apart from the soliton, dominates. Thus, the effects of the soliton are negligible when monitored from the bottom and at ranges far from the soliton.

3.2 EFFECTS ON BACKSCATTER

Finally, the effects of the soliton on the backscattered acoustic field were simulated using the FFRAME model. A measured of the backscattered field from the soliton was obtained by: (1) calculating the total coupled 2-way acoustic complex pressure field, with and without the presence of the soliton; (2) differencing these two complex pressure fields, point-for-point, to obtain the complex field due to the presence of the soliton; and, (3) converting the differenced complex pressure field to dB. This result is shown in Fig. 11 where the acoustic source was at a depth of 25 m and at range zero, the soliton was at a mean depth of 15-20 m and began at range 500 m (Fig. 4) and the vertical receivers were located in the backscattered direction at range 496 m. The results are not surprising -- the backscattered field due to the soliton is a very weak field and is strongest in the backscattered downward angle (since the ensonifying source is below the soliton).

4. SUMMARY

The cumulative effects on propagation of multiple resonances of soliton packets had been postulated and examined by others. Therefore, in this numerical experiment we have focussed the study on the refractive and backscatter effects of a single soliton wave packet embedded in a shallow water waveguide. The waveguide was allowed to contain a bottom region that had small roughness on the water-bottom interface and that supported shear in the bottom and subbottom. The effects of the waveguide dominated the effects due to the soliton packet to the extent that discrimination with sensors near the shallow water bottom and at long ranges was highly unfavorable. The effects of the soliton were more recognizable in the backscattered field since there were no propagation effects to clutter this field. However, the weakness of the backscattered field suggests that only near-field sensor placements are likely to produce favorable discrimination.

5. ACKNOWLEDGEMENTS

This work was supported by the Office of Naval Research (ONR), the Naval Research Laboratory, and the ONR Acoustic Reverberation Special Research Program. J. E. M. was also supported by the Research Program Division of ONR. Computations were performed on the Cray Y-MP8/8128 at the Naval Oceanographic Office's POPS Supercomputer Center in Hancock County, MS. NRL Contribution No. PR 0090:21. Approved for public release; distribution is unlimited.

6. REFERENCES

[1] Haury, L. R., Melbourne, M. G., and Orr, M. H. (1979), "Tidally generated internal wave packets in Massachusetts Bay," Nature 278, 312-317.
[2] Zhou, J. X., Zhang, X. Z. and Rogers, P. H. (1991), "Resonant interaction of sound wave with internal solitons in the coastal zone," J. Acoust. Soc. Am 90, 2042-2054.
[3] Murphy, J. E. and Chin-Bing, S. A. (1988), "A finite element model for ocean acoustic propagation," Mathematical and Computer Modeling, 11, 70-74.
[4] Murphy, J. E. and Chin-Bing, S. A. (1989), "A finite element model for ocean acoustic propagation and scattering," J. Acoust. Soc. Am. 86, 1478-1483.
[5] Chin-Bing, S. A. and Murphy, J. E. (1990), "Acoustic wavefield distortion by sea-mounts: A finite element analysis," in D. Lee, A. Cakmak and R. Vichnevetsky (eds.), Computational Acoustics, Vol. 1, Elsevier, North-Holland, 37-52.
[6] Murphy, J. E. and Chin-Bing, S. A. (1991), "A seismo-acoustic finite element model for underwater acoustic propagation," in J. M. Hovem, M. D. Richardson, and R. D. Stoll (eds.), Shear Waves in Marine Sediments, pp. 463-470, edited by J. M. Hovem, M. D. Richardson, and R. D. Stoll (Kluwer Press, 1991).
[7] Chin-Bing, S. A. and Murphy, J. E. (1991), "Long-range ocean acoustic-seismic propagation modeling using hybrid finite element and parabolic equation models," in Computational Acoustics edited by D. Lee, A. Robinson, and R. Vichnevetsky (Elsevier Publishers, North-Holland, 1992).
[8] Chin-Bing, S. A. and Murphy, J. E. (1991), "Shear effects on ocean acoustic propagation due to step-periodic roughness along the ocean bottom interface," Proc. of the Eight International Conference on Mathematical and Computer Modeling, 1-4 April 1991, University of Maryland, College Park, MD. (In press).
[9] Collins, M.D. (1988), "FEPE user's guide," NORDA TN-365, Naval Ocean Research and Development Activity, Stennis Space Center, Mississippi.
[10] Evans, R. B. (1983), "A coupled mode solution for acoustic propagation in a waveguide with stepwise depth variations of a penetrable bottom," J. Acoust. Soc. Am., 74, 188-194.
[11] Collins, M.D. (1989), "A higher-order parabolic equation for wave propagation in an ocean overlying an elastic bottom," J. Acoust. Soc. Am. 86, 1459-1464.
[12] Schmidt, H. and Jensen, F. B. (1985), "Efficient numerical solution technique for wave propagation in horizontally stratified environments," in M. H. Schultz and D. Lee (eds.), Computational Ocean Acoustics, Pergamon, New York, 699-715.
[13] Lee, O. S. (1961), "Effect of an internal wave on sound in the ocean," J. Acoust. Soc. Am. 33, 677-681.

REVERBERATION MODELING WITH THE TWO-WAY PARABOLIC EQUATION

M.D. Collins, G.J. Orris, and W.A. Kuperman
Naval Research Laboratory
Washington, DC 20375
United States of America

ABSTRACT. The two-way parabolic equation and separation of variables are applied to develop an efficient method for solving acoustic reverberation problems involving a point source in an ocean that varies with both depth and one of the horizontal Cartesian coordinates. Although this reverberation problem is idealized, it is currently the most complex long-range reverberation problem that can be solved routinely with a full-wave acoustic model.

1. INTRODUCTION

In modeling acoustic propagation in the ocean, it is more difficult to account for back-scattered energy than outgoing energy. For many problems, the outgoing field may be obtained with a simple two-dimensional approximation [1]. One of the difficulties of modeling reverberation is that there is no two-dimensional problem that is physically meaningful. The line-source problem in plane geometry is nonphysical. For a point source in cylindrical geometry, back-scattered energy focuses at the source range. If back scattering occurs only in a small range interval (i.e., the scatterer is localized in range), the solution of this problem (with a spreading loss correction) is representative of certain scattering problems for the monostatic case (i.e., back scattering toward the source).

A simple boundary reverberation problem that is physically meaningful involves a point source in an environment that varies with range (one of the horizontal Cartesian coordinates) and depth but is independent of cross range (the other horizontal coordinate). Problems of this type that involve range-localized scattering features embedded in range-independent waveguides have been solved with separation of variables and the boundary integral method, including problems involving perfectly-reflecting boundaries [2] and problems involving elastic bottoms [3].

The parabolic equation (PE) method is the most efficient approach for solving range-dependent propagation problems [4]. In this paper, the two-way PE [5,6] is extended to handle reverberation problems involving a point source and range and depth dependence. This approach is applicable to large-scale reverberation problems involving complicated range and depth dependence, including scattering features separated by long ranges. A simple example is presented to illustrate three-dimensional reverberation.

2. THE SPECTRAL PE

We work in Cartesian coordinates, where x is range, y is cross range, and z is depth below the ocean surface. The complex wavenumber k and density ρ are assumed to depend on x and z. The complex pressure p satisfies

$$\rho \frac{\partial}{\partial x}\left(\frac{1}{\rho}\frac{\partial p}{\partial x}\right) + \frac{\partial^2 p}{\partial y^2} + \rho \frac{\partial}{\partial z}\left(\frac{1}{\rho}\frac{\partial p}{\partial z}\right) + k^2 p = \delta(x)\delta(y)\delta(z-z_0) , \qquad (1)$$

where z_0 is the source depth. We Fourier transform (1) in y to obtain

$$\rho \frac{\partial}{\partial x}\left(\frac{1}{\rho}\frac{\partial \hat{p}}{\partial x}\right) + \rho \frac{\partial}{\partial z}\left(\frac{1}{\rho}\frac{\partial \hat{p}}{\partial z}\right) + \left(k^2 - h^2\right)\hat{p} = \delta(x)\delta(z-z_0) , \qquad (2)$$

where $\hat{p}$ is the Fourier transform of p and h is the separation parameter. Since (2) is the two-dimensional wave equation for a line source with the wavenumber $\sqrt{k^2 - h^2}$, it may be solved with a modified version of the two-way PE of [5]. Some of the existing PE starting fields may be generalized to handle the case $h \neq 0$. Since these PE starters break down as $h \to k$, however, we have developed a new PE starting field that is valid in this limit [7]. To avoid singularities in $\hat{p}$, the Fourier transform is inverted by numerical integration along a contour that is shifted off the real h axis [2].

The environment is approximated by a sequence of range-independent regions. In each range-independent region, (2) factors for $x > 0$ as follows:

$$\left(\frac{\partial}{\partial x} + ik_0 L\right)\left(\frac{\partial}{\partial x} - ik_0 L\right)\hat{p} = 0 , \qquad (3)$$

$$L = \sqrt{1+s} , \qquad (4)$$

$$s = k_0^{-2}\left(\rho \frac{\partial}{\partial z}\frac{1}{\rho}\frac{\partial}{\partial z} + k^2 - k_0^2 - h^2\right) , \qquad (5)$$

where k_0 is an average wavenumber. The solutions of (3) satisfy

$$\frac{\partial \hat{p}}{\partial x} = \pm ik_0 L \hat{p} , \qquad (6)$$

where the positive sign corresponds to outgoing solutions and the negative sign corresponds to incoming solutions.

PE derivations are based on approximations of the operator square root in (4) for small s, which corresponds to small differences between k and k_0 and propagation nearly parallel to the x-axis. With higher-order Padé approximations, these smallness restrictions may be relaxed significantly. Substituting the approximation,

$$L \cong 1 + \sum_{j=1}^{m} a_{2j-1,m}\left(1 + a_{2j,m}s\right)^{-1} s , \qquad (7)$$

into (6), we obtain the spectral PE

$$\frac{\partial \hat{p}}{\partial x} = \pm i k_0 \left[1 + \sum_{j=1}^{m} a_{2j-1,m} \left(1 + a_{2j,m} s \right)^{-1} s \right] \hat{p} \ . \tag{8}$$

Values for the Padé coefficients are given in [8]. The numerical solution of (8) is based on a splitting method [9]. The spectral PE is used to propagate the outgoing and incoming fields through the range-independent regions.

3. A PE-BASED METHOD FOR SCATTERING

In this section, we derive an approach based on the PE method for handling the vertical interfaces between range-independent regions. The following conditions for continuity of pressure and the normal component of velocity are to be satisfied at the vertical interface between range-independent regions A and B:

$$\hat{p}_i + \hat{p}_r = \hat{p}_t \ , \tag{9}$$

$$\frac{1}{\rho_A} \frac{\partial}{\partial x} (\hat{p}_i + \hat{p}_r) = \frac{1}{\rho_B} \frac{\partial}{\partial x} \hat{p}_t \ , \tag{10}$$

where $\hat{p}_i$, $\hat{p}_r$, and $\hat{p}_t$ are the incident, reflected, and transmitted fields and the subscripts A and B denote evaluation in regions A and B. Evaluating the range derivatives of $\hat{p}_r$ and $\hat{p}_t$ directly would require information of these quantities in a neighborhood of the vertical interface. Since this information is not available, we use (6) to replace the range derivatives in (10) with the depth operator L using the subscripts A and B for the two regions and the appropriate sign for the reflected field to obtain

$$\frac{1}{\rho_A} L_A (\hat{p}_i - \hat{p}_r) = \frac{1}{\rho_B} L_B \hat{p}_t \ . \tag{11}$$

Since a range operator appears in (10), (9) may not be applied to eliminate $\hat{p}_t$ from (10). Since L_A and L_B are depth operators, however, $\hat{p}_t$ may be eliminated from (11) using (9) to obtain

$$\left(\frac{1}{\rho_A} L_A + \frac{1}{\rho_B} L_B \right) \hat{p}_r = \left(\frac{1}{\rho_A} L_A - \frac{1}{\rho_B} L_B \right) \hat{p}_i \ . \tag{12}$$

Solving (12) directly is not an efficient approach for computing $\hat{p}_r$ because non-banded matrices are involved. The following iteration formula, which is obtained by rearranging (12), only requires the solution of tridiagonal systems of equations:

$$\hat{p}_r = \alpha^{-1} \left(1 - L_A^{-1} \frac{\rho_A}{\rho_B} L_B \right) (\hat{p}_i - \hat{p}_r) + \alpha^{-1} (\alpha - 2) \hat{p}_r \ , \tag{13}$$

where α is a convergence parameter. The initial iterate is taken to be $\hat{p}_r = 0$. For most problems, the iteration formula converges rapidly (a single iteration is sufficient in many

Figure 1. The outgoing field in a horizontal plane at $z = 50$ m for $0 \leq x, y \leq 12$ km. The transmission loss levels vary from 55 to 75 dB. The origin is at the lower left corner.

cases) for $\alpha = 2$, which corresponds to the one-part iteration formula of [5]. If a strong scattering feature is encountered, it may be necessary to take $\alpha > 2$ and perform several iterations. This approach is an alternative to the two-part iteration formula of [5].

We have implemented the iteration scheme using the following Padé approximations:

$$L \cong \prod_{j=1}^{n} \left(1 + b_{2j,n} s\right)^{-1} \left(1 + b_{2j-1,n} s\right), \tag{14}$$

$$L^{-1} \cong \prod_{j=1}^{n} \left(1 + b_{2j,n} s\right) \left(1 + b_{2j-1,n} s\right)^{-1}, \tag{15}$$

which are equivalent to the sum representation given by (7) if $m = n$. Values for the Padé coefficients are given in [5].

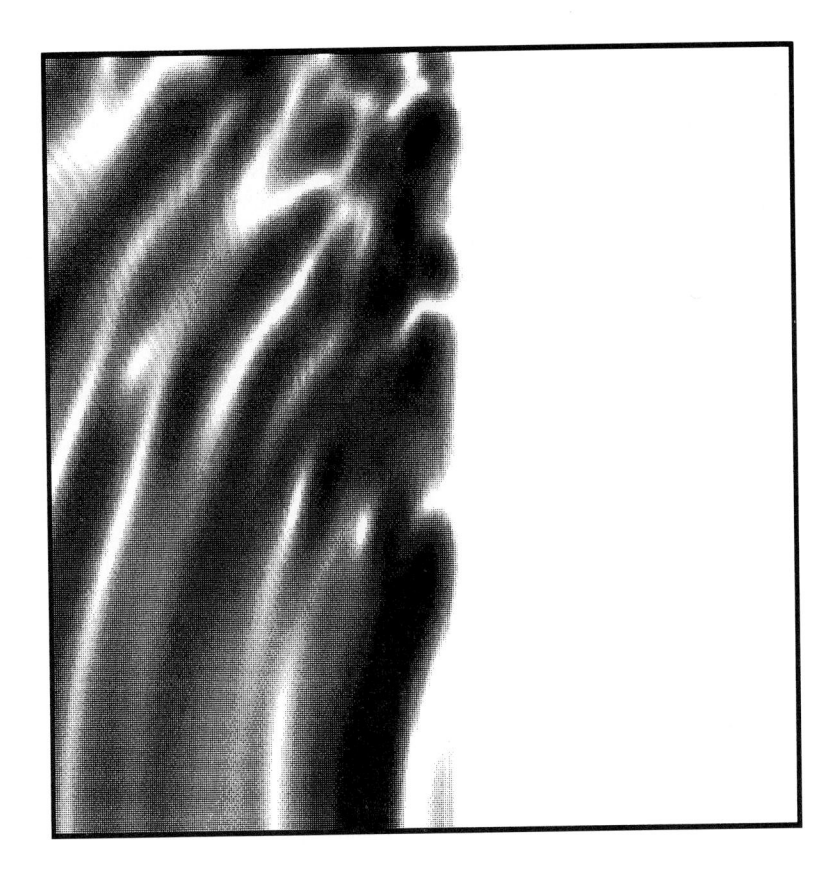

Figure 2. The back-scattered field in a horizontal plane at $z = 50$ m for $0 \leq x, y \leq 12$ km. The transmission loss levels vary from 75 to 95 dB. The origin is at the lower left corner.

4. THE STAIR-STEP PROBLEM

In this section, we apply the two-way PE to solve a problem involving a discontinuity in ocean depth. A 25-Hz source is placed at $z = 50$ m in an ocean that is 500 m deep for $x < 7$ km and 250 m deep for $x > 7$ km. The speed of sound is 1500 m/s in the water column and 1700 m/s in the sediment. The other sediment parameters are $\rho = 1.5$ g/cm^3 and $\beta = 0.5$ dB/λ.

A horizontal slice of the three-dimensional outgoing field is displayed in Figure 1. Although the solution was generated by solving 500 line-source problems, it correctly exhibits cylindrical wavefronts in the range-independent region $x < 7$ km. The solution breaks down near the y-axis (the left edge of the display), which corresponds to energy propagating at nearly 90 deg from the preferred direction (the x-axis). Since a five-term Padé approximation was used for the calculation, this forbidden pie slice is narrow. Although the solution exhibits three-dimensional structure for $x > 7$ km, the wavefronts are roughly cylindrical in this region.

The back-scattered field, which appears in Figure 2, also exhibits three-dimensional structure and roughly cylindrical wavefronts (with curvature in the opposite direction). Since the back-scattered field is weak directly above the stair step, Figure 2 suggests that the stair step is located less than 7 km from the source. Similarly, Figure 1 suggests that the stair step is located more than 7 km from the source. This behavior can be understood by studying a vertical slice of the field in the direction of the x-axis. Appearing in [5] for this environment is a display of this type for the line-source field, which is qualitatively similar to the point-source field for the monostatic case. Since the apparent translation distance depends on the angle of incidence, the fields in Figures 1 and 2 also suggest that the stair step is slightly skewed (in opposite directions for the two fields) relative to the y-axis.

5. CONCLUSION

We have applied the two-way PE and separation of variables to develop a model for solving three-dimensional reverberation problems. Problems that involve a point source and both range and depth dependence are currently the most complex type of long-range reverberation problem that can be solved routinely with a full-wave model.

6. REFERENCES

[1] J.S. Perkins and R.N. Baer, "An approximation to the three-dimensional parabolic-equation method for acoustic propagation," J. Acoust. Soc. Am. 72, 515-522 (1982).

[2] J.A. Fawcett and T.W. Dawson, "Fourier synthesis of three-dimensional scattering in a two-dimensional oceanic waveguide using boundary integral equation methods," J. Acoust. Soc. Am. 88, 1913-1920 (1990).

[3] H. Schmidt and H. Fan, "Numerical modeling of three-dimensional reverberation from elastic bottom facets," J. Acoust. Soc. Am. 90, 2277 (1991).

[4] F.D. Tappert, "The parabolic approximation method," in Wave Propagation and Underwater Acoustics, edited by J.B. Keller and J.S. Papadakis (Springer, New York, 1977).

[5] M.D. Collins and R.B. Evans, "A two-way parabolic equation for acoustic back scattering in the ocean," J. Acoust. Soc. Am. 91, 1357-1368 (1992).

[6] M.D. Collins, "A two-way parabolic equation method for elastic media," J. Acoust. Soc. Am. 91, 2463 (1992).

[7] M.D. Collins, "A self starter for the parabolic equation method," J. Acoust. Soc. Am. (to appear).

[8] M.D. Collins, "Higher-order Padé approximations for accurate and stable elastic parabolic equations with application to interface wave propagation," J. Acoust. Soc. Am. 89, 1050-1057 (1991).

[9] M.D. Collins, "Applications and time-domain solution of higher-order parabolic equations in underwater acoustics," J. Acoust. Soc. Am. 86, 1097-1102 (1989).

CALCULATIONS OF OCEAN BOTTOM AND SUB-BOTTOM BACKSCATTERING USING A TIME-DOMAIN FINITE-DIFFERENCE CODE

Dale D. Ellis, Nikolaos A. Kampanis[1], and Ralph A. Stephen[2]
SACLANT Undersea Research Centre,
Viale San Bartolomeo 400, 19138 La Spezia (SP), Italy

ABSTRACT. Conventional measurements of acoustic backscatter from the ocean bottom are often attributed to, and modelled by, scattering from a rough water-bottom interface. However, there can be contributions due to reflections from sub-bottom layers and scattering from sub-bottom heterogeneities. All these effects can be handled by the FINDIF model, which is a finite-difference code developed at Woods Hole Oceanographic Institution to solve the two-way elastic wave equation in the time-domain. Here, two basic scattering models are constructed: a rough-bottom model, where the interface between the water and sediment is a randomly-generated surface; and a volume-heterogeneity model, where the sediment contains randomly-distributed scatterers with specific acoustic properties. Using an initial pulse as the source, the time series of pressure is computed, and output at a number of receiver depths. The computed pressure is also used to extract a conventional measure of acoustic backscattering strength versus grazing angle. The inclusion of a sub-bottom layer illustrates how the apparent backscatter can be corrupted at steep angles by sub-bottom reflections, as can happen in the analysis of experimental measurements.

INTRODUCTION

During 1990–91 a version of the Woods Hole Oceanographic Institution finite difference model FINDIF was implemented at the SACLANT Undersea Research Centre [1]. Though developed as a low-frequency propagation model [2], it can be used to study two-way propagation and scattering [3]. Our specific interest was scattering, so we applied it to some hypothetical bottom types to simulate backscatter from a rough ocean bottom and scatterers buried within the bottom. We also included a sub-bottom layer to be realistic, and to illustrate how sub-bottom reflections can corrupt the scattering at steep angles.

The FINDIF model is unique in that it produces a time series of the backscattered energy. The model, as installed at the Centre, produced only synthetic seismograms – time series of unnormalized pressure versus time at selected receiver depths. We extended it to extract dB plots of intensity versus time, and performed a "ray-like" analysis to extract an apparent backscattering strength versus grazing angle for a quantitative comparison of results.

---

[1] *Present address: Department of Mathematics, University of Crete, Iraklion, Crete, Greece; and Institute of Applied and Computational Mathematics, F.O.R.T.H., Iraklion, Crete, Greece.*
[2] *Present address: Woods Hole Oceanographic Institution, Woods Hole, MA 02543, USA.*

To keep the computer run times low we chose a relatively small problem – very short range in shallow water at 100 Hz. We used rather large scattering inputs, and our results indicate that rather large backscatter is indeed observed.

ENVIRONMENTS

Three environments were chosen, see Figure 1. The basic environment consists of 130 m of water overlying 10 m of low-speed sediment, all over a homogeneous half-space of semi-consolidated material. The acoustic parameters include elastic properties, though not intrinsic attenuation, and were chosen to be representative of a shallow water – see Table 1.

Two modifications of this environment were chosen. The first is a rough water-sediment interface intended to simulate a Gaussian surface with rms roughness of 1 m and correlation length of 10 m. The second variation has sediment inhomogeneities – "buried rocks" – imbedded in the transition layer. The rocks were of uniform size, and for the calculations shown were 4 m square, in 10% concentration, randomly distributed. They were allowed to overlap, but to remain at least 2 m below the water-sediment interface, so there would be no interface roughness or impedance. Their acoustic properties are the same as the basement.

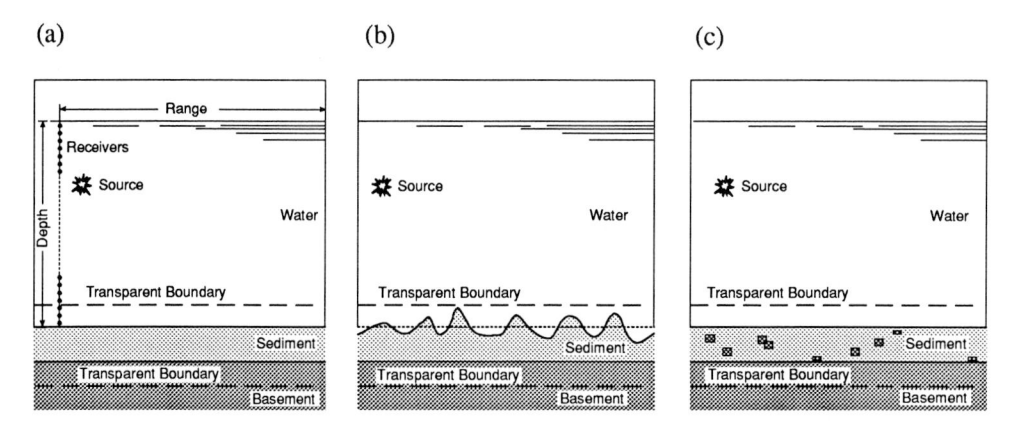

Figure 1. Plots of the three environments: (a) basic environment, (b) rough bottom, (c) buried scatterers.

Material	Thickness (m)	Density (t/m^3)	Sound Speed (m/s) Compressional	Shear
Water	130	1.0	1500	–
Sediment layer	10	1.8	1600	200
Basement	∞	2.0	2000	500
Buried rocks		2.0	2000	500

Table 1: Acoustical parameters used in the study.

THE FINDIF MODEL

Here we provide a brief, qualitative description of the FINDIF model, and describe the inputs used in solving our problem. The FINDIF model solves the elastic wave equation in the time domain in heterogeneous media, using an explicit second-order finite-difference scheme on a uniform grid. It has two unique properties that are especially valuable for studying backscattering: it solves the problem in the time domain, and it includes two-way propagation. It is a full-wave calculation, including all shear and interface waves. Limitations include: it is a 2-D model, and it does not handle intrinsic attenuation in the media.

The environmental model consists of three parts: an upper region of homogeneous water, a transition region that can be completely arbitrary in two dimensions, and a homogeneous elastic basement. The left boundary is an axis of symmetry, the upper boundary is a perfectly reflecting surface, and, to simulate the radiation condition and avoid spurious reflections from the right and bottom boundaries, a telegraph equation region with 180 grid points is used. In the transition region each grid point can have a unique density, P-wave, and S-wave velocity. Boundary conditions at interfaces are handled implicitly by spatial derivatives of the acoustic parameters.

For our problem the source pulse (in pressure) was the third derivative of a Gaussian with a peak frequency at 100 Hz and half-power points at 68 and 136 Hz. The computational grid had 601 points in range, 501 points in depth, and 2501 points in time, with a grid size of 0.5 m and a time step of 0.0001 s. The problem was 300 m by 250 m by 0.25 s, though the actual box is 90 m smaller due to the attenuating boundaries. The CPU time for each run was typically 35 minutes on a VAX 9000-210, or 110 minutes on a FPS 264 array processor.

SCATTERING CALCULATIONS

Figure 2a shows the basic output from the model – plots of time series at a number of receiver depths for a source at 65 m depth. (Snapshots of the field on the spatial grid are also available, but not currently implemented at SACLANTCEN.) Note the reflections at the surface and bottom near 0.06 and 0.15 s, and some subsequent reflections due to the sub-bottom layer. Figure 2b shows the corresponding synthetic seismogram for the rough bottom. Note the differences due to the scattering, at times later than the first bottom reflection. The scattering from the sediment inhomogeneities (not shown) is similar. In both cases the scattering is quite strong, noticeable even on a linear plot.

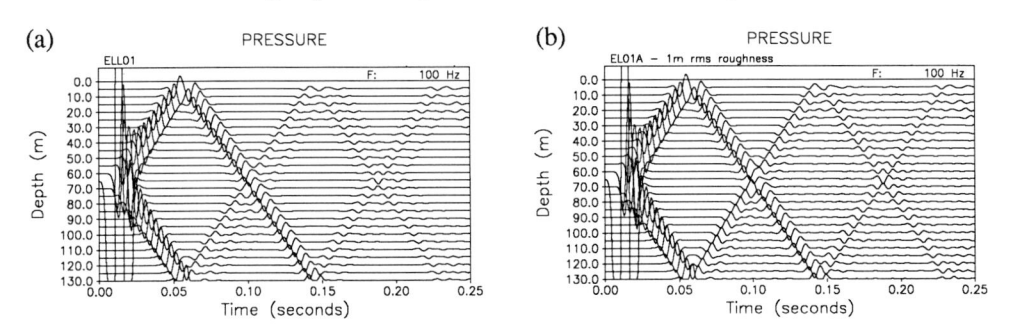

Figure 2. Pressure time series for: (a) the basic environment, (b) the rough bottom.

To get a more quantitative picture we extracted the (unnormalized) time series for a receiver near the source. The intensities from the basic model and the rough-bottom model are illustrated in Figure 3. The large peaks at 0.09 s and 0.19 s are the fathometer returns and surface reflections. For the basic model the level between the peaks is due to sub-bottom multiples which die out to what is essentially numerical noise. The scattering from the rough surface fills in the levels between the peaks, and is 30–50 dB higher than for the basic model. There are differences in detail, but no obvious qualitative differences, between the rough-interface scattering and scattering from sub-bottom inhomogeneities.

DERIVED BACKSCATTERING STRENGTH

To allow a more quantitative comparison between these calculations and the results from other calculations, a ray-like backscattering strength can be extracted. It is analogous to the bottom scattering strength obtained in conventional analysis of deep-water measurements. The scattering is assumed to occur at the water-sediment interface, and the received intensity for a single path is given by

$$R(t) = \int_0^{\tau_0} I_0(\tau) \, H_1(t_1) \, H_2(t_2) \, S(\theta_1, \theta_2) \, dA,$$

where I_0 is the source intensity, τ_0 its duration, H_1 and H_2 are respectively the propagation losses from the source to the scattering "area" dA and from the scattering area to the receiver, t_1 and t_2 are the corresponding travel times, S is the scattering function, and θ_1 and θ_2 are the grazing angles at the bottom. The times are connected by the relation $t = t_1 + t_2 + \tau$. The equivalent sonar equation (cf. [4]) for monostatic backscattering (source and receiver at the same location) due to a short pulse is

$$SS = RL - SL + 2TL - 10 \log A,$$

where $SS = 10 \log S(\theta,\theta)$, $SL = 10 \log I_0$, $\theta = \sin^{-1}[2h/c(t-t_0)]$, $TL = 10 \log [c(t-t_0)/2]$, and $A = c\tau_0/(2 \cos \theta)$. Note that due to the 2-D geometry, the transmission loss is cylindrical spreading, and A is a length rather than an area. The $(\cos \theta)^{-1}$ area dependence requires that $S(\theta,\theta) = 0$ at $\theta = 90°$. For our calculations the source and receiver were at mid-depth $h = 65$ m, the mid-point of the initial pulse was at $t_0 = 0.0135$ s, the source level SL was 50 dB, and for the area calculation a duration of $\tau_0 = 0.02$ s was used. The source intensity was obtained by extrapolating backwards to 1 m from the source using cylindrical spreading and a fit to the pressure at several receiver depths.

Figure 4 shows the derived backscattering strength for the case when there are no scatterers – only reflections from the surface, bottom, and sub-bottom. The solid line shows the result for the basic model; the dashed line shows the result without sub-bottom multiples, i.e., when the basement properties are made identical to the sediment layer. Since time and angle are simply related here, the sub-bottom multiples show apparent scattering from 90° down to about 50°, whereas in fact all arrivals are from the vertical. In deeper water the fathometer peaks would be narrower, but the calculation illustrates that interpreting the results must be done with some care. Measurements of backscattering are often interpreted in the same way, by assuming that all of the scattering occurs at the interface. The steep-angle scattering at low frequencies may be sub-bottom reflections – though bottom attenuation plays a role in damping them out. Below 35° the second fathometer return and other scattered paths complicate the picture, so for this geometry the scattering strength cannot be simply obtained. The dotted envelope is for comparison with the results shown in Figures 5 and 6.

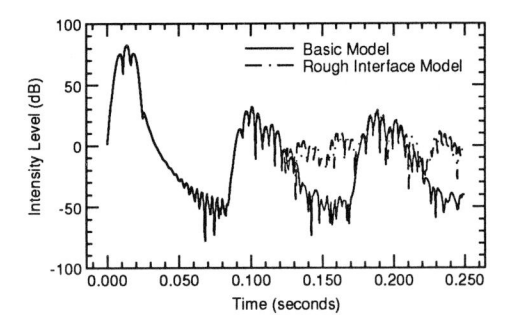

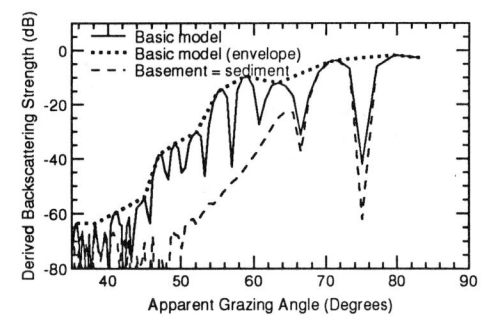

Figure 3. Intensity vs time for the basic model and the rough bottom.

Figure 4. Derived backscattering strength due to sub-bottom reflections.

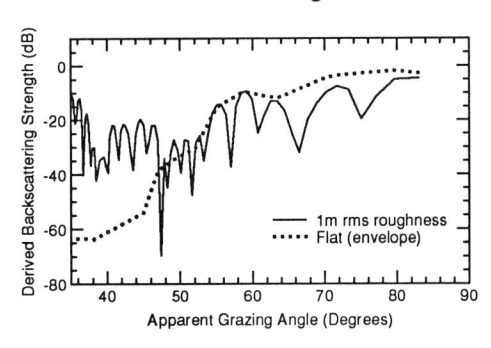

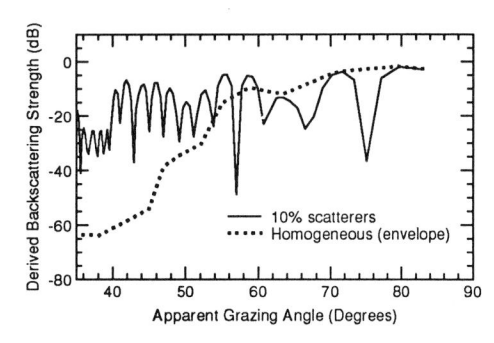

Figure 5. Derived backscattering strength for rough bottom.

Figure 6. Derived backscattering strength for volume inhomogeneities.

Figures 5 and 6 illustrate the backscattering derived from single realizations of the rough interface and buried scatterers' environments respectively. The "scattering" at steep angles is dominated by the sub-bottom reflections – as can be seen by comparing with the dotted envelope from Figure 4. At lower angles the scattering is relatively independent of angle and quite strong – ~30 to 50 dB above the sub-bottom multiples at the lowest angles shown. Whereas the apparent grazing angle in Figure 4 is an artifact of the ray analysis, the grazing angles here should be closely related to the actual angle of the scatterers. One should take care in making quantitative comparisons with measured backscatter data or Lambert's law, since this is a 2-D model and the inputs were chosen to produce large scattering.

DISCUSSION AND CONCLUSIONS

The FINDIF model provides a powerful, and in many ways unique, tool for investigating the physics of bottom and sub-bottom scattering at low frequencies. In particular it includes the full-wave solution, including shear and interface wave effects, and the transition layer allows a completely arbitrary 2-D grid to be specified. Its time-domain formulation means that the time series for a pulse can be simulated.

We have performed a feasibility study and extracted a backscattering strength, much as one would do in the analysis of measurements. The model can equally well be applied to the study of bistatic scattering. To obtain more meaningful scattering strengths the problem would have to be scaled up and multiple runs performed to extract statistical scattering strengths. This is possible to a limited extent, though the problem size increases as the cube of the box size. Long-range calculations and low angles would require a hybrid model – a propagation model combined with the scattering chamber idea of references [5] and [6]. Another possible way of extracting the angular dependence is by beamforming with a vertical array of hydrophones – or the equivalent F-K diagram of seismologists.

We believe the real potential of the model lies in its time-domain formulation. Though it is only 2-D in space, since reverberation is intrinsically a time-dependent problem, the model treats (the most important?) three of the four relevant dimensions. For example, some recent measurements have indicated that reverberation is dominated by features, and our modelled time series suggest a similar effect. The model output would be useful in generating synthetic input for data analysis schemes, or clutter rejection algorithms. The time snapshots, not shown here (but see Refs. [5] and [6]) would provide insight into the scattering process, especially when combined in a movie sequence.

ACKNOWLEDGEMENT

This work was partly supported by ONR Contract N00014-89-J-1012 and ONR Grant N00014-90-J-1541.

REFERENCES

1. Stephen, R.A. (1992). "User's Guide for FINDIF at SACLANTCEN," Woods Hole Oceanographic Institution, Technical Memorandum WHOI-07-92.

2. Stephen, R.A. (1988). "Finite difference methods for bottom interaction problems," in *Computational Acoustics: Wave Propagation,* D. Lee, R.L. Sternberg and M.H. Schultz (eds), pp. 225–238.

3. Dougherty, M.E. and Stephen, R.A. (1991). "Seismo/acoustic propagation through rough seafloors," J. Acoust. Soc. Am. **90**, 2637–2651.

4. Urick, R.J. (1983). *Principles of Underwater Sound* (McGraw-Hill, New York), 3rd edition, Chapter 8.

5. Stephen, R.A. (1992). "A numerical scattering chamber for studying reverberation in the seafloor," elsewhere in these proceedings. [*Ocean Reverberation,* D.D. Ellis, J.R. Preston, and H.G. Urban (eds), Kluwer, Dordrecht, NL, 1993].

6. Levander, A., Harding, A. and Orcutt, J.A. (1992). "Numerical scattering results for a rough unsedimented seafloor," elsewhere in these proceedings. [*Ocean Reverberation,* D.D. Ellis, J.R. Preston, and H.G. Urban (eds), Kluwer, Dordrecht, NL, 1993].

TIME AND ANGLE SPREADING FROM ROUGH SEDIMENTS
by

Diana F. McCammon
The Pennsylvania State University
Applied Research Laboratory
P. O. Box 30
State College, PA 16804

ABSTRACT

Employing the Fresnel and Kirchhoff approximations in the Helmholtz Integral equation and assuming a Gaussian correlation function, an equation can be derived that predicts the time and spatial distribution of the acoustic intensity from a pulse propagated over a rough bottom. Inputs to this equation include an "effective" slope of the acoustic basement obtained from echo half duration time estimates. [McCammon and Burns, JASA,90 No4, Pt2, 1991] Numerical evaluations of this expression have produced angle and time distributions whose statistical characteristics have been extracted. In this paper, the relationships between the statistical descriptors (variance, skew and kurtosis) of the time and angle distributions will be described. It will be shown that at low frequencies over rough thin sediment regions where significant reflection for the acoustic basement is encountered, there is a clear relationship between time and angle spreading.

INTRODUCTION

The relationship between time and azimuthal angle spreading of pulses that have interacted with the sediment is examined in this paper. Specifically, we wish to show their inter-connectedness by discussing a theoretical expression for intensity with functional dependence on these two variables, and by illustrating the behavior of this expression with variations in grazing angle and basement slope.

The theories that will be employed assume that the scattering is being accomplished by the layer roughness of a few specific sediment layers, the surficial, an intermediate conformal layer, and the basement. Volume scattering by localized inhomogeneities and reinforced resonance between layers are not considered in this theory, thus limiting the applicability of the results to relatively thin sediment regions of the oceans where the overlying sediment is thin or nearly non-existent acoustically. The theory assumes all scattering is a result of a Gaussian distribution of surface elevations above a mean depth, and that the actual RMS surface slopes are relatively small.

THEORETICAL BASIS OF RELATIONSHIP

The theoretical basis for the intensity distribution to be presented comes from a solution to the Helmholtz integral equation using the Kirchhoff approximation for the surface velocity field and Fresnel phase approximation. Many of the basic analysis techniques have been discussed in reference [1], but in this development, a pulse of arbitrary shape and duration has been introduced and the spatial average of the product of the pressures at two separated receivers has been found following the development of Berry [2]. A thumb nail sketch of the derivation follows. Let θ be the grazing angle and ϕ be the azimuthal angle. Let source and receiver be at equal shallow depths in a deep ocean. Then

$$\alpha = \cos\theta_s \cos\phi_s + \cos\theta_r \cos\phi_r, \quad \beta = \cos\theta_s \sin\phi_s + \cos\theta_r \sin\phi_r, \quad \gamma = \sin\theta_s + \sin\theta_r \tag{1}$$

$$M = \frac{1}{2}\left(\frac{1}{|r_r|} + \frac{1}{|r_s|}\right), \quad r_{r,s} = \sqrt{x_{r,s}^2 + y_{r,s}^2 + z_{r,s}^2}, \text{ and } c = \text{medium sound speed}$$

Recast the Helmholtz integral equation for the surface pressure from a stiff, locally reacting surface in polar form, and employ the Fresnel phase approximation and the Eckart approximation for small surface slopes. Introduce the pulse via Fourier transform by assuming

$$p(r_r,t) = p(r_r) a(t)e^{-i\omega_o t} \quad \text{with} \quad a(t)e^{-i\omega_o t} = \int_{-\infty}^{\infty} d\omega\, A(\omega-\omega_o)e^{-i\omega t} \tag{2}$$

These steps produce an expression for the scattered pressure

$$P_{(r_r,t)} = \int_{-\infty}^{\infty}\int d\omega d\vec{R}\, A(\omega-\omega_o)e^{-i\omega t}\, \frac{e^{i\omega(|r_r|+|r_s|)/c}}{(|r_r||r_s|)}\, \frac{\omega \sin\theta_r}{c}\, e^{-i\omega\gamma z/c}\, e^{i\omega M|\vec{R}|^2/c}\, e^{-i\omega\left(\frac{\alpha^2+\beta^2}{4Mc}\right)} \tag{3}$$

Now form the spatial average of the product of pressures $<p(r,t)\, p^*(r-Q,t)>$ where Q represents a horizontal separation from the receiver position. The spatial average will require the evaluation of the characteristic equation for two variables, and will result in an expression involving the surface correlation function, C. This development will assume a Gaussian correlation function with isotropic slopes ζ, and further assume it can be expanded in a small slopes approximation to give

$$C = e^{-R^2\zeta^2/4h^2} \approx 1 - R^2\zeta^2/4h^2 \tag{4}$$

Following the application of these approximations and several integrations, the expression for the average pressure product is normalized by the total intensity (no receiver separation) from a flat surface from which the returns come from the specular spot, denoted sp. Finally, the normalized intensity from a point in the plane at time t, to a pair of closely spaced receivers at a distance Q is

$$<I(r_r,t,Q)> \geq \frac{\sin^2\theta_r}{\sin^2\theta_{sp}}\, \frac{(|r_{sp}|)^4}{(|r_r||r_s|)^2}\, \frac{M_{sp}}{M}\, \frac{e^{-\left(\frac{\alpha^2+\beta^2}{\zeta^2\gamma^2}\right)}}{2\pi\zeta^2} \int_0^\infty d\tau\, \frac{|a(t-\tau)|^2}{|a(t_{sp})|^2}\, e^{-4Mc\tau/\zeta^2\gamma^2}\, J_o\!\left(2Q\omega_o\left(\frac{M\tau}{c}+\frac{(\alpha^2+\beta^2)}{4c^2}\right)^{1/2}\right) \tag{5}$$

If we assume three layers, a surficial water/sediment interface, an intermediate conformal layer within the sediment, and a basalt basement, then Equation (5) becomes a sum over these layers with the appropriate layer reflectivity. Three layers have been chosen because this configuration has produced good agreement with measured angle spreading from the Pacific Echo experiments of 1988. [3]

In the work that follows, we will concentrate on intensity distributions with time and angle for zero receiver separation, that is for Q=0, thus removing the Bessel function factor of spatial coherence. The remaining unknowns are the reflective and transmissive properties of the sediment layer and the RMS slopes of the layers. The sediment density and sound speed can be expected to be well approximated by typical deep ocean sediment characteristics, in fact the sensitivity of the result for most typical sediment compositions is very slight. However the determination of appropriate values for layer slopes is more difficult. There is evidence that the surficial sediment slopes are of the order of 2-4° everywhere, and the sensitivity of the result to this value is also very slight. Thus, if we assume that the slope of the intermediate conformal layer varies linearly from the surficial to deeper layer, then the only true unknown left is the slope of the basalt basement.

Berry [2] has suggested a method by which the "effective" slope of the basement can be measured. This method, presented in more detail in Reference [4], involves using a very short pulse at normal incidence to the surface so that $r_r = r_{sp}$, etc, and $\alpha = \beta = 0$. Equation (5) reduces to
exp $[-ct/\zeta^2 H]$, H being the depth from receiver to bottom. Then, integrating this energy over all time, an expression for the echo half duration time $T_{1/2}$ is derived.

$$\int_0^{T_{1/2}} e^{-ct/\zeta^2 H} \, dt = 1/2 \int_0^{\infty} e^{-ct/\zeta^2 H} \, dt \qquad (6)$$

From this, an equation is obtained for the one dimensional slope $\zeta = \tan\theta_{slope}$,

$$\zeta = (-cT_{1/2}/H \, \ell n.5)^{1/2} \qquad (7)$$

When this expression is evaluated using $T_{1/2}$ from the Pacific Echo experiment [5], an estimate of "effective" slope is found to be 29°. Algebra gives a basement slope of 25° from this measurement. We can readily see that this is an "effective" slope because the approximations within the theory restrict it to single scatter with no volume effects or resonance. In addition, actual basement slopes are thought to be in the neighborhood of 8-10°. However, this within the theory restrict it to single scatter with no volume effects or resonance. In addition, actual basement slopes are thought to be in the neighborhood of 8-10°. However, this measurement does produce the appropriate value for use in this theory that enables it to predict spreading comparable with that measured.

In the discussion that follows, the basement slope will be varied from 8 to 30°, with emphasis placed on the 20-25° values, since these seem to correspond best to the measurements in the Pacific Echo experiment and others in thin sediments.

NUMERICAL PREDICTIONS OF SPREADING

Equation (5) has been evaluated as a function of time and receiver azimuthal angle for several values of effective basement slope and for a span of bottom grazing angles. The resulting intensity distributions are shown in figure 1 for a 0.03sec pulse, a basement slope of 25° and grazing angles of 20 and 30°. These plots show the normalized intensity in 10 dB contours as a function of time and azimuthal angle at an omnidirectional receiver. Notice that there is a fairly linear increase in angle with time. In the 30° case, the contours show this same increase until the sound begins to echo from behind the receiver (>2sec) when the angle distribution becomes spread over the entire circle.

In order to quantify these distributions, the output is summed over one of the variables to form two dimensional representations of the total angle integrated time spread or time integrated angle spread. Figure 2 displays the integrated distributions for the same two grazing angles of 20 and 30° with a variety of effective basement slopes ranging from 8 to 30°. There is a

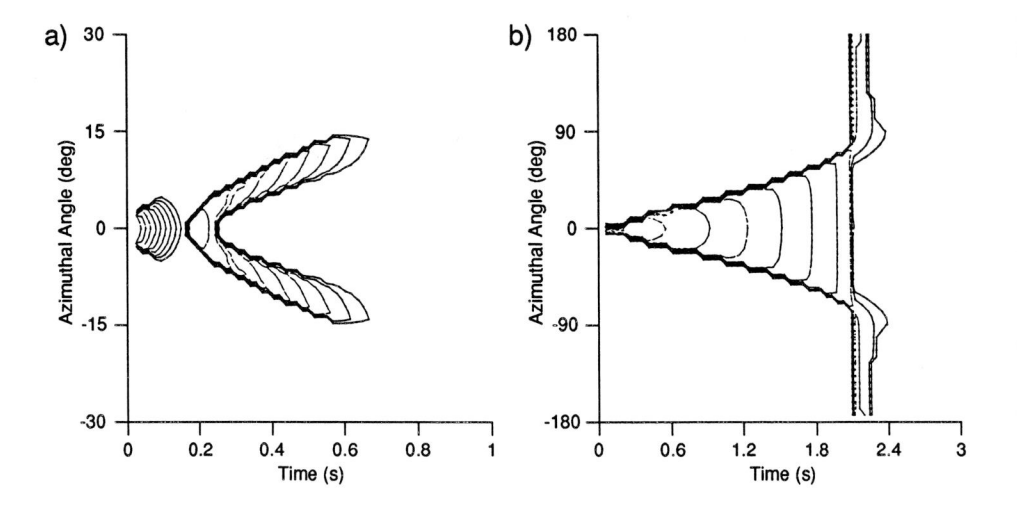

Figure 1. Intensity distributions at a) 20° and b) 30° grazing

steady increase in spreading with increasing basement slope. There are several convenient ways of quantifying the amount of spreading that has occurred. For example, the time or angle where the level drops by 3dB, 6dB or 10dB can be a very useful descriptor. Even more important for the theory of sediment scattering may be the use of descriptors that quantify the fundamental shape of the curves, since these distributions are usually assumed to be Gaussian. The statistical moments of the time and angle spreading curves can provide important clues on the Gaussian-ness of the spreading, and with the use of a Gram-Charlier series, the distributions can be theoretically recovered from just a knowledge of its moments. In this paper, we present the square root of the second moment, the standard deviation of these curves, in order to demonstrate the linear relationship between these two types of spreading. The question of the Gaussian-ness of the spreading and the generation of higher order moments will be considered in the future.

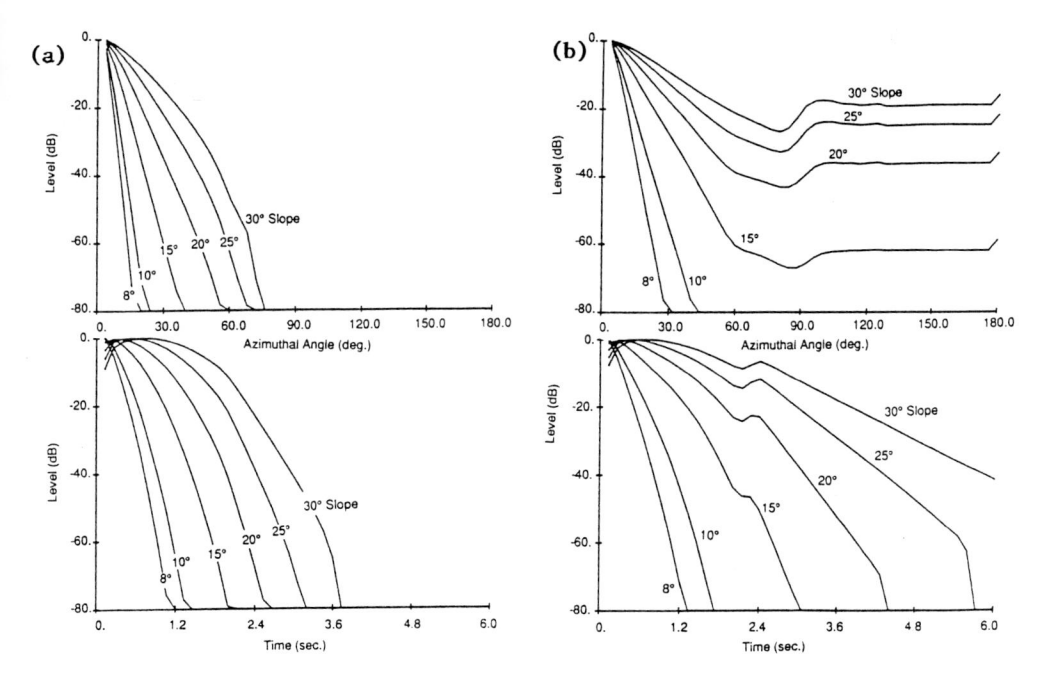

Figure 2. Integrated levels for parameter changes of basement slope
a) 20° grazing b) 30° grazing

Figure 3 displays a plot of the standard deviation of the integrated time curves vs. the standard deviation of the integrated angle curves for parameter changes of the basement slope, the grazing angle and the pulse length. Each line traces the σ_t and σ_a values as the effective basement slope is increased. Grazing angle changes result in an intercept change and sometimes a increase in the steepness of these curves, and pulse length changes cause an

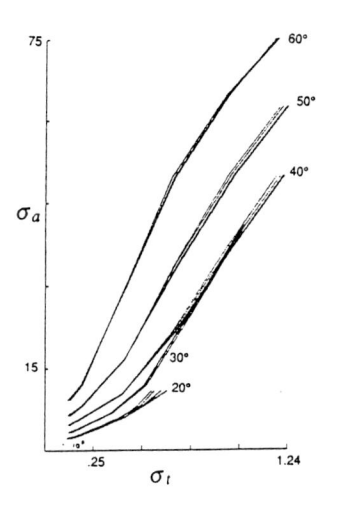

Figure 3.
Standard deviation of angle spread vs. standard deviation of time spread for changes in slope, grazing angle and pulse length.

inconsequential variation about σ_t. There is a clear fairly linear increase in σ_a for changes in σ_t at a specific grazing angle. The nearly parallel curves for grazing angles 40-60° indicate larger angle spreads for the same time spreads. This is shown more clearly in Figure 4, a plot of the integrated levels at the effective slope of 20°, with parameter changes in grazing angle. There is a saturation in the time spreading from 40° grazing on up, while the angle spread continues to increase. This saturation effect is a simple consequence of the changes in the shape of the equal time contours on the bottom from highly elliptical at low grazing angles to nearly circular at high grazing angles.

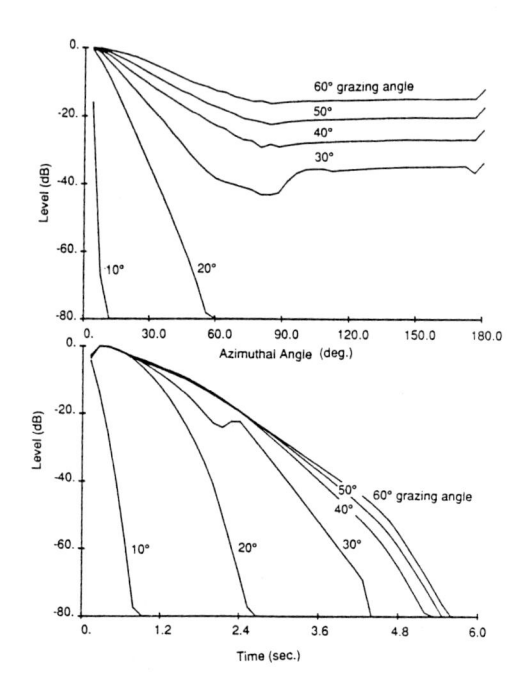

Figure 4. Integrated levels at 20° slope for grazing angles from 10 - 60°

SUMMARY

An expression has been presented for the received intensity of sound that has reflected from and refracted within a thin sediment overlying a rough basement. The theories used were the Helmholtz integral, the Fresnel phase approximation and the Eckart approximation for small slopes. In addition, the correlation function was assumed to be Gaussian and was approximated for small slopes. This work differed from previous derivations by the addition of a pulse representation and by including a small receiver separation in the average product of pressures. The resulting expression for intensity has been examined as a function of travel time and receiver azimuthal angle in order to explore the relationships between the spread of energy in these two ways. We have shown that the effective basement slope, which controls spreading through the probability of slope orientation function, can be obtained by a simple normal incidence measurement of echo duration. This effective slope can be used to predict time and angle spreading distributions. We have also shown that this theory predicts a nearly linear relationship between time and angle spreading at a given grazing angle.

The impact of these findings, of the linear relationship between angle spreading and time spreading from rough thin sediments at low frequencies, is that one can now be predicted from a measurement of the other. Angle integrated time spread measurements should be fairly easy to make, whereas angle spread measurements require large arrays capable of forming many narrow beams. With these findings, it may be possible to gain a reliable estimate of angle spread from a knowledge of time spreading.

ACKNOWLEDGEMENT

This work was sponsored by the ONR/AEAS program, E. Estalote, program manager.

REFERENCES

[1]   McCammon, D.F. "A sediment time-angle spreading model", J. Acoust. Soc. Am., 87(3), p1126-1133 (1990).

[2]   Berry, M.V. "The Statistical Properties of Echos Diffracted from Rough Surfaces", Phil. Trans. Roy. Soc., A273, p611-658 (1973).

[3]   McCammon, D. F. "Comparison between Sysloss and the 1988 Pacific Echo Experiment", ARL TM 91-26, Applied Research Lab (1991).

[4]   McCammon, D.F. and T. Burns, "Estimating bottom slope from pulse duration", J. Acoust. Soc. Am.,90 No4, Pt2, (1991).

[5]   Dr. Ronald Dicus, personal communication

ROUGH SURFACE SCATTERING AS SEEN THROUGH THE RENORMALIZATION GROUP

Gregory J. Orris
U.S. Naval Research Laboratory, Code 5160
Washington, DC 20375-5000
Roger F. Dashen
Physics Department, University of California, San Diego
9500 Gilman Drive, La Jolla, CA 92093-0319

Abstract: *A Renormalization Group Equation can be developed to address the ensembled averaged acoustic pressure field scattered by a random rough surface. It has been shown that there is no solution to the Renormalization Group Equation for fractal-like scatterers that obey Dirichlet or pressure-release boundary conditions, in the long wavelength limit. This leads us to develop an argument based upon dimensional analysis that suggests several fixes that would make this scattering theory finite under these conditions.*

1.1 Introduction

Anyone who has solved a scattering problem knows that even a deterministic surface can lead to a very difficult technical solution, from which, little insight can be gained. Instead of working with deterministic systems in which one knows the shape and the characteristics of the scatterer, it is much more common in the real world to experiment; go to a test site and measure the scattering of sound from the ocean bottom or surface, and then try to develop insight into a theory based upon measurement, without *a priori* knowledge of the surface. One assigns a probability that the scatterer has a specific shape, but it is not surprising that this is exceedingly difficult to accomplish given the complexity of real world problems. Instead one invokes ideas of statistical mechanics, and uses an ensemble average to introduce surface statistics into the problem. In this context, the scatterer is assumed to be an object, the shape of which is determined by some random process, so that an overall structure is created and the surface is modeled after the statistics of the real scatterer, with the modeling now being the art of the problem, where the various statistical moments of the functions describing the surface are important quantities that determine the scattering process.

The statistics of fractals are characterized by a power spectrum with power-law dependence on the fourier modes used to describe the surface. One problem with scattering from fractal-like surfaces is that there is no scale for which the scattering is geometrically optical, because the wavelength of the incoming radiation is never shorter than all length scales of the scatterer. Renormalization Group techniques come into play at this point, because one wishes to sum up the effects of all scales that are smaller than some length scale, and to put all of the effects into an effective set of parameters. In this scheme chosen parameters of the problem become dependent upon the degrees of freedom removed.

Here we demonstrate that in calculating the scattered field and the cross section of a scatterer there is a hidden caveat of boundary perturbation theory, such that it remains valid at all orders. The surface must be at least twice differentiable, thus we offer the conjecture that no scattering problem involving fractal-like rough surfaces is a purely classical perturbative problem until it is cut off from length scales not accessible to the incoming wave. [9] [3] [8] [4] We believe that without specifying a cutoff these problems are asymptotic and at some point a cutoff on the minimum scale probed by the scattering must be defined. The exact positioning of this cutoff and the physical reasons for choosing it are what makes the theory of rough surface scattering a finite perturbative theory.

2.1 The Scattering System

Consider a wave obeying the Helmholtz equation
$$\partial^2 \psi + k^2 \psi = \delta(\mathbf{x} - \mathbf{x}_0),$$
with a general impedance boundary condition
$$\left[\gamma \frac{\partial \psi}{\partial n} + \psi\right]_S = 0, \tag{2.1}$$
incident upon a rough surface. The perturbation consists of nothing more than making the surface height a function of the two transverse coordinates. A point on the surface is represented as
$$\mathbf{x}|_S = (x, y, h(x,y)) = \left(\mathbf{x}_{||}, h\left(\mathbf{x}_{||}\right)\right).$$
This choice of representation manifests itself in the metric of the surface element and in the normal derivative as,
$$dS = \sqrt{g}\, dx\, dy = \sqrt{1 + \nabla h \cdot \nabla h}\, dS \Rightarrow \frac{\partial}{\partial n} = \mathbf{n} \cdot \nabla = \frac{(-\nabla h(\mathbf{x}), 1)}{\sqrt{1 + \nabla h \cdot \nabla h}} \cdot \nabla. \tag{2.2}$$
Our scatterer will be assumed to have a rough surface that is gaussian in nature, implying that it has no statistical moments beyond the second. The height of the surface of the scattering object is assumed to be given by $h(\mathbf{x})$, and a mean value of h is defined to be the zero of the coordinate system. Any particular representation of the scattering surface is assumed to be random, so that translational invariance of the ensemble averaged moments is assumed, which yields
$$\langle h(\mathbf{x})\rangle = \langle \tilde{h}(\mathbf{k})\rangle \equiv 0 \tag{2.3}$$
$$\langle h^*(\mathbf{x}) h(\mathbf{x}')\rangle = \xi(|\mathbf{x} - \mathbf{x}'|) \Rightarrow \langle \tilde{h}^*(\mathbf{k}) \tilde{h}(\mathbf{k}')\rangle = \delta(\mathbf{k} - \mathbf{k}') W(\mathbf{k}) \tag{2.4}$$
$$W(\mathbf{k}) = \begin{cases} 0 & |\mathbf{k}| > q_{max}; \\ C|\mathbf{k}|^{-\beta} & q_{max} \geq |\mathbf{k}| \geq q; \\ 0 & q > |\mathbf{k}| \end{cases} \tag{2.5}$$
We assume that for each realization of the surface, $h(\mathbf{x}_{||})$ is an $\mathcal{L}^2$ function, allowing us to use the fourier decomposition
$$h(\mathbf{x}) = \int d\mathbf{k}\, \exp(i\mathbf{k} \cdot \mathbf{x}) \tilde{h}(\mathbf{k}). \tag{2.6}$$
Upon substitution of equations 2.2 and 2.6 into the boundary condition of equation 2.1, a series solution in powers of the expansion parameter can be developed as a recurrence relation. Using this tedious method, the solution is worked out in reference [6] through order $\mathcal{O}(\epsilon^6)$. The last bit of information necessary to complete the perturbation series is conservation of momentum, which forces the outgoing

wave to behave as
$$\psi(\mathbf{x}) = \int d\mathbf{k}' \exp i(\mathbf{k}' \cdot \mathbf{x} + \sqrt{k^2 - k'^2} z) \tilde{\psi}(\mathbf{k}'), \qquad (2.7)$$

defining the wave at $z = 0$. As is standard the dimensionless parameter $\epsilon = h_0 k$ is introduced so that the perturbative solution is formally written out as,

$$\psi = \psi_0 + \epsilon \psi_1 + \epsilon^2 \psi_2 + \cdots = \sum_{n=0}^{\infty} \epsilon^n \psi_n, \qquad (2.8)$$

where ψ_0 is the solution to the system with the perturbation turned off and where h_0 is the root-mean-square of the height function $h(\mathbf{x})$. For our purposes we will assume that implicitly an h_0 is included in every factor of $h(\mathbf{x})$. We assume that in the expansion of equation 2.2, h_0 can be made small enough for any surface considered, so we may expand the square root using the binomial expansion.

The scattered field in the far-field limit is usually calculated using the Green's function method presented in reference [5]. To paraphrase, one first considers two boundaries S and S_0. The Helmholtz Green's function for the un-perturbed problem (the surface S_0), yet still satisfying the boundary condition

$$\left[\gamma \frac{\partial G}{\partial n} + G \right]_{S_0} = 0, \qquad (2.9)$$

is used for the integral equation inside the perturbed surface S. The information about the perturbed surface boundary conditions is then inserted into the usual Green's function surface integral, giving

$$\psi_{Scattered}(r_0) = -\frac{1}{4\pi} \int dS \frac{\partial \psi(\mathbf{x})}{\partial n} \left(\gamma \frac{\partial G(\mathbf{x}|\mathbf{x}_0)}{\partial n} + G(\mathbf{x}|\mathbf{x}_0) \right)_{\mathbf{x} \text{ on } S}$$

3.1 The Renormalization Scheme

The perturbative solution defines the field scattered by a rough surface to any arbitrary order in ϵ. This solution is then averaged over all possible realizations of the surface. Inherently, the averaged solutions satisfy the Helmholtz equation and some boundary condition on the unperturbed surface S_0. So, we define γ_0 to be the new impedance constant satisfying

$$\left[\gamma_0 \frac{\partial \langle \psi \rangle}{\partial n} + \langle \psi \rangle \right]_{S_0} = 0 \qquad (3.1)$$

Assuming that the series solution for $\langle \psi \rangle$ is valid one uses the binomial theorem and expands this equation for γ_0 to any order of accuracy desired, although to find the largest contributions to γ_0 need only be calculated to $\mathcal{O}(h^2)$. Now all of the degrees of freedom we are trying to remove from the system reside in the definition of γ_0 and as such γ_0 is a function of the choice of both the maximum and minimum scales of the surface. Now suppose that we move this lower bound by an amount $\delta q > 0$. This removes the roughness of scales between $1/q$ and $1/(q + \delta q)$ from the problem, thus rendering the surface physically smoother. This is exactly the transformation desired. Formally, the Renormalization Group Equation is then defined as the solution to the ordinary differential equation

$$\frac{d\gamma_0}{dq} = -\frac{d}{dq}\frac{\langle\psi\rangle}{\frac{\partial\langle\psi\rangle}{\partial n}} \quad (3.2)$$

with the inital condition

$$\gamma_0(q = q_{max}) = \gamma.$$

The solution to equation 3.2 has similar meaning as the running coupling constant in Quantum Field Theory [1], implicitly expressing γ_0, as a function of q, where γ_0 represents an effective impedance parameter with the average effect of the wavenumbers $\{k|q < k < q_{max}\}$ included in the new parameter. As applied to the surface this is equivalent to solving the system of equations for a less rough surface. When $q \to q_{max}$ we are back at the original fundamental surface, S, explaining our initial condition on γ.

It has been shown in reference [2], that to $\mathcal{O}(h^2)$

$$\gamma_0 = \gamma + \gamma \int d^2q \left\{ \left(-\frac{1}{2}q^2 + \mathbf{k}\cdot\mathbf{q}\right) W(\mathbf{q}) \right.$$

$$\left. + \left\{\gamma\mathbf{k}\cdot(\mathbf{k}-\mathbf{q}) - i(k^2-q^2)^{\frac{1}{2}}\right\} \times \frac{1}{\gamma}\frac{\gamma^2\mathbf{k}\cdot(\mathbf{k}-\mathbf{q})+1}{1+i\gamma(k^2-q^2)^{\frac{1}{2}}}W(\mathbf{q}-\mathbf{k}) \right\} \quad (3.3)$$

Taking the further limiting condition $\mathbf{k} \to k\hat{\mathbf{z}}$ and keeping only the lowest non-zero terms,

$$\gamma_0 = \gamma + \gamma \int_0^\infty dq\, 2\pi q W(q) \left\{-\frac{1}{2}q^2 + \frac{q}{\gamma - \gamma^2 q}\right\} \quad (3.4)$$

To second order equation 3.2 gives

$$\frac{d\gamma_0}{dq} = \gamma q W(q) \left\{\frac{1}{2}q^2 - \frac{q}{\gamma - \gamma^2 q}\right\}, \quad (3.5)$$

where a factor of 2π has been absorbed into the definition of $W(q)$. Seemingly, this is a fairly benign equation. But, the solution to this problem is only desired to $\mathcal{O}(h^2)$, so we may make the substitution on the right-hand-side, $\gamma \to \gamma_0$, giving,

$$\frac{d\gamma_0}{dq} = \gamma_0 q W(q) \left\{\frac{1}{2}q^2 - \frac{q}{\gamma_0 - \gamma_0^2 q}\right\}, \quad (3.6)$$

As a special case we consider Dirichlet conditions, requiring the pressure to be zero on the boundary. When one considers the different values of β for equation 2.5, it can be proven that the smallest value of β that still offers a solution consistent with Dirichlet boundary conditions if $q_{max} \to \infty$ is $\beta_{Cr} = 4$. The Renormalization Group technique has many advantages, one of which is the ability to sum up many terms of a perturbation theory, by solving a differential equation or transformation law. These summations are then expressed in terms of a new parameter. The fact that there is a limit to the ability to find a solution for the Dirichlet problem without cutoff, is strongly suggestive that the averaged fields become infinite under these conditions. As the next section will demonstrate, this is not only true, but much more disturbing is that the average cross section per unit area is also divergent under these conditions.

4.1 Dimensional Analysis

One main thrust of this work is that dimensional analysis gives an easy power counting argument when one considers the Dirichlet boundary value problem ($\gamma \to 0$ in equation 2.1) in the limit that the wavelength of the incoming wave is much larger that the longest wavelength of the surface, represented here as $k \ll q$ with q being the lowest mode on the surface representing the largest feature of the surface. It is under these limiting conditions that the integrands for the scattering cross section are simple polynomials in **q**. To start we need to find the dimensionality of a few of the important parameters.

$$[W] = L^{1+d}$$

where d is the dimension of the space the system is embedded. For the inverse power-law spectrum considered here,

$$W = C\Lambda^{p-1-d} q^{-p} \qquad (4.1)$$

with $[\Lambda] = \frac{1}{L}$. Then the ensemble averaged cross section per unit area will be proportional to

$$\langle \sigma_n \rangle \propto k^4 \int_{q_{min}}^{q_{max}} \left\{ \prod_{i=1}^{\frac{n}{2}} d^{d-1} q_i W(q_i) \right\} P_m(\{q_i\}) \qquad (4.2)$$

where P_m is a polynomial of order m. Using the fact that this is a dimensionless form, m must satisfy,

$$4 + (d-1)\frac{n}{2} - (d+1)\frac{n}{2} + m = 0 \Rightarrow m = n - 4 \qquad (4.3)$$

If $\langle \sigma_n \rangle < \infty$ as $q_{max} \to \infty$, then

$$\frac{n}{2}(d-1-p) + n - 4 < 0 \Rightarrow n < \frac{8}{d+1-p}, \qquad (4.4)$$

implying for the case $d = 3$ and the boundary being two dimensional,

$$\frac{8}{4-p} = n_\sigma.$$

Thus at order $n > \frac{8}{4-p}$, σ_n is no longer finite. For $p = \frac{7}{2}$, $n = 16$, which implies that we need to calculate the field to $\mathcal{O}(h^{16})$, which is a considerable task indeed. Similarly the averaged field can be shown to be finite for the condition [6], [7]

$$2n - 1 + \frac{n}{2}(d-1) - \frac{n}{2}p < 0 \Rightarrow n < \frac{2}{d+1-p} = n_c < n_\sigma. \qquad (4.5)$$

For the case $d = 3$ and $p = \frac{7}{2}$, $n_c = 4$ or $\mathcal{O}(h^4)$.

5.1 Conclusion

Considering the solution to this equation under the two extreme conditions $\gamma = 0$ and $\gamma^{-1} = 0$, which represent Dirichlet and Neumann boundary conditions respectively, it can be readily seen the case of Neumann conditions is a fixed point of the RGE theory, suggesting that the impedance parameter remains of the Neumann variety. On the other extreme, the Dirichlet problem has no fixed point, and furthermore there is no solution for the condition of a divergent mean-square-slope, implying that there could be a fundamental problem with the region of parameter space considered to be valid for finding the fields. All of this suggested to us that there might also be a problem with the scattering cross section. This result seems to be in direct contradiction of the earlier work of Thorsos [10], but is actually not. The difference lies in the fact that no one calculates terms of a perturbative expansion up to $\mathcal{O}(h^{16})$, because of the ridiculous amount of work involved. Since most of the previous work in rough surface scattering has tended to concentrate on relatively low orders of the scattering cross section, it has only been seen that a superficial divergence in the cross section is removed by considering the cross terms in the limit of large $|\mathbf{q}|$ [10]. Yet it has been communicated to me by Thorsos that the "exact" integral solution explored in [10] does not suffer from the type of divergence presented in this work, as it should not, since any numerical realization cannot have a divergent mean-square-slope, i.e. any exact numerical technique will have an upper cutoff q_{max}, and thus not be subject to these divergences. There seem to be three reasonable explanations: 1) The case of rough surface scattering is in reality an asymptotic theory. 2) All media is lossy, and thus after several interactions with the surface the divergent part is exponentially cut off. 3) There must be an upper cutoff placed on any scattering problem, which seems reasonable, in the sense that no one expects that a macroscopic theory, such as Ocean Acoustic Reverberation should depend on the microstructure of the scatterer.

References

1. Daniel J. Amit. *Field Theory. The Renormalization Group, and Critical Phenomena.* World Scientific, Singapore, second edition, 1984.
2. Roger F. Dashen and Gregory J. Orris. Rough Surfaces and the Renormalization Group. *J. Math. Phys.*, 31(10):2352-2360, 1990.
3. Bertrand Duplantier. Can one "hear" the thermodynamics of a (rough) colloid? *Physical Review Letters*, 66(12):1555-1558, 1991.
4. Thomas C. Halsey. Double-layer impedance at a rough surface: A perturbative approach. *Physical Review A*, 36(12):5877-5880, 1987.
5. Philip M. Morse and Herman Feshbach. *Methods of Theoretical Physics,* Part II. McGraw-Hill, 1953.
6. Gregory J. Orris. *Rough Surfaces and the Renormalization Group with the Impedance Boundary Condition.* 1991. Doctoral Thesis at the University of California, San Diego.
7. Gregory J. Orris and Roger F. Dashen. Renormalization Groups Analysis of Rough Surface Scattering., *J. Acoust. Soc. Am.*, 1992, Submitted for publication February 1992.
8. Adriana I. Pesci, Raymond E. Goldstein and Victor Romero-Rochin. Electric Double Layers near Modulated Surfaces. *Physical Review A*, 41(10):5504-5515, 1990.
9. Sergei Lvovich Sobolev. *Partial Differential Equations of Mathematical Physics.* Pergamon, New York, 1964.
10. Eric I. Thorsos and Darrel R. Jackson. The Validity of the Perturbation Approximation for Rough Surface Scattering using a Gaussian Roughness Spectrum. *J. Acoust. Soc. Am.*, 86(1):261-277, 1989.

LONG RANGE 3-D REVERBERATION AND SCATTERING MODELING METHODOLOGY

Robert Bruce Williams, code 7104,
Homer P. Bucker, code 541,
Angela D'Amico, code 731,
NRaD Division, NCCOSC, and
David F. Gordon, Computer Sciences Corporation
San Diego, Ca 92152-5000

ABSTRACT. A methodology is being implemented for low frequency surface and bottom reverberation and scattering that uses 2-D and 3-D beam propagation similar to Gaussian beam (Porter & Bucker JASA 82(1987)) calculations in determining the source to scattering areas and the scattering areas to receiver paths. The approach makes use of source-receiver coupling tables. Temporal, acoustic intensity and direction information are filled into these tables by the propagation calculations from one or more sources to the scattering areas. Next, propagation from the scattering areas to the receivers are calculated, using a scattering function to couple to the receiver the various sources contributing to the acoustic energy impacting that area. Full ocean basins can be accommodated but at the expense of large tables and lengthy runs. The environmental factors used for the propagation calculations include five minute grided bathymetry and sediment thickness, thirty minute grided sound speed profiles and a variety of wind speed sources including historical, near real-time and simulations. For the bottom scattering, measured bottom roughness (Fox & Hays Rev. Geophys. 23(1985)) is used. Several scattering functions have been studied: the composite model (Backman (JASA 54(1973)), Berman, et. al. (JASA 88(1990), and the recent "full spectrum" formalism (Dashen, et. al. JASA 88(1990)). The Rayleigh method, described by Berman, is currently being used, due to application to both surface and bottom. Model runs are compared to low frequency bistatic reverberation data.

1. INTRODUCTION

Full simulation of scattering and reverberation involves not only the local scattering processes, but the accurate propagation from the sources to the scattering areas and then the propagation from the scattering areas to the receivers. For low frequencies, large areas are involved, and therefore environmental databases are required. We have taken the approach of using propagation code similar to Gaussian beam (Porter and Bucker (1987)) for the source to scattering area and the scattering area to receiver paths for two reasons: 1) 3-D approaches are possible, and 2) the time dimension can be easily added. The ocean surface and bottom are divided up into small scattering areas and, in principal, each can have its own roughness or deterministic surface characteristics. The scattering function that relates the incoming energy to the scattered energy is modular, so that different theories can be used and compared. This work is intended to take a global approach to this three tiered problem. It is still in the development stage, so that this report can be considered a progress report. In section 2.1, we describe the source to scattering propagation, section 2.2 details the scattering approaches taken, 2.3 discusses the surface simulations, while 2.4 describes the scattering areas to receiver propagation. Section 3 presents the comparison to some real bistatic low frequency data.

2. METHODOLOGY

2.1 SOURCE TO SCATTERING AREA PROPAGATION

The software module that propagates beams from a source to the scattering areas uses a fan of beams in the vertical whose density is higher (20 per degree) in the near horizontal angles, while those above 20 degrees in magnitude are reduced. This was found to be necessary in the long distance range-dependent environment in order to have enough beams distributed on the scattering surfaces within the small scattering areas of 1km x 1 degree azimuth. The beam density in the azimuth is, however, uniform at one per degree. When a beam encounters either the ocean surface or bottom, the intensity, time of arrival and the pair of directions (azimuth and vertical arrival angles) are recorded in temporary files. A database is then built up from the files of each source run. The information from each file is sorted in range, and then time within a range. Although adjustable, the entries are currently rounded into units of degrees in azimuth and vertical arrival angle, seconds in time and kilometers in range. Indexing is done by range, so that one entry of time, intensity, azimuth and vertical angle require only 6 bytes, the two angles being packed into a 2 byte integer. Indexing files are generated for accessing the data randomly. These databases, one for each scattering surface, currently completely fit into normal sized workstation memory (2-5 Mbytes for our runs to date), and are used for the scattering area to receiver propagation part. The maximum range from the source is a parameter, which we have varied from 400km to 1000km, although a full basin model run could be made. Multiple source propagations can be used, although again for simplicity only data from one will be described here.

This propagation, although labeled as Gaussian beam-like, uses an Airy function approximation for the calculation of the acoustic field that couples a beam's energy that vertexes below the surface to the surface. In the ray sense, the surface would be in a complete shadow, so that this approximates the field refracted into the shadow zone. Likewise, beams vertexing above the bottom couple energy to the bottom in the same manner. Perhaps this technique more appropriately should be labeled Airy beams. It is our feeling that, while Gaussian beam techniques should be used for acoustic intensity calculations interior to the surfaces, the Airy coupling should be more accurate for the vertexing beam to surface (and bottom) coupling, either for the surfaces being sources or receivers of intensity. This is therefore used both for the source to scattering surfaces and the scattering area to receiver propagation. Insight can be gleaned from the examination of these databases.

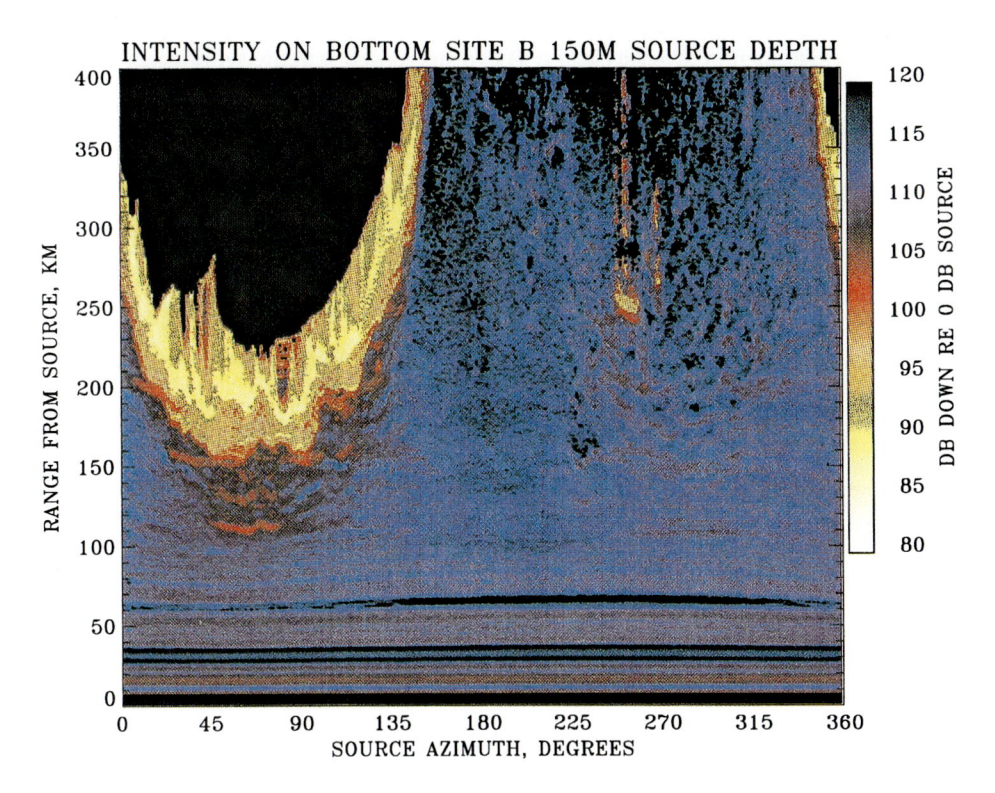

Figure 1. Intensity on the bottom for a Pacific site, which was used to compare to actual data.

2.2 SCATTERING METHODS

We consider a small local volume near a scattering surface (bottom or ocean surface) where the sound speed can be considered constant. Propagation to and from the local area are handled elsewhere. We consider a single incoming wave. Let **r** and **x** be vector space variables with three dimensions (x,y,z), **x2** be a vector space variable with two dimensions (x,y), **k** be a vector wavenumber variable with three dimensions (k_x, k_y, k_z), **q** be a vector wavenumber variable with two dimensions (k_x, k_y) and t be time. The early development here follows closely that of Dashen, et al (1990). The outgoing field from the scattering area is given by

$$P(r, t) = \int A(k) P_k(r) e^{-\omega_k t} d^3 k \quad \text{(EQ 1)}$$

$$P_k(r) = e^{ik \cdot r} - \int_S (4\pi)^{-1} G_k(r-x) \hat{n} \cdot \nabla P_k(x) \, dS \quad \text{(EQ 2)}$$

where the wavefunction of pressure, P(r,t) is decomposed into the set of functions $P_k(r)$, each satisfying the boundary conditions and obeying the Helmholtz equation:

$$(\nabla^2 + k^2) P_k(r) = 0 \quad \text{(EQ 3)}$$

$$\omega_k = c|k| = ck \quad \text{(EQ 4)}$$

and the normal to the local surface pointing into the ocean is

$$\hat{n} = (-\nabla h(x2), 1) / \sqrt{1 + (\nabla h(x2))^2} \quad \text{(EQ 5)}$$

and h(x2) is the surface height. The Green's function G_k is, as yet, unspecified.

Taking the two-dimensional Fourier transform of the scattered portion of the field (the second right hand term in eq. 2),

$$\Phi_k(q, z) = (2\pi)^{-2} \int e^{-iq \cdot x2} d^2x2 \int -(4\pi)^{-1} G_k(r-x) \hat{n} \cdot \nabla P_k(x) \, dS \quad \text{(EQ 6)}$$

Letting

$$q_z = \frac{\sqrt{k^2 - q^2}}{i\sqrt{k^2 - q^2}} \quad \text{(EQ 7)}$$

for k > q and q > k respectively, and q = |q|. Defining a scattering amplitude T(**k**,**q**)

$$\Phi_k(q, z) = i \frac{e^{iq_z z}}{8\pi^2 q_z} T(k, q) \quad \text{(EQ 8)}$$

$$T(k, q) = -\int_S e^{-iq \cdot x} \hat{n} \cdot \nabla P_k(x) \, dS \quad \text{(EQ 9)}$$

Experimentally, T(**k**,**q**) has been sampled, although in the ocean it impossible to simultaneously obtain T for all of the **q**, due to the requirement of having enough receiver elements in the body of water near the scattering surfaces to adequately sample the scattered field. Usually, some average of T(**k**,**q**) is reported by a series of experiments with different geometries.

Theoretically, for the evaluation of eq. 9, the normal derivative of the pressure on the surface is needed. This can be obtained by operating on eq. 2 by $\hat{n} \cdot \nabla$ and taking the limit as $r \to x$.

$$\hat{n} \cdot \nabla P_k(x) = 2ik \cdot \hat{n} e^{ik \cdot x} - (2\pi)^{-1} \int_{S'} \hat{n} \cdot \nabla G_k(x-x') \hat{n} \cdot \nabla' P_k(x') dS' \quad \text{(EQ 10)}$$

In the past, successive iterations are used to solve eq. 10 and therefore eq. 9. By various expansions, two-scale models (Bachman 1973)) have been used, breaking the problem into scales much smaller and those much larger than an acoustic wavelength, leaving a part of the spectrum unaddressed.

Dashen, et al (1990) supply a new formalism using a generalized **n**, which appears to obviate the iteration procedure, and supply scattering through a full spectrum. In these approaches a Green's function must be used. We are unaware, though, that this as yet has been applied to bottom scattering. We therefore will postpone the implementation of this approach, although apparently more rigorous, for the future.

Berman, et al (1990), on the other hand, point to a simulation approach which uses the Rayleigh method, and does not require knowing the Green's function. For this paper, we have adopted that method, which uses an over-determined set of linear equations to satisfy the boundary conditions. It is a simulation, requiring a realization of the scattering surface, but can deal with pressure release, fluid-fluid and fluid-solid boundaries. Only Bragg angles are considered for the scattered field, which would require the size of the scattering surface to be infinite. However, through the use of periodic surface replication, and quasiperiodic Green's functions, he re-formulates the problem on a finite surface. Given the incoming field

$$e^{i(k_I \cdot r)}, where \quad k_I = (k_{Ix}, k_{Iy}, k_{Iz}) = (k_h, k_z). \quad \text{(EQ 11)}$$

The scattered Bragg field is given by

$$\Phi_k(q, z) = \Sigma_i \Sigma_j a_{ij}(k_h + q_{ij}) e^{i(q_{ij} \cdot x2)} e^{iq_z z} \quad \text{(EQ 12)}$$

where

$$q_{ij} = i\Delta k_x + j\Delta k_y \quad i,j = 0, \pm 1, \pm 2, ..., \pm N$$
$$\Delta k_x = 2\pi/L_x \quad \Delta k_y = 2\pi/L_y \quad \text{(EQ 13)}$$

The L are the size of the local scattering area. N is chosen so that we include the largest wavenumber that can propagate:

$$N \leq \frac{\omega L}{2\pi c} \quad \text{(EQ 14)}$$

We have used only square areas from 250m to 500m on a side to date. There are some evanescent waves near the corners of the k_x, k_y space. We next make a spatial grid such that there are at least twice the number of points than Bragg waves. Using eq. 11 and 12 with boundary conditions, a least squares problem can be formulated. For example, for the pressure release, we minimize the difference of the two in the least squares sense, since the pressure must go to zero. The resulting scattering amplitudes, a_{ij}, are then specified, which specifies the scattered field.

2.3 SIMULATION OF THE SURFACES

For the ocean surface, a two dimensional random surface is constructed to match a Pierson-Moskowitz wind driven surface after propagating a given distance from an area of generation. Multitudes of such waves can be superimposed. From this, the amplitude components of a two dimensional spectrum are specified, random phases are inserted, and an inverse FFT produces a realization. The time evolution of the surface waves can be obtained using an appropriate surface wave dispersion, which then allow realistic simulation of ping-to-ping fluctuations.

The ocean bottom is quite variable in many parts. Fox and Hayes (1985) have mapped roughness for many areas, and presents the data as either a single power law or multiple segment power law spectra. In principal, a sample realization is straight forward by assigning random phases and taking an inverse FFT. However, as with many geophysical phenomena, there is no natural low wavenumber cutoff. For this work, we use the roughness for wavelengths less than 1km, superimposed on a plane tilted in two dimensions, obtained from the grided bathymetry.

2.4 SCATTERING AREA TO RECEIVER PROPAGATION

The software module that propagates beams from the scattering area are very similar to those that propagate from the source to scattering area. However, reciprocity is used, so that the beams are actually propagated from the receiver. The basis for this and the modules of section 2.1 stem from the range independent work of Talham (1964), and are extended for bottom and sound speed range dependence. The database information from the source runs are used, together with the scattering functions of section 2.2 to obtain the receiver reverberation directionality, which is a function of time.

3. COMPARISON TO REAL DATA

A bistatic active experiment was performed off the northwestern coast of the United States using low frequencies. A bottomed line array was used as a receiver. Using the experimental geometry, the output from the reverberation directionality run can be multiplied and summed with a theoretical response for each beam to simulate the array outputs as a function of time. This simulation can be compared to the actual array data.

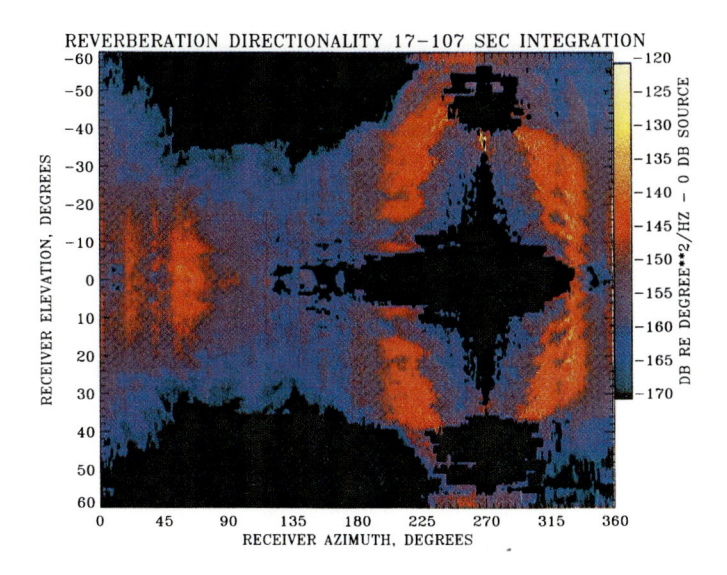

[Figure: Reverberation directionality plot, 17-107 sec integration, showing receiver elevation (degrees) vs receiver azimuth (degrees), with dB re degree**2/Hz - 0 dB source color scale from -120 to -170.]

Figure 2. Simulated reverberation directionality, Site B, integrated for 90 seconds, starting 17 seconds after the direct arrival. Site B is a Pacific slope area northwest of the United States.

The real data, as seen in figure 3 show a time granularity of about 0.1 seconds, near the theoretical minimum for the transmission bandwidth. The simulations, on the other hand, are using only 1 second granularity, so that the outputs will not be able to simulate the fine structure. However, the overall time history of the simulated data (figure 4) has a similar beam-time history when viewed in coarser time resolution. Much of the reverberation is coming from the local shelf structure where the slope is large.

4. SUMMARY AND CONCLUSIONS

A methodology has been presented for yielding a global approach to a large scale, low frequency reverberation simulation. A comparison to real data show some correspondence, but the simulations will require a smaller time resolution for comparing some of the fine structure found in the data. However, it may well be that other factors of the simulation such as the coarseness of the bathymetry may also cause the simulations not to have the fine structure in time. The intent of this paper is to lay down the methods used, however, and not to draw conclusions on the output. Future work will continue the development using 3-D propagation, increase the time resolution and continue with data comparisons.

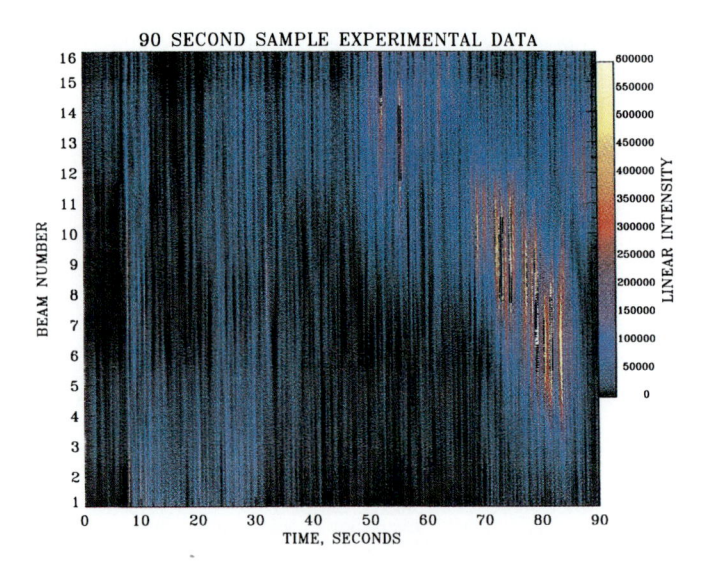

Figure 3. Beam level intensity vs. time for the experimental data. Note the trending to lower numbered beams as time progresses.

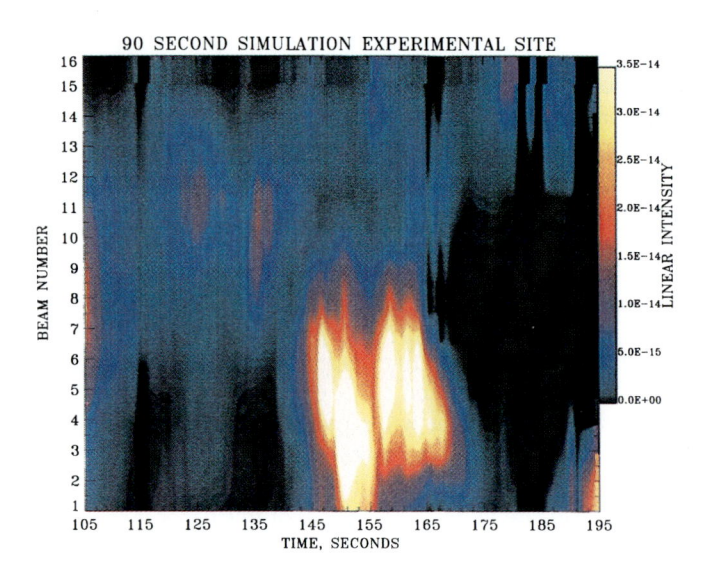

Figure 4. Simulated beam level time history. Due to the coarseness in time of one second, the fine structure of the real data does not appear. A trend of higher intensity to the lower beam numbers with the same time is seen. The strong reverberation levels from 145 to 170 seconds correspond to the high levels in the real data from 70 to 80 seconds (figure 3).

5. REFERENCES

Bachman, W. "A theoretical model for the backscattering strength of a composite roughness sea," J. Acoust. Soc. Am. **54**, 712-716 (1973).

Berman, D.H. and J.S. Perkins "Rayleigh method for scattering from random and deterministic interfaces," J. Acoust. Soc. Am. **88**(2) 1032-1044 (1990).

Dashen, R., F.S. Henyey and D. Wurmser "Calculations of acoustic scattering from the ocean surface," J. Acoust. Soc. Am., **88**(1) 310-323 (1990).

Fox, C.G. and D.E. Hayes "Quantitative methods for analyzing the roughness of the seafloor," Rev. Geophys. **23**(1) 1-48 (1985).

Porter, M.B. and H.P. Bucker, "Gaussian beam tracing for computing ocean acoustic fields," J. Acoust. Soc Am., **82**(4), 1349-1359 (1987).

Talham, R.J. "Ambient sea noise model," J. Acoust. Soc. Am., **36**(8) 1541-1544 (1964).

RANGE-DEPENDENT, NORMAL-MODE REVERBERATION MODEL FOR BISTATIC GEOMETRIES

David Meloy Fromm, B. J. Orchard, and Stephen N. Wolf
Acoustics Division, Naval Research Laboratory
Washington, D.C. 20375-5000
USA

ABSTRACT. An acoustic normal-mode reverberation model for bistatic source and receiver configurations in multi-layered, range-dependent underwater environments has been developed. Adiabatic normal-mode theory is used to propagate the signal field (with dispersion) to and from the scattering regions. The model is designed such that scattering from the interfaces and volume inhomogeneities may be easily included. Currently, scattering from the ocean bottom is treated by using a three-dimensional Lambert's-law/facet reflection model. Sources and receivers may be either omnidirectional or vertical/horizontal line arrays. An explicit description of the coupling between incident and scattered modes and their contribution to reverberation is obtained.

1. INTRODUCTION

Low frequency reverberation models developed to date have generally treated the propagation from the source to the scatterer and from the scatterer to the receiver via a ray-trace formulation. An exception is the normal-mode reverberation model developed by Ellis for a homogeneous waveguide[1].

Since many shallow water areas have acoustic environments which change over regions of dimension tens of kilometers, it is desirable to have a model which determines the influence of environmental range dependence. The extension to a range-dependent waveguide described here was developed by combining the WRAP normal-mode propagation code[2,3] for a range-dependent environment and the BiRASP bistatic-geometry area-integration code[4] previously used with ray-trace propagation routines.

2. THEORY

An acoustic source in a range-dependent, multi-layered medium emits a sinusoidal signal with angular frequency ω. The acoustic signal propagates from the source to a scattering region on the ocean surface, in the water or sediment layers, or at the interface between layers. The ocean environment is assumed to be sufficiently slowly varying in range that the assumptions of adiabatic normal mode theory are satisfied. The pressure $P(x|x_o)$ at position $x = (r, z)$ due to an acoustic source located at position $x_o = (r_o, z_o)$ in a medium with sound speed $c(r, z)$ is:[1,5]

$$P(x|x_o) = \sum_n A_n(r|r_o) \, U_n(r_o, z_o) \, U_n(r, z) \,, \tag{1a}$$

where

$$A_n(r|r_o) = \sqrt{\frac{2\pi}{k_n(r)|r - r_o|}} \, \exp\left(\int_{r_o}^{r} [ik_n(r') - \delta_n(r')]dr'\right) \,, \tag{1b}$$

and the local eigenfunctions $U_n(r_o, z_o)$ and $U_n(r, z)$ are evaluated at the source and receiver, respectively. Attenuation coefficient $\delta_n(r')$ and the horizontal wavenumber $k_n(r')$ of the nth mode depend only upon the position r' between r_o and r.

We assume the normal modes propagate and scatter independently and add incoherently. The reverberation component $R_{n,m}$ due to the transmission of the nth modal component of the field from a source at position x_o to the scattering area at x_b, scattering by a unit area, and propagation of the mth modal component of the scattered field to the receiver at position x is:[1]

$$R_{n,m}(x|x_b|x_o) = U_m^2(r, z) \, |A_m(r|r_b)|^2 \, S_{n,m}(x_b, \theta) \, |A_n(r_b|r_o)|^2 \, U_n^2(r_o, z_o) \,, \tag{2}$$

where $S_{n,m}(x_b, \theta)$ is the scattering coefficient that couples the field of the nth incident mode with the field of the mth scattered mode. The bistatic angle θ is the angle measured in plan view between the vectors $(r_b - r_o)$ and $(r - r_b)$, where $\theta = 180°$ corresponds to direct backscatter.

For bottom reverberation, $S_{n,m}(x_b, \theta)$ is obtained by assuming that the downward propagating component $U_n^{(-)}$ of the pressure field is incident upon the bottom with grazing angle $\phi_n = \cos^{-1}[c(r_b, z_b)k_n(r_b)/\omega]$, and that diffuse and near-specular three-dimensional scattering can be described by a combination of Lambert's law and Kirchhoff's approximation.[6,7] With these assumptions,

$$S_{n,m}(x_b, \theta) = |U_n^{(-)}(r_b, z_b)|^2 \, Q(x_b, \theta; \phi_n, \phi_m) \, |U_m^{(-)}(r_b, z_b)|^2 \tag{3}$$

where $U_n^{(-)}$ is obtained with a WKB asymptotic method.[8] The bistatic scattering function $Q(x_b, \theta; \phi_n, \phi_m)$ is discussed in Ref. 7.

For a narrowband signal pulse of duration τ, the reverberation time series $\mathcal{R}(t)$ is:

$$\mathcal{R}(t) = \int_{\mathcal{A}} \sum_n \sum_m R_{n,m}(x|x_b|x_o) I(t - t_{nm}) \, d\mathcal{A} \,, \tag{4}$$

where the integral is over the scattering region. We assume that the signal has a sufficiently narrow bandwidth that each modal term is frequency independent. The function $I(t - t_{nm})$ is the signal's envelope function and is nonzero only when $0 \leq t - t_{nm} \leq \tau$, where t_{nm} is the total travel time.

Although Eq. (4) is correct only for omnidirectional sources and receivers, the effect of arrays may be readily incorporated.

Table 1 – Geoacoustical Properties of the Sample Wedge Environment

Medium	Speed of Sound (m/s)	Density Ratio wrt Water	Attenuation (dB/kHz-m)
Water	At Right	1.0	0.00
Sand	1750	2.0	0.50
Basement	5000	2.0	0.68

Water Depth (m)	Speed of Sound (m/s)
0	1500
25	1500
35	1490
200	1490

3. RESULTS

In this section we present and discuss a sample calculation. The environment consists of a 3 layer ideal wedge with the water depth increasing linearly from 0 to 200 m over a range of 100 km. The sediment extends 200 m below the water depth at all ranges, and then an infinite half-space basement. The inclusion of the hard, half-space basement permits the approximation of the continuous modes by a discrete but highly attenuated set of modes. The geoacoustic properties are given in Table 1.

We consider bistatic reverberation from a 1-sec gated pulse projected from an omnidirectional source with an acoustic frequency of 350 Hz and source level 220 dB re 1 μPa@1m. The source and receiver are at depths 12.5 and 65 m, respectively, and are located over the 100 m depth contour with the source 10 km to the left of the receiver. The receiver is assumed to have a idealized beam pattern with 0.5° azimuthal resolution and no left/right ambiguity. The intensity normalization is such that summing the reverberation over azimuth gives the result for an omnidirectional receiver. For this example, there are 35 propagating modes that can make a significant contribution at ranges greater than approximately a kilometer. To show dependencies on mode order, Fig. 1 presents the received reverberation in plan view with respect to the receiver for 5 different partial sums over the contributions by source mode, and for a sum over the 35 propagating modes. The reverberation intensity is represented by color according to the scale at the bottom. Each color corresponds to a change of 10 dB and variations of individual color brightness indicating changes within the 10 dB interval. The calculation does not include the direct path; the high intensity (i.e., yellow and white) region between the source and receiver is due to the facet term in the scattering function.

There are several different phenomena represented in the plots of Fig. 1. The most evident is the mode order cutoff due to the 0.11° slope of the wedge. In the up-slope direction, a given mode will propagate until it can no longer be supported by the water column. In general, the lower the order of the mode, the shallower the water column can be and still support propagation. The result is the sharp drop-out of the reverberation at the bottom of each plot. The exact location of the mode cutoff cannot be determined from these plots since attenuation also impacts the contribution to the reverberation displayed.

Mode-dependent attenuation is most evident in Figs. 1(c–e). There are two effects to be noted: First, the higher order modes experience a higher attenuation. This is best illustrated by comparing the reverberation on the isobath located beneath the source/receiver where the environment is range independent. Second, for a given mode, the attenuation up-slope is greater than that down-slope.

In Figs. 1(a) and (b) the we note that for the first 50 km, the reverberation in the up-slope direction is actually greater than in the down-slope direction. This is not a contradiction with respect to the earlier comments on attenuation. In general, as a given mode propagates in the up-slope direction its wave number decreases, and thus, its effective grazing angle increases. The low order modes have effective grazing angles on the order of a few degrees, and the increase in the scattering strength due to the increase in grazing angle dominates the attenuation of the mode. The higher order modes have effective grazing angles on the order of 25° and the although the effective angle (and scattering strength) increases, the attenuation dominates. The difference in behavior is due to the $\sin\theta$ dependence of the Lambert's law scattering strength function which is most sensitive at small angles.

4. SUMMARY

A bistatic, normal-mode, range-dependent acoustic reverberation model has been described. Examples of its analysis capabilities/outputs have been shown. The detailed information provided by the model was used to explain the behavior of the received reverberation field for a sample calculation. Currently, the model only addresses bottom scattering using a Lambert's law/facet reflection description of the scattering process. Surface and volume scattering will be incorporated in the near future.

ACKNOWLEDGEMENTS

This work was supported by the Office of Naval Technology, Code 234. The authors are grateful to Mr. John Perkins of the Naval Research Laboratory for advice and assistance in the use of the WRAP propagation code and to Mr. Kenneth Luther, also of NRL for assistance in running the model. Dr. Gary Gibian and Mr. Gary T. Murphy of Planning Systems, Inc., and Mr. Kenneth Luther made substantial contributions to the development of the color displays shown here. Mr. Terrell Dossey, also of PSI, was instrumental in moving the model to a color workstation.

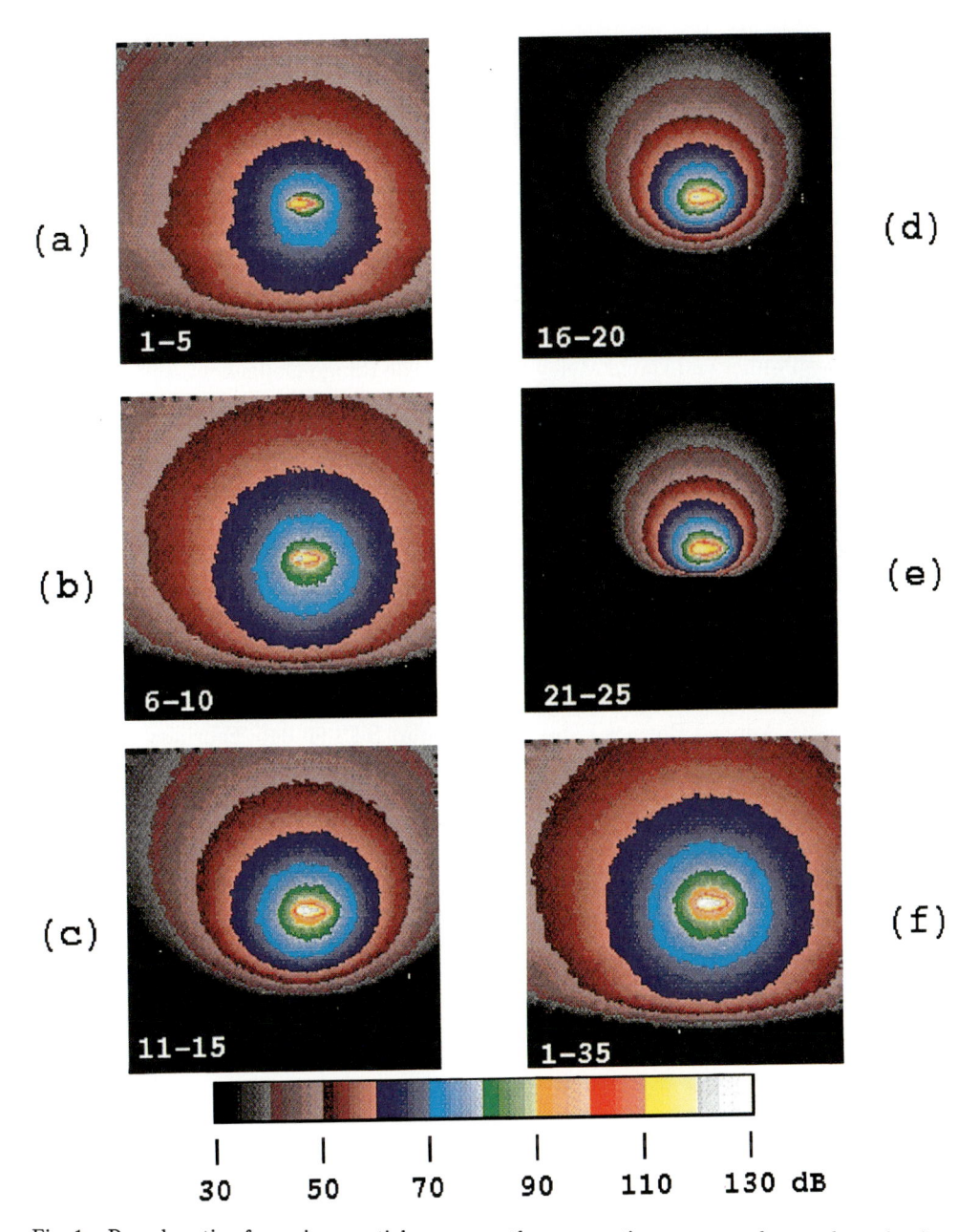

Fig. 1 – Reverberation for various partial sums over the propagating source modes are shown in plan view. The source modes used are indicated in the lower left corner of each plot. Each displayed area is 100 by 100 km with depth going from 0 m at the bottom to 200 m at the top. The reverberation time series were mapped into range with respect to the receiver using 1500 m/s as the assumed sound speed.

REFERENCES

1. Dale D. Ellis, "Shallow Water Reverberation: A Heuristic Ray-Mode Model," J. Acoust. Soc. Am. **80**(S1), S116(A) (1986); and informal communications.

2. M. B. Porter, "The KRAKEN normal mode program," SACLANT Undersea Research Centre Memorandum SM-245, September 1991.

3. Kuperman, W. A., M. B. Porter, J. S. Perkins, and R. B. Evans, "Rapid Computation of Acoustic Fields in Three-Dimensional Ocean Environments," J. Acoust. Soc. Am. **89**(1), 125–133 (1991).

4. L. Bruce Palmer and David Meloy Fromm, "The Range-Dependent Active System Performance Prediction Model (RASP)" Naval Research Laboratory Report 9383, 1992. A report documenting the bistatic version is in preparation.

5. L. Brekhovskikh and Yu. Lysanov, *Fundamentals of Ocean Acoustics* (Springer-Verlag, New York, 1982), Chap. 7, pp. 141–144.

6. Ref. 5, Chap. 6, pp. 187–197.

7. Ellis, D. D. and D. R. Haller, "A Scattering Function for Bistatic Reverberation Calculations," J. Acoust. Soc. Am. **82**(S1), S124(A) (1987).

8. K. J. McCann, M. E. Ladd, and C. S. Hayek, "Upward and Downward Propagating Modes for a Normal-Mode Treatment of Reverberation," The John Hopkins University Applied Physics Laboratory, Report STS-90-046, February 1990.

MEASUREMENT, CHARACTERIZATION, AND MODELING OF ICE AND BOTTOM BACKSCATTERING AND REVERBERATION IN THE ARCTIC OCEAN

by

T. C. Yang and T. J. Hayward
Naval Research Laboratory, Washington, DC 20375

ABSTRACT. Acoustic reverberation data from SUS explosions collected during the CEAREX 89 Arctic experiment are analyzed to estimate both under-ice and bottom backscattering strengths at very low frequencies (below 50 Hz). Short range direct-path returns are analyzed to measure the backward scattering strengths of the under-ice surface for scattering angles between 2.5° and 40° grazing and frequencies between 24 Hz and 105 Hz. Medium range (5–20 km) ice and bottom reverberation returns received on a 1200-m vertical array are separated using the time-arrival angle relationship to measure ice and bottom backscattering strengths at 23 Hz for grazing angles of 15°–60°. Long range reverberation is modeled by combining a normal-mode based scattering description with grazing angle dependent scattering functions based on small-ka (low frequency) approximations. Reverberation data at 23 Hz are successfully modeled for four cases of different bottom bathymetry and source depth.

1. INTRODUCTION

The 1989 Coordinated East Arctic Experiment (CEAREX 89), sponsored by the Office of Naval Research, was conducted in April 1989 in an area northeast of Greenland. At one of the ice camps, a large aperture (1200-m) vertical array and a cross-shaped horizontal planar array were deployed to record, among other data, reverberation from explosive charges. Objectives included measurement of scattering strengths of the ice and the bottom at low frequencies (15–50 Hz), development of a model for low-frequency, long-range reverberation and testing reverberation imaging (imaging of boundary features using reverberation data). The experimental measurements and modeling results are summarized here. Detailed analyses and results will be reported in separate publications.

Reverberation data were collected using 1.8-lb SUS charges detonated at depths of 91 and 244 m, approximately 350 m away from the 1200-m vertical array. The vertical array contained 31 hydrophones, 21 of which were spaced 30 m apart at nominal depths of 30–630 m, and 10 of which were spaced 60 m apart at nominal depths of 690–1230 m. Data from 26 channels were used in the analysis. For the analysis of long-range reverberation, we used a 14-element subarray of the cross-shaped horizontal array having an aperture of approximately 160 m.

2. DIRECT-PATH MEASUREMENTS

Previous measurements of under-ice backscattering strengths were reported by Milne using data collected in April 1962 near the Canadian Arctic Archipelago[1]. In that experiment, shots were deployed in close proximity to a single receiver to measure the monostatic backscattering strength from direct-path returns. In the CEAREX 89 experiment, a vertical array was used to measure the backscattering strength as a function of the scattering angle using reverberation data from SUS charges deployed at a depth of 244 m. The data for the direct-path returns covered an annular area of ice of approximately 1–3 km radius. Both this and the previous experiment were conducted at deep-water (~3000 m) sites so that reverberation returns from the ice could be separated from those from the bottom.

The measurement of scattering strengths was limited at both low and high frequencies by the decay of the SUS spectrum levels below the ambient noise. The scattering strength results are presented in Fig. 1, covering a frequency range of 24–105 Hz and scattering angles of 2.5°–40° grazing. The scattering strengths are averaged over the span of incident angles (10° to 15° grazing) covered by the data.

The scattering strength results for scattering angles of 10–15° grazing provide an estimate of the monostatic backscattering strength. Our results are approximately 10 dB lower than Milne's. We note that the nominal ice roughness in the CEAREX 89 experiment site is ≤ 2 m versus a typical value of 3–4 m in the area of the Milne experiment.

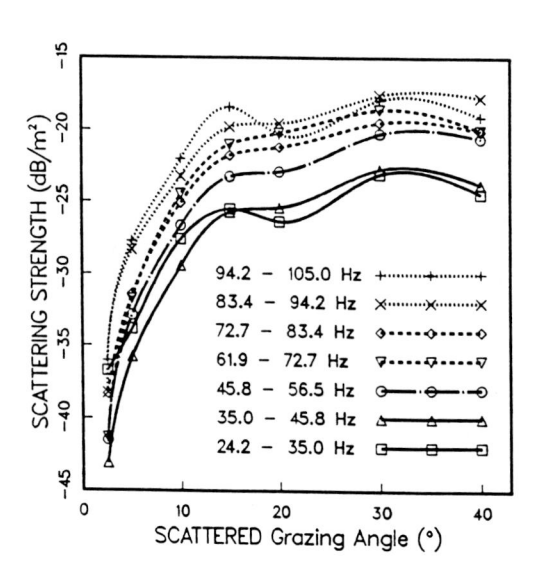

Fig. 1
Backscattering strength as a function of back scattered angle for 7 frequency bands.

3. MEASUREMENTS OF ICE AND BOTTOM BACKSCATTERING STRENGTHS FROM MEDIUM-RANGE DATA

The measurement of ice backscattering strength using short-range, direct-path returns is inherently limited by the small area (< 3 km radius) ensonified during the period before multiple-bounce returns arrive; hence the statistical sample is small unless the measurements are repeated many times at different ice floes. To obtain an estimate of scattering strength from statistically large samples, one resorts to long range reverberation data. However, the returns from long range ice and bottom features arrive simultaneously and are usually not separable. We find that at medium ranges surface and bottom backscattered returns encountering one or two bottom reflections can be distinguished in the reverberation data using vertical-array processing. This approach allows measurement of the surface and bottom backscattering strengths separately over a large ensonified area (up to 20 km radius).

To separate the surface and bottom returns, we use two-way travel time versus arrival angle (T-θ) contours calculated for the CEAREX 89 environment. The vertical array data are beamformed as a function of arrival angle and time; data located on the T-θ contours are used to deduce the surface and bottom scattering strengths.

The measured monostatic backscattering strengths from surface (ice) and bottom are plotted in Fig. 2 compared with some model predictions. The sources were four SUS shots detonated at 91 m depth. The scattering strengths deduced above have an estimated uncertainty of ~ 5 dB due to the small number of data sets available.

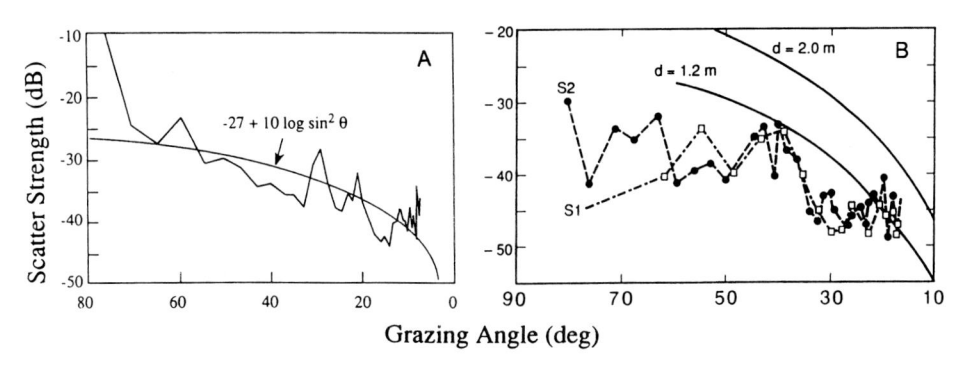

Fig. 2 (A) Measured bottom backscattering strength as a function of grazing angle compared with the Lambert's law. (B) Measured surface backscattering strengths associated with the S1 and S2 rays. Solid curves are Burke-Twersky model predictions for ridge depths of 1.2 and 2.0 m.

4. LONG-RANGE REVERBERATION MEASUREMENTS AND MODELING

Reverberation from a moderate charge in the central Arctic Ocean basin has been observed to last more than 30 minutes. Analysis of long-range reverberation at 9 Hz indicated returns from features of the ocean bottom [2]. At this frequency, the under-ice surface is effectively "smooth" from the standpoint of reverberation, except possibly for scattering from the very largest ice keels. At frequencies above 50 Hz, it is believed that backward scattering from under-ice roughness (particularly keels of typical size) will become dominant. In the frequency range 15–50 Hz, how much reverberation is due to bottom versus ice is not known a priori and has not previously been investigated quantitatively.

At long ranges reverberation returns from the under-ice surface and bottom arrive simultaneously at the receiver and cannot easily be identified in the data. Thus the interpretation of the reverberation data is much more difficult. To model low-frequency (15–50 Hz) reverberation, a field-theoretical model is required. We use a (rigorous) normal mode model of reverberation based on a generalization of Ingenito's earlier work [3]. The normal-mode scattering description is combined with the adiabatic normal mode representation of propagation to and back from the receiver. The backscattered acoustic field at a point (r, z) lying in the plane of propagation from a CW source at (r_s, z_s), of 0 dB source level, to a protuberance at $(r_0, z_0 = 0)$ on a pressure release surface is given by

$$P(\mathbf{r}, z) = \frac{1}{4k(0, r_0)} \sum_{m,n} \frac{\bar{\alpha}_m \bar{\bar{\alpha}}_n}{\sqrt{\rho \rho_s}} e^{i(\bar{k}_m \rho_s + \bar{\bar{k}}_n \rho)} \psi_m(z_s, r_s) \psi_n(z, r) S(\theta_m, \theta_n) \frac{\frac{\partial \psi_m}{\partial z}(0, r_0)}{\gamma_m} \frac{\frac{\partial \psi_n}{\partial z}(0, r_0)}{\gamma_n}, \quad (1)$$

where $S(\theta_{inc}, \theta_{scat})$ is the plane-wave scattering function (far-field scattering amplitude) of the protuberance, $\rho_s = |r_0 - r_s|$, $\rho = |r_0 - r|$; $\bar{k}_m = \frac{1}{\rho_s} \int_{r_s}^{r_0} k_m(r) \, dr$ and $\bar{\bar{k}}_m = \frac{1}{\rho} \int_r^{r_0} k_m(r) \, dr$ are the range-averaged mode wavenumbers for the forward and return propagation paths; $\psi_m(z, r)$ is the mode depth function for the m-th mode at range r; $\gamma_m = \sqrt{k(z_0, r_0)^2 - k_m(r_0)^2}$, $\bar{\alpha}_m = \sqrt{2\pi/\bar{k}_m}$, and $\bar{\bar{\alpha}}_m = \sqrt{2\pi/\bar{\bar{k}}_m}$. The incident and scattering grazing angles, denoted by θ_m and θ_n, are related to the mode wavenumbers by $\cos \theta_m = k_m(r_0)/k(z_0, r_0)$.

The formula for scattering from a protuberance on a rigid bottom is obtained by replacing $\frac{\partial \psi_m}{\partial z}(0, r_0)/\gamma_m$ by $\psi_m(z_0, r_0)$ in Eq. (1), where z_0 is now the bottom depth.

For the scattering function S, we assume small-ka (low frequency) approximations, where a is the dimension of the protuberance. These approximations take the form $S(\theta, \alpha) = A \sin \theta \sin \alpha$ for a protuberance on a pressure-release surface and $S(\theta, \alpha) = A + B \cos \theta \cos \alpha$ for a protuberance on a rigid bottom [4]. These scattering functions were used to calculate the reverberation level as a function of range via Eq. (1). We find that for the cases considered below, both terms in the bottom scattering function yield the same range dependence (within 1 dB) and hence can be adequately represented by the first term. The two coefficents of the scattering functions for surface and bottom are determined by fitting the data.

We now compare our model calculations with reverberation data received on the horizontal array. Data from SUS charges at 91 and 244 m were beamformed at two bearings to measure the reverberation level versus range along the two different bottom bathymetries shown in Fig. 3. Modeled and measured reverberation levels for the 91 m SUS are shown in Fig. 4 for both the flat bottom and the upslope propagation. We find good agreement

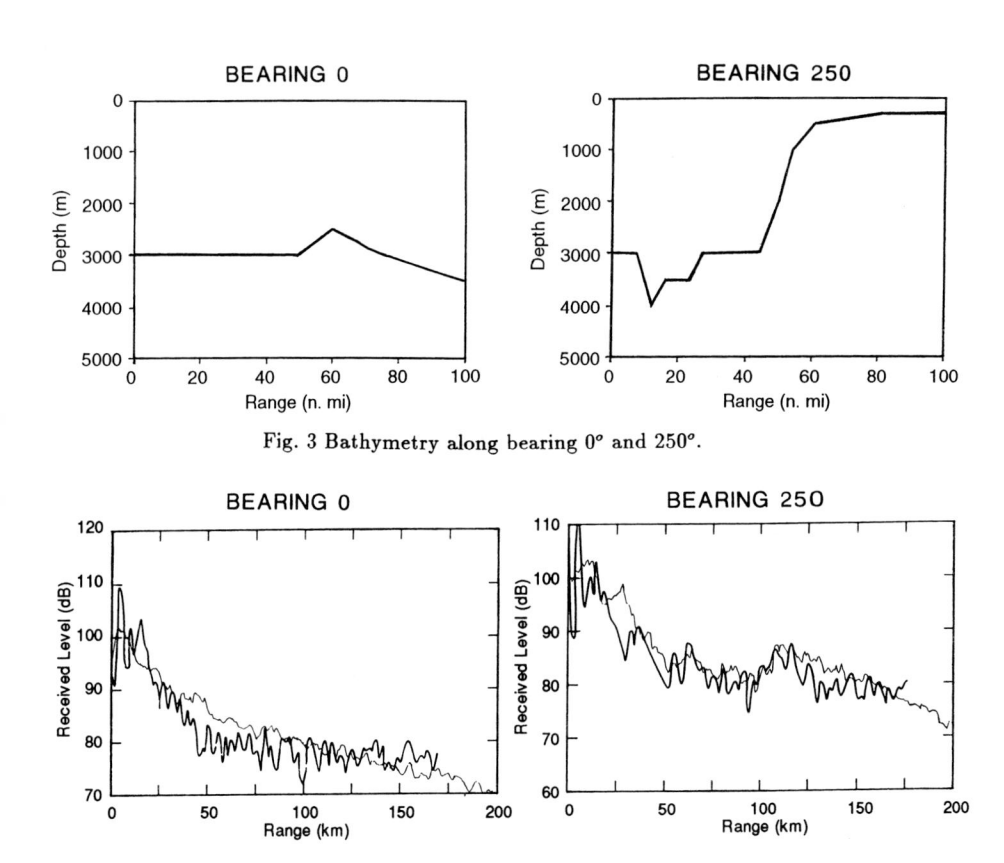

Fig. 3 Bathymetry along bearing 0° and 250°.

Fig. 4 Modeled reverbertion levels (thin lines) compared with measured data for source depth 91 m along bearings 0° and 250°.

between the model and the data for each of these two contrasting bottom bathymetries. We have assumed a bottom scattering strength $10\log(|A/k|^2\sigma) = -29$ dB/m^2, where σ is the density of the bottom scatterers, and a surface scattering strength of -52 dB/m^2 at 10°. Using these values for the scattering functions given above, we are able also to fit the reverberation versus range for the 244 m SUS case for both bathymetries. (The results are not shown here.) The modeled reverberation is smaller than the data for ranges > 125–150 km, where we speculate the data may be ambient noise limited. The differences in reverberation levels for the different source depths result from a combination of the source depth dependence of mode excitations and the angular dependence of the scattering function.

5. SUMMARY

Vertical array data from the CEAREX 89 experiment were used to measure the backward scattering strength of the ice as a function of scattering angle using short-range (< 3 km), direct-path returns from the under-ice surface. Backscattering strengths of ice and bottom at grazing angles between 15° and 60° were estimated at 23 Hz using medium-range (5–20 km) data. Long range reverberation data received on the horizontal array were beamformed at different bearings to measure the reverberation level versus range and investigate the effects of bathymetry and source depth on the reverberation. A normal mode model using small-ka approximation for the surface and bottom scattering functions successfully models the reverberation data at 23 Hz.

ACKNOWLEDGEMENT. This work was supported by the Office of Naval Research and the Office of Naval Technology.

REFERENCES

1. Milne, A. R. (1964) "Underwater backscattering strengths of Arctic pack ice," J. Acoust. Soc. Am. 36(8), 1551-1556.

2. Dyer, I., Baggeroer, A. B., Zittel, J. D. and Williams, R. J. (1982) "Acoustic Backscattering from the basin and margins of the Arctic ocean," J. Geoph. Res., 87, 9477-9488.

3. Ingenito, F. (1987) "Scattering from an object in a stratified medium," J. Acoust. Soc. Am. 82, 2051-2059.

4. Bowman, J.J., Senior, T. B. A. and Uslenghi, P. L. E. (eds), (1987), "Electromagnetic and Acoustic Scattering by Simple Shapes", Hemisphere Publishing Co., New York.

A REVERBERANT-UNDERSEA MODEL FOR BOTTOM-LIMITED ENVIRONMENTS

HOMER P. BUCKER, JOSEPH A. RICE and NEWELL O. BOOTH
Naval Command, Control and Ocean Surveillance Center
Research, Development, Test and Evaluation Division
(NRaD)
San Diego, CA 92152-5000
United States

Abstract. A Reverberant-Undersea Model for Bottom-Limited Environments (RUMBLE) has been developed. RUMBLE estimates component and total reverberation levels as functions of time for bistatic-sonar geometries. General projector and receiver arrays are specified, along with desired target scenarios for the event. Influential environmental parameters such as bathymetry, sound speed, surface-scattering mechanisms, bottom loss and volume-scattering layers may be modeled as spatially-dependent variables. The projected signal is propagated as a fan of Gaussian beams through the modeled undersea medium. Acoustic pressure at any point in space is computed as a sum of the contributions of individual beams. Numerical efficiency is achieved by one-time calculation of sound propagation as intermediate results to be subsequently accessed via look-up tables. Retained in the tables are arrival times, angles and levels for a specified number of lowest-loss propagation paths from projector and receiver to discrete ranges. The reverberation calculation proceeds by numerically integrating over all scattering layers. For each area differential or "cell" of a scattering layer, all applicable projector-to-cell paths are combined with all applicable cell-to-receiver paths and weighted by the angle-dependent scattering strength. The energy contribution from each propagation-path combination is added at the appropriate time, resulting in reverberation time series for surface, bottom and volume scattering. Total received sound is calculated by summing the reverberation components with specular and target-echo components. Theoretical case studies have been formulated to verify the accuracy of the numerical method. Modeling of numerous sites for which measured data are available has confirmed the validity of using RUMBLE to model ocean reverberation.

1. Introduction

The Reverberant-Undersea Model for Bottom-Limited Environments (RUMBLE) has been developed. RUMBLE is a numerical algorithm implemented in FORTRAN and presently maintained on both UNIX and VAX computer systems. A spin-off of an early RUMBLE release is a C-language version developed by Castile (1992) and presently maintained on UNIX and Macintosh computer systems. Both versions have been successfully applied to an assortment of sonar problems.

RUMBLE estimates component and total reverberation levels as a function of time for bistatic-sonar geometries. General projector and receiver arrays are modeled. Situations may include multiple targets and scattering layers. Such environmental characteristics as bathymetry, sound-speed profile, bottom loss, and surface-scattering coefficients may be modeled as space-dependent variables.

The key to RUMBLE's numerical efficiency is the one-time calculation and storage of sound-propagation information. These intermediate data are later accessed in the form of sound-angle-level-time (SALT) tables, named for the information stored therein. The arrival angle, sound-intensity level, and arrival time for a user-specified quantity of lowest-loss-propagation paths are retained in the SALT tables as discrete functions of range.

2. Propagation

2.1. Gaussian Beams

Propagation is modeled using the Bucker and Porter (1986,1987) concept of Gaussian beams. The projected sound is approximated by a fan of beams numerically propagated through the modeled medium. Acoustic pressure at a given location is computed as a sum of the contributions of each of the individual beams. The central ray of the beam is traced by integrating the usual ray equations. A beam is then constructed about each ray by integrating a pair of equations governing arclength-dependent beamwidth and curvature. The resulting beam-pressure field diminishes in a Gaussian fashion as a function of normal distance from the central ray. Gaussian-beam tracing avoids undesirable artifacts common to ray tracing. The solutions are free of singularities at caustics and abrupt discontinuities at shadow-zone boundaries.

2.2. SALT Tables

The undersea medium is modeled as a set of scattering layers. For each scattering layer, a full set of multipaths are traced for discrete ranges from the projector. Each multipath is indexed according to the boundary reflections or inversions it has experienced.

For every range r considered, a SALT table is created for each scattering layer. For multipath i from the projector, sound-intensity level $L_{P \to r}(i)$ is computed from the contributing Gaussian beams. If the level exceeds the minimum threshold, it, along with the cosine of the vertical ray angle and the multipath index, is saved as a SALT-table entry.

SALT tables containing propagation of scattered sound from scattering layers to the receiver are compiled similarly. Propagation reciprocity justifies using the receiver position as the origin of the Gaussian beams for these calculations.

2.3. Propagation Loss

The entries to the projector SALT tables represent sound intensities equivalent to the source level less vertical-beampattern strength, propagation loss, scattering loss and absorption loss suffered en route to a given range for a particular layer. The entries to the receiver SALT tables, on the other hand, represent propagation losses for various paths from the receiver to the scattering elements and include vertical beampattern gain. Horizontal beampatterns of projector and receiver are separately accounted for by azimuth-dependent sector definitions.

3. Reverberation

3.1. SCATTERING LAYERS

The sea surface and sea bottom are treated as two-dimensional scattering layers, A_S and A_B, divided into contiguous-area differential "cells", denoted ΔA. Similarly the volume is modeled as a collection of scattering layers. Volume layers at each target depth and a volume layer A_P at the projector depth are included.

On the projector-depth volume layer A_P, the rectangular cells are tiled on a horizontal Cartesian grid originated at the projector location and oriented with the x-axis toward the vertical projection of the receiver location. The cells have a 1-m^2 area at the projector and receiver locations. Cell size increases geometrically with distance from these points, roughly in inverse proportion to the significance of the cell on the reverberation effect. The cell grid is superimposed upon each of the scattering layers and the resulting cells form the scattering-layer differentials, ΔA.

3.2. INTEGRATION

Numerical integration over the scattering-layer areas proceeds, one cell at a time. For a given cell, the projector SALT table corresponding to the range of the cell from the projector is referenced. Likewise, the receiver SALT table containing data for the receiver-to-cell range is accessed. For each cell, all applicable projector-to-cell SALT entries are paired with all applicable receiver-to-cell SALT entries and weighted by the angularly-dependent scattering strength. The reverberation from a scattering layer A is

$$R_A(t) = \sum_{\Delta A} \sum_{i=1}^{I} \sum_{j=1}^{J} L_{P \to \Delta A}(i) \; L_{R \to \Delta A}(j) \; f(t-\tau_i-\tau_j) \; \sigma_{ij} \; \Delta A \qquad (1)$$

For a given scattering cell ΔA, the SALT tables contain I propagation paths from the projector and J propagation paths from the receiver. $L_{P \to \Delta A}(i)$ is the sound that travels to ΔA via path i from the projector. $L_{P \to \Delta A}(i) \, \sigma_{ij} \, \Delta A$ is the sound that is scattered at ΔA in such a direction that it travels to the receiver via path j. The scattering coefficient σ_{ij} depends upon the elevation angles associated with paths i and j and the relative geometry of the projector, receiver and cell locations. $L_{R \to \Delta A}(j)$ is the factor that accounts for the propagation loss on path j and for the receiver beampattern. The function $f(t)$, the correlator response to the transmitted signal, is zero for $t < 0$ and falls off rapidly for $t > 1/\Delta f$, where Δf is the bandwidth. The times τ_i and τ_j are the respective travel times for paths i and j.

3.3. TIME SERIES

Equation (1) registers the time dependence of reverberation power by determining when each packet of acoustic energy arrives at the receiver. This is merely the sum of the two SALT-table time entries, τ_i and τ_j, for each projector-cell-receiver combination. The energy contribution from each propagation-path combination is added at the appropriate time, resulting in reverberation time series for the surface, bottom and volume. The projector-to-receiver specular (direct, unscattered multipath) propagation response and projector-to-target-to-receiver echo response are also maintained.

The onset of the signal produced by the projector is arbitrarily assigned to be time zero. Commonality of the time scale allows for direct summation of these component time series to obtain total received level. At any time t, the total received sound $T(t)$ is calculated by summing the reverberation components with the direct (specular) and target-echo components, $D(t)$ and $E(t)$:

$$T(t) = R_S(t) + R_B(t) + R_V(t) + D(t) + E(t) . \qquad (2)$$

In some cases, an ambient noise term, $N(t)$, is included in equation (2).

4. Input

In it's present form, RUMBLE is executed as a batch program. The user completely specifies the sonar system, the environment, and run parameters in a single input file.

4.1. SONAR-SYSTEM PARAMETERS

The projector location is treated by RUMBLE as the origin of the spatial coordinate system. User-specified projector parameters are depth, signal frequency, source level, signal duration, beampattern, and orientation. RUMBLE treats the receiver location as colinear with the positive x-axis (0° polar). User-specified receiver parameters include horizontal range from projector, operating depth, beampattern, and orientation.

4.2. ENVIRONMENTAL PARAMETERS

Correct modeling of propagation through the ocean medium and of boundary interactions at the sea surface and bottom requires an accurate environmental description. Some parameters, especially physical processes occurring at the sea surface, are assumed to be constant at all ranges and azimuths. Others, particularly bottom characteristics and temperature-influenced parameters, may be specified by the user to be spatially dependent.

4.2.1. SPATIALLY-INDEPENDENT PARAMETERS

Boundary roughness and inhomogeneities at the sea surface are a significant source of reverberation. These nonstationary processes have small time constants and are assumed to be uniformly distributed over the entire region of interest. In addition to specifying a base surface-scattering factor, the user may specify the average wind speed and average peak-to-trough wave height. The Chapman and Harris (1962) equations employ the wind speed to estimate surface scattering factors. Surface-reflection loss is computed from either wave height or wind speed, at the discretion of the user. Additional surface-reflection loss due to bubbles and other mechanisms may be user-specified as a function of surface-grazing angle.

The volume-attenuation factor typical for the basin is user supplied. If scattering layers exist within the medium, the layer depth and volume-scattering coefficients must be specified. A horizontal-scattering function to control the off-axis spread of energy at boundary interactions is required. The user may select a standard function or provide an angle-dependent function.

4.2.2. SPATIALLY-DEPENDENT PARAMETERS

With the projector as the origin of a polar coordinate system, the user may define azimuthally-dependent, wedge-shaped sectors for which the spatially-dependent parameters are separately specified. A separate sector geometry is defined for the receiver.

A range-dependent bathymetry profile is required for each sector. Other user-specified bottom parameters are bottom loss and bottom scattering, both functions of grazing angle. While the user may fully specify either function, options exist for invoking standard bottom-loss functions such as Weinberg's (1980) implementation of the Marine Geophysical Survey (MGS) curves or selection of standard bottom-scattering functions described by Wong and Chesterman (1968) and Mackenzie (1962) for various bottom compositions. To model reflection losses from a layered bottom, the user may optionally include a phase-shift-vs-grazing-angle function to augment the bottom-loss function.

For each sector the user provides range-dependent sound-speed profiles. RUMBLE grids these and then smoothly interpolates over range.

To allow realistic representation of a given environment, RUMBLE allows adjacent sectors to overlap. For a given location within the region of intersection, the spatially-dependent parameters are a weighted average of the environments associated with the two overlying sectors. The weighting provides linearly smooth transition within the intersection from one sector to another. RUMBLE does not allow gaps between sectors.

4.3. RUN PARAMETERS

Execution parameters are required to orchestrate RUMBLE's usage of input data.

For each segment of a target track, the user provides range, bearing, depth, volume-scattering strength, and target strength.

Acoustic propagation modeling is controlled by specifying angular and horizontal increments and limits, transitions between ray theory and beam theory, and maximum number of boundary reflections. SALT-table creation is controlled by the number of tables permitted, and by the specified range increment.

Reverberation computation is controlled by the initial cell dimensions and their geometric growth factor, the time-averaging period, the reverberation time increment, and the maximum event time.

Use of previously computed SALT tables or reverberation data may be specified if the user wishes only to vary the bistatic geometry or target track.

5. Output

Separate time series for surface reverberation, bottom reverberation, volume reverberation, direct-path specular, target echoes and total received sound are the raw output products of RUMBLE. Also available as a raw output are signal gains due to Doppler processing for nonstationary targets.

Processed output includes intensity plots and echo-to-reverberation ratios plotted as functions of space and time. Such information is presented as a two-dimensional horizontal slice at the intended receiver depth.

6. Validation

The first step in validating a computation-intensive algorithm such as RUMBLE is to apply it to a set of simplified problems which can be analytically solved. Monostatic cases are fairly easy to develop. Cox (1988) provides a good start for extending the analysis to bistatic geometries. The model results must agree with the analytical solutions for these simple problems before it can be trusted with more complex scenarios.

Several such elementary problems have been formulated and comprise a diagnostic kit for verifying the correct behavior of RUMBLE following a new installation or after modifying the source code. Each test case exercises a specific property of the algorithm. As an example, one such test case places the projector near the surface of an infinitely deep ocean to isolate surface-reverberation effects.

Next, ocean environments from which active sonar data have been obtained are modeled. Any differences between the measured and predicted results must be reconciled by improving the model inputs or refining the model assumptions. Several such iterations have been performed, resulting in the reliable prediction tool RUMBLE is today.

As with any model of a complex physical process, improvements are always possible. Indeed, several major enhancements are planned. Most notably, incorporation of the three-dimensional-Gaussian-beam algorithm will allow RUMBLE to directly use a standard or user-generated bathymetric data base for the computation of bottom-interacting propagation.

7. Applications

Applications for RUMBLE include designing sonar projectors and receivers, analyzing ocean environments, and planning active-sonar experiments. The usual approach is to specify a base case as the standard against which subsequent cases are compared. The analyst thereby obtains a measure of sensitivity to each parameter of the sonar configuration of interest.

Results of a simple parametric study are presented in the figure below. Two environments with differing bottom-loss characteristics are considered; all other parameters are identical. The plotted functions were computed by RUMBLE according to equation (2).

Advanced applications use RUMBLE results as input for more complex calculations. Examples include engagement modeling and simulation.

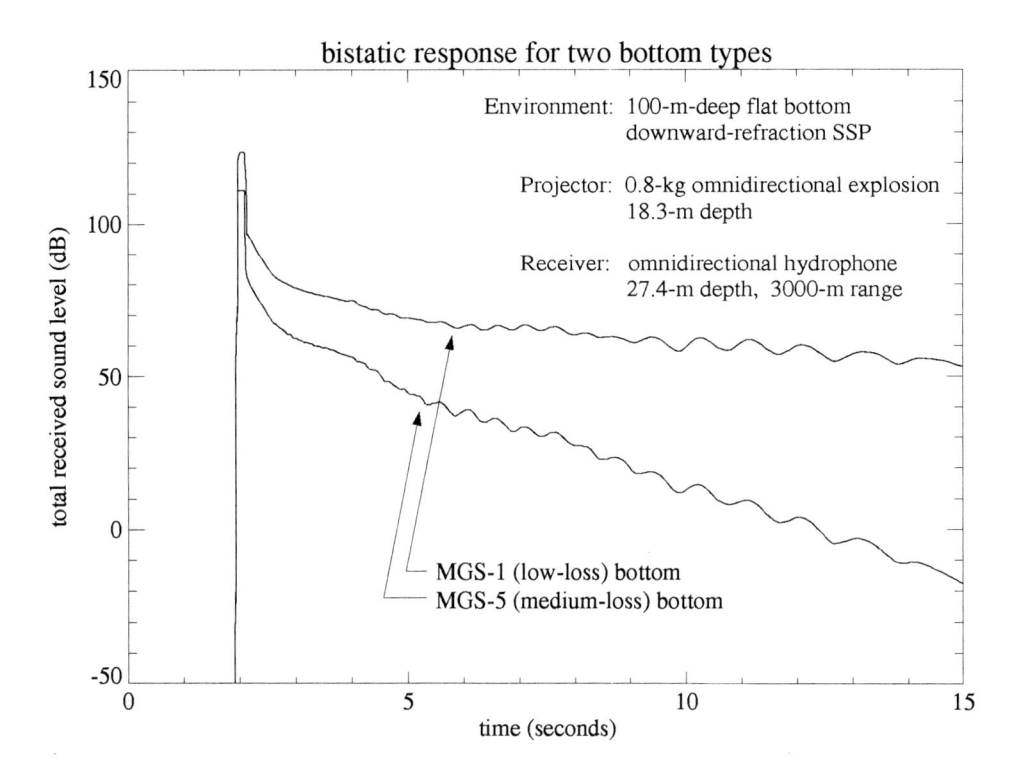

bistatic response for two bottom types

Environment: 100-m-deep flat bottom
downward-refraction SSP

Projector: 0.8-kg omnidirectional explosion
18.3-m depth

Receiver: omnidirectional hydrophone
27.4-m depth, 3000-m range

MGS-1 (low-loss) bottom
MGS-5 (medium-loss) bottom

8. Acknowledgements

The authors are grateful for assistance from Brett Castile of Orincon Corporation and Paul Hursky of Lockheed Aeronautical Systems Company in validating RUMBLE calculations and providing processed results. We also acknowledge the contribution to RUMBLE development by Douglas Kewley of Australian Defence Science and Technology Organization.

9. References

Bucker, H. P. and M. B. Porter, "Gaussian Beams and 3-D Bottom-Interacting Acoustic Systems," in Ocean Seismo-Acoustics, edited by T. Akal and J. Berkson (Plenum, New York, 1986).

Castile B., "Software for Bistatic-Sonar Predictions over Planar-but-Tilted Ocean Bottoms," unpublished NRaD/Orincon report (1992).

Chapman, R. P. and J. H. Harris, "Surface-Backscattering Strengths Measured with Explosive Sound Sources," *J. Acoust. Soc. Am.*, V.34 p.1592 (1962).

Cox, H., "Fundamentals of Bistatic Active Sonar," presented at NATO Advanced Study Institute, Underwater Acoustic Data Processing, Royal Military College of Canada, Kingston, Ontario, Canada (1988).

Mackenzie K. V., "Long-Range Shallow-Water Bottom Reverberation," *J. Acoust. Soc. Am.*, V.34 p.62 (1962).

Porter, M. B. and H. P. Bucker, "Gaussian-Beam Tracing for Computing Ocean Acoustic Fields," *J. Acoust. Soc. Am.*, V.82 p.1349 (1987).

Weinberg, H., "Generic Sonar Model," Naval Underwater Systems Center, Technical Document 5971-A (1980).

Wong, H. - K. and W. D. Chesterman, "Bottom Backscattering Near Grazing Incidence in Shallow Water," *J. Acoust. Soc. Am.*, V.44 p.1713 (1968).

SHALLOW WATER REVERBERATION MODELING

James K. Fulford
Naval Research Laboratory Detachment
Stennis Space Center, MS 39529

Abstract

Shallow water reverberation in the high sub-kHz to low-kHz range is modeled using ray theoretic algorithms. The algorithms assume multiple-forward/single-backscatter along with reciprocity in the propagation paths. The environments modeled are assumed to be range dependent in bathymetry, bottom loss, and bottom scattering strengths, and range independent water column, wind velocity, and surface scattering strengths. The model is capable of simulating bistatic, and monostatic geometries. The output of the model is reverberation as function of time for specified beams (either idealized look directions, or realizations of the beam pattern of an array). Comparisons of predicted reverberation versus time with measured data for shallow water sites off the south coast of the United States are presented.

Introduction

Prediction (or estimation) of reverberation for near kHz frequencies in shallow water (where shallow water is defined as water bounded on the one side by the location at which the continental shelf transitions to the continental slope and the other side by the 100 ft depth contour nearest the coast line) is typically made using range independent ray codes which assume gradually changing vertical gradients in the sound speed fields, simple bottom loss structures, and relatively simple relations between travel time and range. These assumptions lead to models which perform well in certain simple, low gradient cases which are not typical of coastal shelf environments.

Within the context of near kHz frequency regime, the environmental considerations suggest that range independent sound speeds with the ability to maintain convexity in high vertical gradient areas, coupled with appropriate geoacoustic models can lead to a conceptually simplistic reverberation model that is adequate. This concept is embedded in a range dependent (bathymetry, bottom loss, and bottom backscattering strength) n x 2 d ray theoretic reverberation model.

Shallow Water Environments

The object of reverberation modeling is to predict the observed reverberation as a function of time, that is the transient response of the scattering bodies in the water column, or along its boundary to an acoustic transmission. Thus the acoustic characteristics of the acoustic system, and the environmental characteristics of the experimental site must be known.

Regardless of the water depth it is necessary to know the geoacoustics of the subaqueous layers, the sound speed structure of the water column, and be able to characterize the interfaces that bound the water column to determine the scattering characteristics.

Of these three factors, the sound speed structure would seem to be the driving influence. Under the influence of solar heating, the upper layers of the water will be heated relative to the near bottom water. Typically low summer wind speeds will lead to the typical three layer summer temperature structure in shallow water: a warm surface, a thermocline, and a cool bottom layer. This leads to an essentially downwardly refracting sound speed profile with a high gradient layer in the thermocline ,thus large gradients in the sound speed regime which must be modeled such that the essential characteristics of the sound speed field are conserved. In winter surface cooling will lead to well mixed layer extending to near the sediment layer.

The sea bed in shallow water environments can be generally characterized by shallow slopes, both in the surface outcrop, and in the subsurface layer, with increasing isopacs toward deep water. Shallow water implies that the thermal mass of the water column is low, thus appreciable change in near bottom sound speed structure can be expected as a function of weather systems. Thus seasonal dependent bottom loss is indicated, suggesting that geoacoustic models would be appropriate. In shallow water the bottom sound speed will be typically lower the the compressional sound speed in the sediments, thus suggesting low loss environments.

Model Fundamentals

Conceptually the model is a relatively simple n x 2 d reverberation code based on ray theoretic concepts. The model is based on the computational technique contained in the RASP (Franchi et al, 1984) model. That is rays are traced from both the source, and receiver locations along bathymetric range dependent great circle tracks. At each intersection with a boundary, or selected intermediate depth the ray parameters - travel time, grazing angle, path length, and bottom loss - are retained. A caustic corrected m (where m ≥ 2) point estimate of the transmission loss is computed for each family of rays, with appropriate vertical beam patterns. The transmission loss estimates are then gridded as a function of range along each track. The source, and receiver gridded vertical arrival structures for the surface, bottom, and volume are then combined with a scattering kernel, horizontal beam pattern and area calculation to arrive at estimate of the reverberant field along each receiver track. Within in the context of ray theory, low resolution bathymetry is adequately modeled, but high resolution bathymetry is poses serious problems to this model.

Fundamentally this is the RASP approach. However environmental considerations lead to three significant differences: the ray trace algorithm, the approach to range dependency in bottom loss, and the reverberation area calculation:

The nature of the shallow water sound speed environment during summer months generated the need to fit curves that had large gradients, with the constraint that the fitted sound speed field must maintain the original sound speed field convexity. This was accomplished using cubic splines under tension (Cline, 1974). The actual ray tracing method was changed to a Runge-Kutta technique.

The geoacoustics of shallow water areas suggest increasing isopac thickness with water depth. Thus rather than sampling a fixed points, and assigning ranges of applicable bottom loss - a curve was fit to the bottom loss as a function of range along a track. Thus bottom loss became a two dimensional structure in the model - the method of fitting was box splines under tension.

The area calculation was modified to use bipolar coordinates in the range, azimuth plane. This modification was implemented to restrict the time width of any incremental area to a value less than the sound speed incremental time quotient.

Examples

Two examples are now presented to demonstrate initial test results. For each of the examples the source was a 0.82 kg SUS charge, and the receiver was an omni directional sonobuoy.

The information for the first test case was found in Urick (1969). The known environmental data is listed in table 1. The sound speed in this case is particularly simple, a insignificant surface duct atop a downwardly refractive layer. The geology of the area is characterized by a quaternary alluvium, atop relic Pleistocene sand. The bottom, and sub-bottom slope gently toward the west (Gulf of Mexico). The resultant bottom loss is similar to, but less than the MGS class 1 curve by about 1 dB for the entire area of interest, the scattering strength was taken from Urick.

Table 1. Environmental information for first test case.

Water Column

depth (m)	sound speed (m/s)
0	1539.5
36.6	1540.1
61.0	1527.4

Sea Surface: calm
Bottom: 1 meter mixture of coral,sand,mud,shell over sand

Figure 1 shows the comparison of the model output with the observed output for the octave band from 400 to 800 Hz. The figure shows that the model reproduces the essential features, and levels of the observed data.

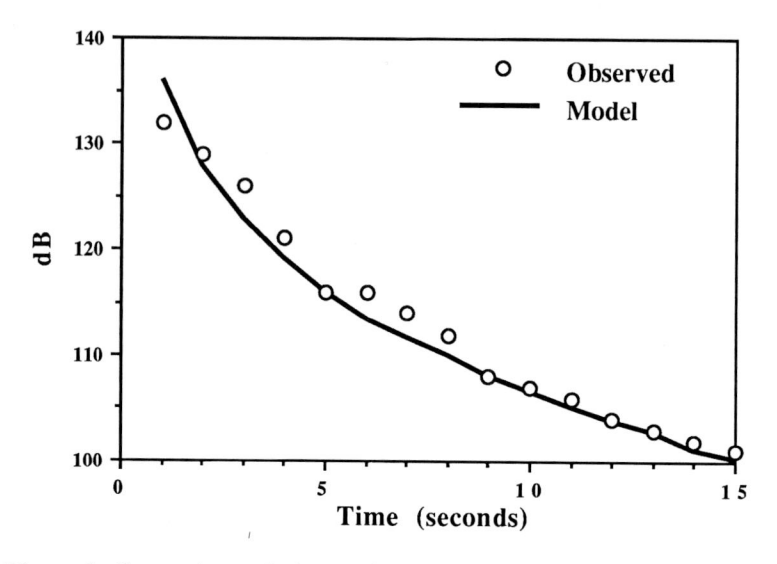

Figure 1. Comparison of observed reverberation data with model prediction for site off the coast of Florida for the 400-800 Hz band.

Table 2. Environment information for second case.

Water Column

depth (m)	sound speed (m/s)
0	1530.3
15.2	1530.5
22.9	1521.9
33.5	1490.5
70.1	1491.1

Sea Surface: waves approximately 3 feet
Bottom: clay silt over sand

The information for the second test case is found in Urick (1971). The known parameters environmental data is listed in table 2. The sound speed listed in the table demonstrates one of the more severe sound speeds that can be found in shallow water, the two isothermal layers separated by a thin high gradient thermocline. In this case the bottom loss, as computed from a geoacoustic model of the area reveals very low bottom loss as grazing angles up to approximately 15-20 degrees (depending on the details of the model).

Comparison of the observed data with the model output is reasonable. The intent of the sampling of the data was to compare the fall off rate of the data with the fall off rates predicted by various models, and as such the comparison of the two rates is favorable.

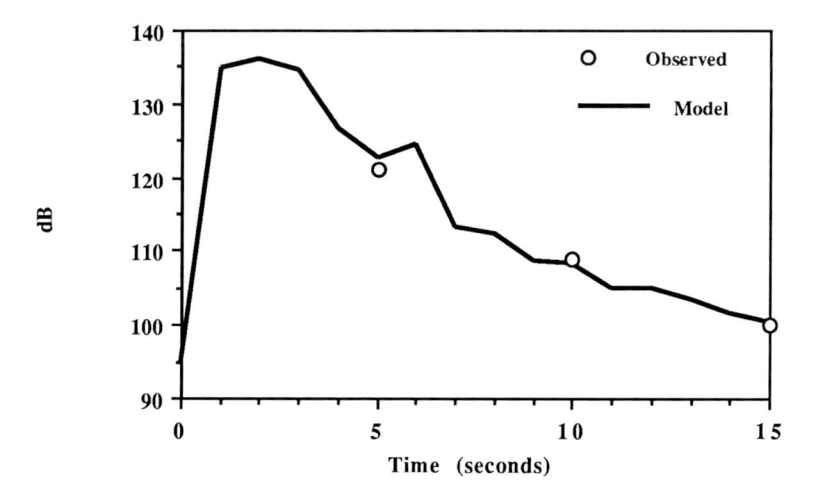

Figure 2. Comparison of observed, and model predicted data for site off eastern coast of the United States for the octave band centered at 1kHz.

Conclusions

The comparison of a conceptually simple model with measured data in continental shelf waters demonstrates that it is possible to obtain reasonable reverberation predictions in hypsometrically shallow water for the typical mid-latitude sedimented regions. Using this basic reverberation modeling approach, typical low latitude environments will be investigated in the future.

Acknowledgements

This work was sponsored by PEO(A) PMA 264, Mr. Bob Davis, program manager, under program element 603254N. This paper is NOARL contribution PR 92:0088:223.

References

1. Franchi, E.R., J.M. Griffin, and B.J. King (1984). NRL Reverberation Model: A Computer Program for the Prediction and Analysis of Medium- to Long-Range Boundary Reverberation. Naval Research Laboratory, Washington, DC, NRL Report 8721.

2. Cline, A. (1974). Scalar- and planar-valued curve fitting using splines under tension. Comm. ACM, 17 218-223.

3. Urick, R.J. (1969). Acoustic Observations At a Shallow Water Location Off The Coast of Florida. Naval Ordance Laboratory, Silver Springs, Maryland, NOL-TR 69-90.

4. Urick, R.J.(1971). Airborne Measurements Of Shallow Water Acoustics At Various Shallow Water Locations Off The Eastern And Gulf Coasts Of The United States. Naval Ordance Laboratory, Silver Springs, Maryland, NOL-TR 71-4.

Section 4

ARSRP Mid-Atlantic Ridge Experiment

AN OVERVIEW OF THE 1991 RECONNAISSANCE CRUISE OF THE ACOUSTIC REVERBERATION SPECIAL RESEARCH PROGRAM [1]

Arthur B. Baggeroer
Depts. of Ocean and Electrical Engineering
Massachusetts Institute of Technology
Cambridge, MA 02139
and
John A. Orcutt
Institute of Geophysics and Planetary Physics
Scripps Institution of Oceanography
Univ. of California, San Diego
La Jolla, CA 92093

ABSTRACT. The Reconnaissance Cruise of the ONR Acoustic Reverberation Special Research Program (ARSRP) was conducted in July - August, 1991. The tracks were west of the Mid-Atlantic Ridge at approximately 26° N from 45° − 53° W. The seafloor in this region is rough with no or very thin sediment cover. The acoustic signals used were centered at 250 Hz with bandwidths up to +/- 40 Hz, the transmitter was a ten element, 5 wavelength, vertical array, the receiver was a 128 element, 64 wavelength, horizontal array. The ARSRP Reconnaissance Cruise was the first of two experiments in this area with the objective of determining the characteristics of low grazing angle bottom reverberation. The second is planned for 1994 and will use *in situ* arrays and sources, towed source and receiver arrays, bistatic receivers and high resolution bathymetry obtained during a 1993 Geophysical Cruise for the ARSRP. This paper provides an overview of the ARSRP Recon Cruise.

1. Introduction

Numerous experiments have demonstrated that backscattered reverberation in the deep ocean basins can persist for a long time. (See Baggeroer and Dyer ([1]) for a summary of the open literature prior to 1985.) Explosive sources have produced echoes which can be correlated with seafloor bathymetry at ranges more than 3000 km distant at frequencies below 50 Hz. While controlled, or modulated, sources operate at higher frequency bands, they also can generate returns backscattered from the seafloor at ranges of 1000 km. The backscattered energy must couple efficiently to the ocean waveguide to be detected at these long ranges, so it is the low grazing angle components which give rise to long range, deep ocean reverberation.

In spite of its importance the mechanisms for low grazing angle backscattering are

---

[1] Work of both authors was supported by the Acoustic Reverberation Special Research Program of the Office of Naval Research.

not well understood. Low grazing angle is difficult to measure experimentally and to predict with simple analytical tools. There are two experimental alternatives: i) Place sources and receivers near the seafloor, or ii) Standoff at ranges of roughly 1/2 convergence zone or more and use the refraction of the sound channel. (i) gives high resolution measurements, but the technology for both sources and array receivers is difficult and coverage is limited because of the fixed nature of the experiment. In (ii) the range offset leads to lower azimuthal resolution, more complicated insonification and the need for more powerful sources, but coverage is more extensive. Modeling and prediction of low grazing angle backscattering is additionally complicated by the low resolution bathymetry normally available. Most theories for backscattering require roughness characterization on the scale of the incident acoustic wavelength or less; however, few locations of the seafloor have been mapped to this scale even at low frequencies. Moreover, slopes, which are felt to be important for many theories, are poorly resolved.

The ONR Acoustic Reverberation Special Research Program (ARSRP) is directed towards a better understanding of the mechanism of low grazing angle reverberation. It combines the use of acoustics systems which are powerful and have high range and angular resolution with detailed bathymetric mapping of the seafloor. The ARSRP Reconnaissance Cruise (ARSRP/RC) during July - August 1991 was the first experiment of this program. Backscattering in a region west of the Midatlantic Ridge at approximately 26° N and between 45° − 53° W was measured using a very high resolution system at ranges greater than 1/2 a convergence zone. Detailed bathymetric mapping in the same region will be done in 1992 and more detailed acoustic experiments which additionally will include bistatic measurements and *in situ* equipment will follow in 1993.

This article gives an overview of the ARSRP/RC and the articles which follow provide some of the initial results and interpretations of the data. A very complete summary is given in the ARSRP/RC Cruise Report [2].

2. Reconnaissance Cruise Objectives

The objectives of the cruise were: i) direct path insonification at offsets of 1/2 convergence zone for low grazing angle incidence with very high range resolution; ii) intermediate scale insonification at offsets of 1/2 to 10 convergence zones in a region of excess depth for determining effects of propagation paths; iii) insonification for ranges greater than 500 km for basin scale reverberation; iv) bistatic insonification using SUS charges dropped from P-3 aircraft; v) long range transmission loss characterization using SUS charges from P-3's. The approach to objectives (i - iii) involved working west of Mid-Atlantic ridge where the bathymetry shoals going from west to east. To the west there is a depth excess where long range propagation occurs without significant bottom interaction, while to the east near the ridge there is no depth excess and signals interact with the bottom. The overall strategy was to insonify first where there is no depth excess for objective (i) and then to explore a region with a transition from depth excess into one without for objectives (ii) and (iii). The rough bathymetry complicates this strategy since there are seafloor features which rise above the excess depth in all regions of the experiment.

The track plan for the ARSRP/RC Recon cruise is indicated in Fig. 1. There were three components: (i) multiple coverage lines providing different viewing angles and ranges of the same sites in eastern regions with no depth excess; (ii) lines transiting from the regions of no excess depth to excess depth providing multiple convergence zone insonification; (iii) a "pentagon" set of lines in the western region for basin wide backscattering. The lines for (i) and (ii) are oriented such that beams in the aft quarter are steered towards the desired target regions; this reduces interference from the bottom bounce, or fathometer, returns. The multiple orientation of the pentagon lines breaks the "right/left ambiguity" of the array geometry. The signals were transmitted at intervals of 15 or 16 min during the eastern tracks and 20 min during the western tracks; this lead to a spatial separation between "ping intervals" of approximately 450 - 480 m and 600 m respectively.

3. Source and receiver arrays

The ARSRP/RC employed a vertical line array (VLA) for a source and a horizontal line array for a reciever (Fig. 2). The source VLA had 10 elements at a spacing of 1/2 wavelength of the carrier frequency. With the uniform shading applied this gives a beamwidth of approximately 12° to the first null. (The bandwidth was wide relative to the carrier, so the beamwidths at the band edges were roughly 15° and 10°.) The beam was steered with a time domain beamformer to generate low grazing angle energy at the seafloor. There was no shading on the array and the on-axis beam level was measured at 232 dB re 1 μPa.

The receiver HLA contained 128 elements with a spacing of 2.5 m. This gives a nominal beamwidth of approximately 2° at broadside. The array was towed at 170 m which was just below a fairly strong sound speed profile gradient.

The product of the VLA and the HLA beampatterns produces the spatial resolution on the seafloor indicated in Fig. 3. The azimuthal resolution has the normal constant wavenumber dependence for a linear array as one goes from broadside to endfire; however, the refraction of the water column in combination with the low grazing angle insonification leads to significant distortion for the vertical resolution which has a fairly sensitive dependence upon the steering of the VLA. In addition, the tilt of the array while underway distorts the ring pattern as one goes from forward to after steering angles.

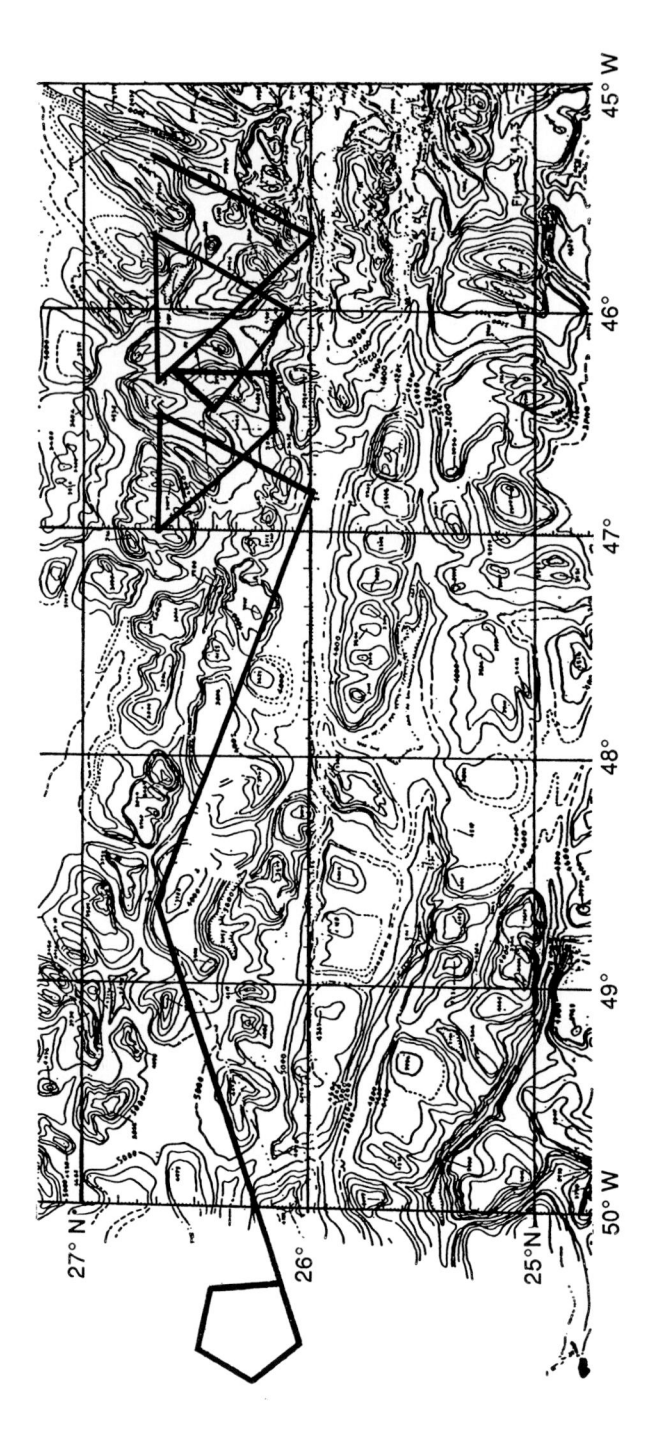

Fig. 1. Track plan of ARSRP Reconnaissance Cruise

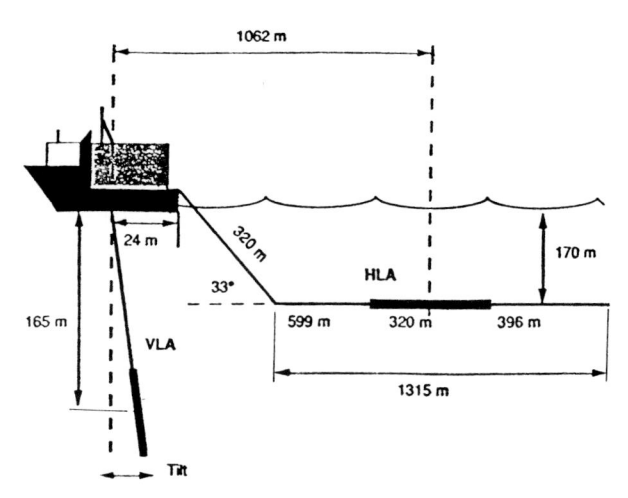

Fig. 2. Configuration of the source VLA and the receiver HLA of the ARSRP/RC

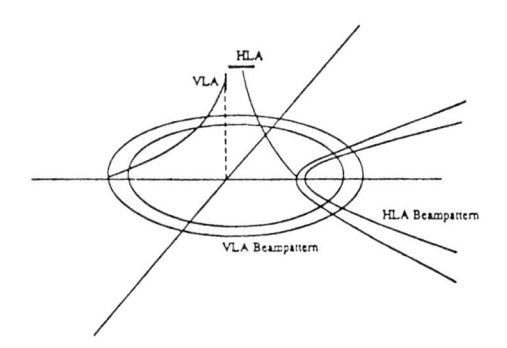

Fig. 3. Spatial resolution pattern for the VLA and HLA of the ARSRP/RC

4. Signal design

Several types of signals were transmitted; each was designed to address objectives in understanding low grazing angle scattering. The important issue *vis a vis* resolving these mechanisms is that the range resolution is given by $\Delta R = c/2W$ where W is the bandwidth of the signal. This resolution can be very high and exceed range resolution on the seafloor provided by the vertical array pattern. Very localized features, many smaller than are indicated by the available bathymetry, can be resolved.

The signals transmitted included:

- Wideband, 210 - 290 Hz, 2 s hyperbolic frequency modulated (HFM) signals with both "up" and "down" chirp. The 80 Hz bandwidth yields the maximum range resolution of approximately 10 m.

- Sequences of 20 Hz, 2 s HFM's and 10 Hz, 10 s HFM's stepped at 20 Hz increments between 210 - 290 Hz. These addressed the frequency dependence of the reverberation over the 1/2 octave band with each HFM yielding a modest range resolution of approximately 40 and 75 m respectively. The longer duration was used for more distant ranges of objectives (ii) and (iii).

- Sequences of windowed, pulsed tones spaced with durations of 2 and 10 s at 10 Hz intervals. These provide data for monochromatic scattering theories with the longer durations providing greater energy for the long range objectives.

5. Scientific questions of the ARSRP

There are a number of scientific issues for the ARSRP. Some pertain to bottom reverberation in general while others are unique to the particular region and frequency band for the ARSRP. Some of these issues are:

1. What are the important mechanisms for rough seafloor reverberation? Can experiments be designed to test these mechanisms?

2. What seafloor features cause distinct, high amplitude backscattering? Is this scattering associated with high slope surfaces such as faults on the seafloor?

3. What is a good characterization of seafloor scattering? Is a combination of diffuse, such as Lambert's law, and frequency selective facet scattering appropriate?

4. What is a good representation of the seafloor? Is the "fractal", or self similar model adequate at the small scales needed for backscattering predictions?

5. How important is the elasticity of the seafloor for reverberation? To what extent does one need to include subsurface propagation? Are Neumann and Dirchlet boundary conditions useful concepts?

6. What is the role of interface waves in the backscattering process?

7. Is the concept of "Scattering Strength" as used in the sonar equation useful for quantifying reverberation with a high resolution system?

8. What is role of the insonification in the near field, at 1/2 convergence zone distances and greater? How stable is it *vis a vis* the environment and array geometry uncertainties?

References

[1] Baggeroer, A.B. and Dyer, I., "Long range, low frequency reverberation: a review," *Seismo-acoustics*, ed. J. Berkson and T. Akal, Reidell Publishing [1986]

[2] ARSRP Reconnaissance Cruise Report, ARSRP Project Office, IGPP, Scripps Inst. of Oceanography

LONG-RANGE MEASUREMENTS OF SEAFLOOR REVERBERATION IN THE MID-ATLANTIC RIDGE AREA

J. M. Berkson, N. C. Makris, R. Menis, T. L. Krout, and G L. Gibian*
Naval Research Laboratory
Washington, D. C. 20375
*Planning Systems Inc., McLean, VA 22102

ABSTRACT. An initial analysis of reverberation measurements made in the Mid Atlantic Ridge area during the Acoustic Reconnaissance Cruise in August 1991 is presented. Sound from a low-frequency source array was scattered by the seafloor at basin-wide ranges and then received by a towed array. Reverberation images show strong returns in the direct-path region followed by returns within concentric bands corresponding to bottom convergence zones. Diffuse and discrete returns occur in both regions. Reverberation images from HFM pulses of differing frequency exhibit similar behavior as do CWs. Discrete features exhibit constant or decreasing relative reverberation level with frequency. Cross-correlation between reverberation beam/time series for different frequency CW subpulses over these features exhibited high but variable correlation throughout the band.

1. INTRODUCTION

During the Acoustic Reverberation Special Research Program Reconnaissance Experiment of August 1991, reverberation measurements were made at hundreds of locations in the Abyssal Hills on the western flank of the Mid-Atlantic Ridge north of the Kane Fracture Zone. Fig. 1 shows the experimental area. In this initial report, we examine long range reverberation measurements at the two locations shown in Fig. 1. The area is characterized by rough ridge topography which often protrudes above the 4000-m conjugate depth for the acoustic source/receiver geometry used during the experiment. Sound from a ten-element acoustic source array at 210 to 290 Hz was scattered by the seafloor at depths 2000 to 5000 m and at ranges up to 600 km and then received by a towed array. After signal conditioning, the data from 64 hydrophone channels were sampled at 2048 Hz and processed by a real-time processor to form 64 beams of 128-Hz bandwidth (186-314 Hz) with hamming shading. These beamformed data were processed and displayed as reverberation images on a map projection. Each point of the spatial grid is assigned the appropriate beam-time data point based on the bistatic travel time in the appropriate receiving beam, using an average sound speed. For grid points where beams overlapped, data were averaged.

2. MEASUREMENTS

At the first location (Fig. 1) the overall characterization of basin-wide reverberation is examined using a 10-s CW transmission. The image in Fig. 2 shows strong returns in the direct-path region followed by returns within concentric bands corresponding to bottom convergence zones predicted by a WRAP model [1] simulation for the experimental area. Diffuse and discrete returns occur in both regions and contain the right/left ambiguity of the line-array. Towship noise appears in the forward beams, to the south. The distinct events on adjacent pings correlate with locations that apparently correspond to specific bottom features. However, the available bathymetry is generally not of sufficient resolution to relate every scatterer with a specific bathymetric feature of the complex ridge-crest topography. A technique is being developed

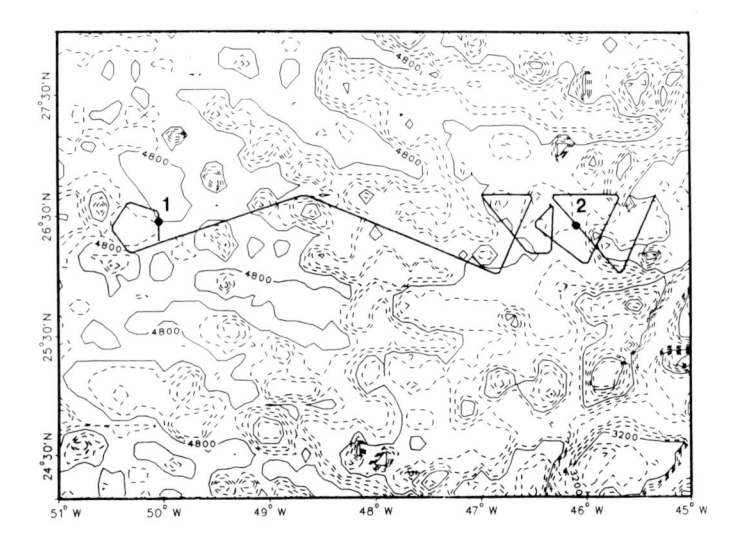

Fig. 1 – Mid-Atlantic Ridge Experimental Area. The two measurement sites are shown in relationship to the overall Acoustic Reconnaissance Cruise tracks. Bathymetry is based on DBDB5 data.

for inverting reverberation data from many array orientations and locations for high-resolution unambiguous images [2]. The absence of returns from surface convergence zones is the result of the very calm seas during the experiment.

At the second location, frequency dependence of long-range reverberation in the band 210-290 Hz was examined using transmission of a sequence of CW and HFM subpulses. Four 1-s, 10-Hz upsweeping HFM pulses with initial frequencies of 215, 235, 255 and 275 Hz were transmitted over an interval of 4 s. The matched filter output was normalized by the signal replica energy. Fig. 3 shows that the images are similar in character over the band. Five minutes later, nine overlapping 2-s time-tapered CWs of different frequency were transmitted over a time interval of 10 s as follows: 210, 250, 290, 220, 260, 230, 270, 240, and 280 Hz. Fig. 4 shows that the images are also similar over the 80-Hz band.

3. ANALYSIS AND DISCUSSION

For further analysis we examined reverberation from specific areas using the nine CW subpulses. Three discrete and one diffuse reverberation features shown in Fig. 5 were selected. By discrete we mean an area characterized by rapid increase in reverberation level over range, while diffuse is gradual and variable over range. On the assumption that the signal loss at long range is due to single scattering from a simplified scattering area defined by the receiving beamwidth and pulse length, the equivalent scattering strength, SS, is estimated:

$$SS = RL - SL + TL_1 + TL_2 - A$$

where RL is the reverberation level, SL is the source level, TL_1 and TL_2 are the transmission losses from the source to the scattering area and from the scattering

Fig. 2 - Reverberation image for 10-s CW pulse at 270 Hz at site 1. Received beam level versus location. Receiving bandwidth 3 Hz. Bathymetry is based on DBDB5 data. Contour interval 100m. Reverberation level (dB re 1μPa) in 3-Hz band. Maximum range of data 300 km.

area to the receiver, and A is $10 log_{10}$ of the scattering area, where all terms are frequency dependent. To obtain RL we average incoherent reverberation, in μPa^2, of between 80-200 not necessarily independent pixels for each region. Since the standard deviation is on the order of the mean, averaging stabilizes the mean, which is then converted to dB as RL. Due to the lack of detailed information on bathymetry and bottom properties, we cannot confidently calculate a transmission loss. [The Geophysical Reconnaissance Cruise scheduled for July 1992 is expected to gather these data]. We therefore arrive at a normalized reverberation level, $RLN = SS - TL_{10} - TL_{20}$, where TL_{10} and TL_{20} are the transmission loss components independent of water-column attenuation.

Figure 5 shows the estimates of normalized reverberation level for the four features as a function of frequency. These narrowband levels exhibit an $f^{-\alpha}$ falloff with frequency, where $0 \leq \alpha \leq 1.8$. This is within $-2.3 \leq \alpha \leq 1.8$, the behavior of scattering strengths reported for distant features using wideband explosive sources over the same band [3-5].

Changes in details of scattering with frequency are also examined using reverberation by the sequence of nine 2-Hz CWs. Since the time interval (1 s) and source/receiver displacement (1.5 m) between each subpulse is small, the experimental geometry is nearly constant. A measure of change in scattering with frequency is the value of the normalized cross-correlation function $\rho(\tau)$ between two beam/time series from different subpulses. An ideal cross-correlation function is of the form

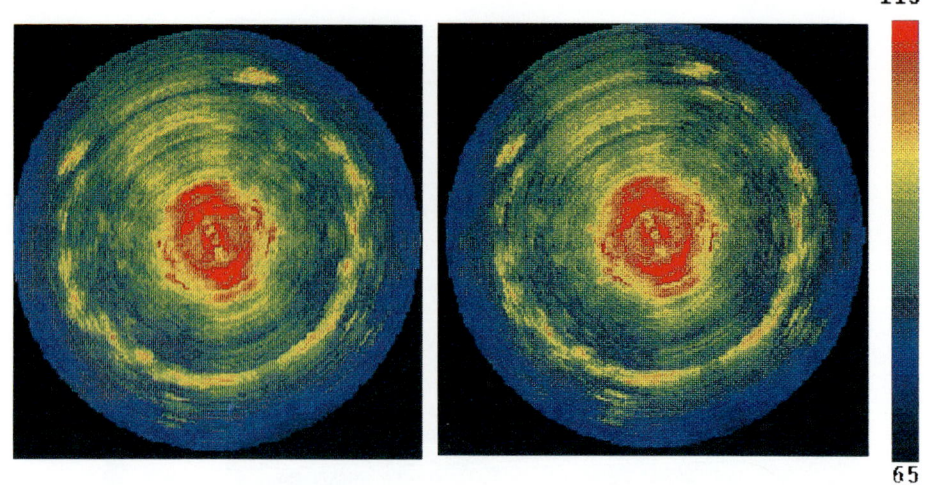

Fig. 3 – Reverberation images for 1-s HFM pulses at 215-225 Hz (left) and 275-285 Hz (right) at site 2. Maximum range of data 120 km. Received beam level versus location. Normalized correlator output of beam level in dB.

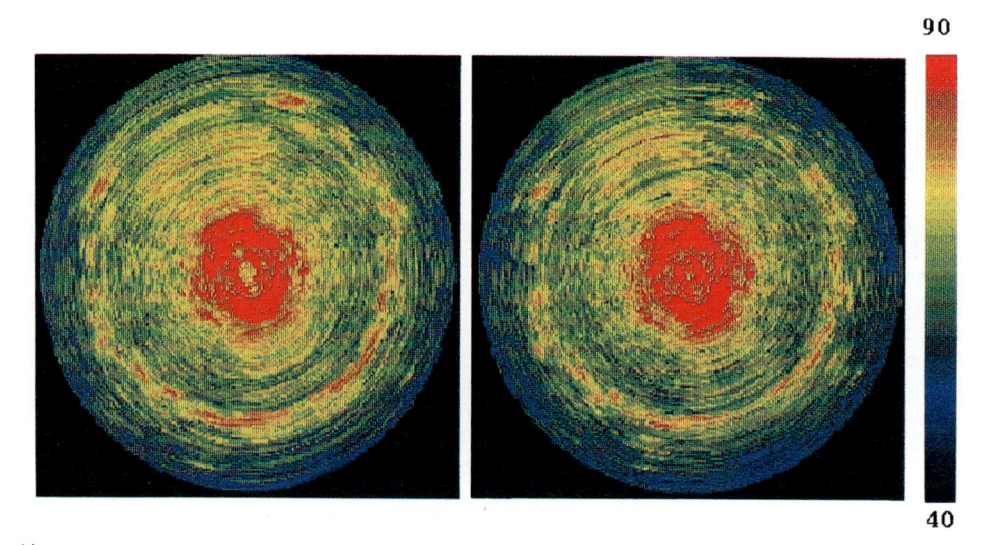

Fig. 4 – Reverberation images for 2-s time-tapered CW pulses at 220 Hz (left) and 280 Hz (right) at site 2. Maximum range of data 120 km. Received beam level versus location. Reverberation level (dB re 1μPa) in 3-Hz band.

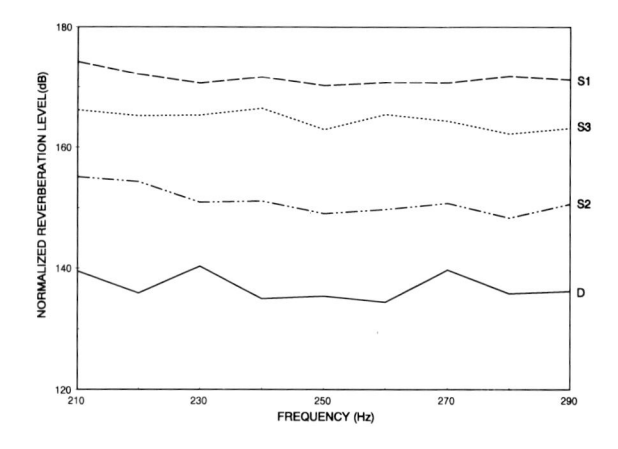

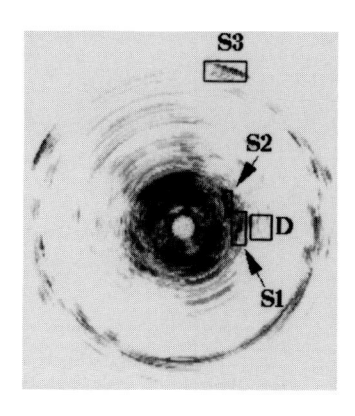

Fig. 5 – Normalized reverberation level, RLN, versus frequency for nine 2-s time-tapered CW pulses for four reverberation features (left). Four features are identified (right).

$\rho(\tau) = C_{12}(\tau)/[C_{11}(0)C_{22}(0)]^{\frac{1}{2}}$, where $C_{12}(\tau) = \langle B_1(t)B_2(t+\tau)\rangle$ and τ=delay; B_1, B_2= beam intensity series at frequencies f_1 and f_2. For this analysis, we used the maximum value of the cross-correlation algorithm from Ref. [6] at the lag corresponding to the subpulse alignment and within a time lag window corresponding to the time lag interval over which the normalized autocorrelation remains above $1/e$. In Fig. 6, the three isolated scattering features exhibit relatively high but variable correlation across most of the 80-Hz band, while the diffuse scatterer is characterized by more rapid decorrelation. Applying the results of Thorsos [7], the former is characteristic of scattering from isolated facets and the latter characteristic of rough surface scattering. The similarity of reverberation images as well as the slow variation of normalized reverberation level within the experimental band suggests that octave-band waveforms at these frequencies could be used for investigating scattering mechanisms in this environment. Finally we note that since the standard deviation is on the order of the mean for beam/time reverberation, there is a tradeoff involved in determining angular and frequency dependence for a particular region. As resolution increases, the number of observations averaged decreases and so does stability of the estimate. We plan to address these issues in the planning of the next acoustic experiment scheduled for 1993.

4. ACKNOWLEDGEMENTS

This work was supported by Office of Naval Research, Code 1125A. We thank our colleagues in the Acoustic Reverberation Special Research Program for planning and carrying out the experiment and M. R. Healey, L. Z. Avelino, and J. C. Richardson for assistance in data processing. We also thank E.I. Thorsos and D.R. Jackson for helpful discussions concerning the stability of scattering strength estimates.

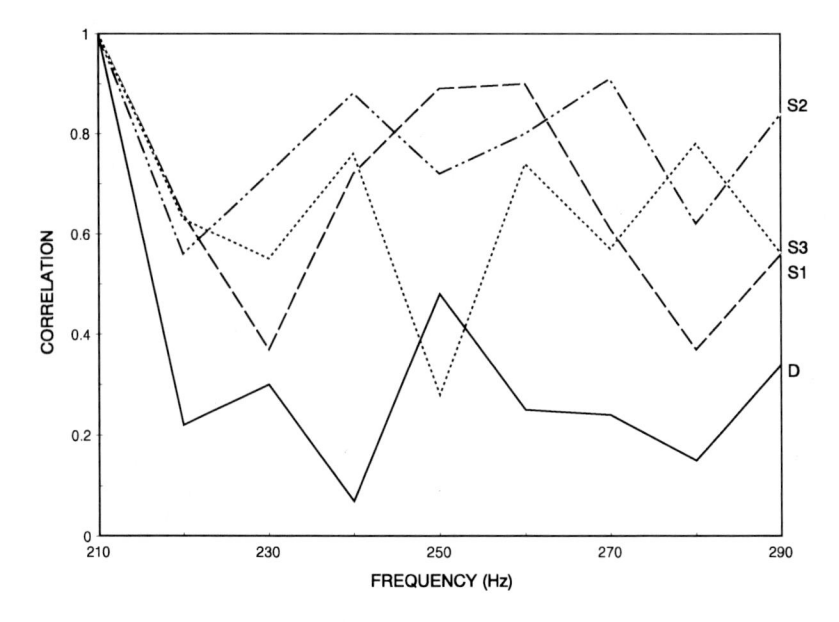

Fig. 6 – Cross correlation of 210 Hz CW subpulse beam/time series with CW subpulses ranging from 210 to 290 Hz for four scattering features using 16-s analysis window.

5. REFERENCES

[1] Kuperman, W.A., Porter, M.B., and Perkins, J.S. (1991) 'Rapid computation of acoustic fields in three-dimensional ocean environments,' J.Acoust.Soc.Am. **89**, 125-133.

[2] Makris, N.C., J.M. Berkson, Kuperman, W.A., and Perkins, J.S. (1992) 'Ocean-Basin Scale Inversion of Reverberation Data,' This Volume.

[3] Berkson, J., Akal, T., and Berrou. J. (1984) 'Techniques for Measuring Backscattering from the Seafloor with an Array,' in H. Urban (ed), Adaptive Methods in Underwater Acoustics, Reidel, Dordrecht, pp. 741-746.

[4] Berkson, J.M., and Akal, T. (1987) 'Simultaneous Reception of Long Range Low Frequency Backscattered Sound by a Vertical and Horizontal Array,' J.Oceanic.Eng. **OE-12**, 362-367.

[5] Preston, J.R., Akal, T., and Berkson, J. (1990) 'Analysis of backscattering data in the Tyrrhenian Sea,' J.Acoust.Soc.Am. **87**, 119-134.

[6] Makris, N.C. and Dyer, I. (1991) 'Environmental correlates of Arctic ice-edge noise,' J.Acoust.Soc.Am. **90**, 3288-3298.

[7] Thorsos, E.I. (1992) 'The Validity of Simple Models for Low Frequency Bottom Scattering,' ARSRP Research Symposium, Scripps Institution of Oceanography, April 1992.

OCEAN-BASIN SCALE INVERSION OF REVERBERATION DATA

N.C. Makris, J.M. Berkson, W.A. Kuperman, and J.S. Perkins
Naval Research Laboratory
Washington, D.C. 20375

ABSTRACT. A technique for imaging large areas of the ocean floor by globally inverting reverberation data is under development. We simultaneously invert beamformed data measured at a variety of towed-array locations and orientations. By optimal use of the data we produce a high resolution scattering strength image of the ocean bottom. Inherent to this global inversion is the removal of the "right-left ambiguity" of the beamformed data. The method has been successfully tested with synthetic data for a simple scattering model. It is being developed further to analyze bottom reverberation data acquired during the Office of Naval Research/Special Research Project Acoustic Reconnaissance Cruise.

1. INTRODUCTION

Our goal is to image the ocean basin using acoustic reverberation data. In particular we are attempting to image portions of the western flank of the Mid-Atlantic Ridge with beamformed towed-array data acquired during the Acoustic Reconnaissance Cruise (ARC) of August 1991. Since this region generally has negative excess depth and rugged bathymetry, reverberation tends to be broadly distributed in horizontal angle at ranges corresponding to bottom-bounce convergence zones, see [1] Fig. 2 in this volume. This poses problems for identifying the location of individual scattering features since beamformed towed-array data has an inherent "right-left ambiguity," and has decreasing resolution for more distant returns and off-broadside beams. Averaging together scattering strength maps generated from individual reverberation observations has proven to be successful in resolving right-left ambiguity [2] [3]. However, this "stacking" technique has only been applied to situations where isolated scattering features, such as seamounts or continental margins, disturb sound channels which otherwise have excess depth. It is not expected to work as well for the broadly distributed backscattering of the ARC data. Right-left angular ambiguity in ambient noise data has been resolved by an iterative optimization procedure [4]. This can be further applied to resolve angular ambiguity in long range reverberation data by collapsing towed-array observations to a single point and performing angular inversions at independent range steps. However, stacking would still be necessary for spatially distributed observations, as in the ARC. We propose an alternative imaging method based upon a global inversion of spatially distributed observations. We test this method with a simple scattering strength model using synthetic data.

2. REVERBERATION MAPPING

We choose a scattering strength representation of the region to be imaged. The ocean basin is discretized into a Cartesian grid with a scattering strength assigned to each grid point. In this simplified problem we assume that the scattering strength is independent of insonification and observation angle in both the horizontal and vertical. This enables us to plot the scattering strength of the region with a single image. It also reduces the number of observations needed for the inversion. Figure 1 shows the water depth measured in hundreds of meters for the region to be imaged, referred to as region A. Here the 8 km sampling of bathymetry has been interpolated to a 1 km grid size spanning a 167x167 km² region centered at 26.35º N 46.04º W. Increasing negative values indicate increasing water depth as measured from the ocean's upper surface. The overlain tracks

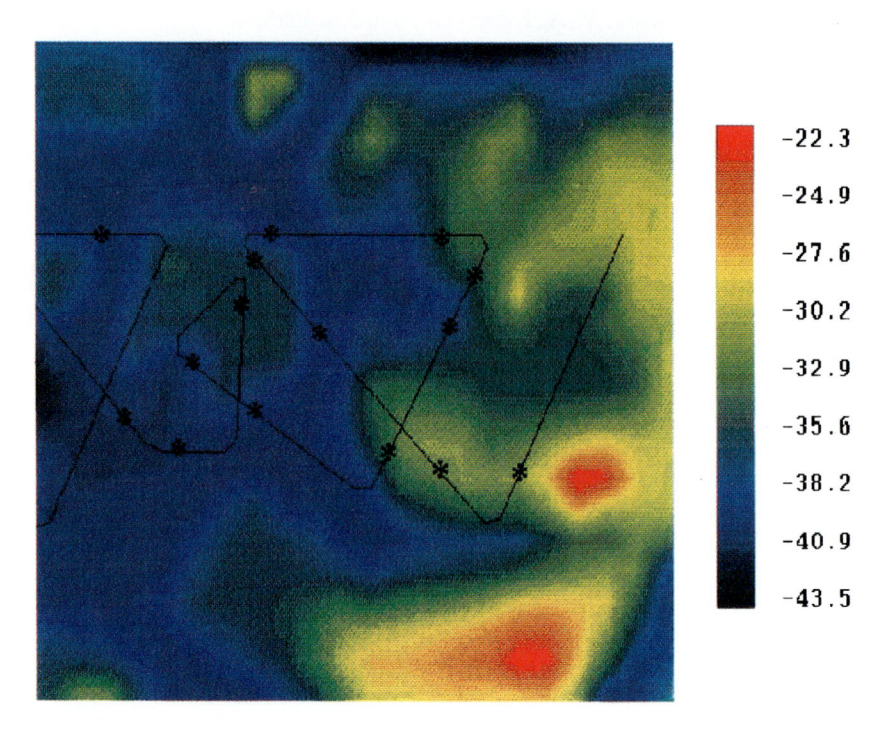

Figure 1. Actual bathymetry and assigned scattering strength of region A with ARC tracks overlain, in hundreds of meters and dB respectively.

indicate the path of the research vessel which towed both a vertical source and a horizontal receiving array during the ARC. Asterisks indicate the position of fifteen reverberation observations to be used in the simulation. The array orientation is roughly parallel to the ship's course. For our simulation, we take the scattering strength to be equal to the numerical value of the depth at each grid point or pixel. This serves two purposes. It defines scattering strength values which fall within the range typically measured by low frequency sources in ocean bottom reverberation experiments. And, it defines a scattering region which has variations related to the bathymetry.

We choose parameters for our synthetic array which are consistent with those used during the ARC. For example, the radial resolution is $\Delta r = c\tau/2$ =1.525 km where c = 1.525 km/s is the sound speed at the array's depth , τ = 2 s is a typical pulse duration for cw signals, λ = 6.1 m is a typical wavelength, and L = 160 m is the array aperture, for the 64 element array used by the onboard real-time-processor. The 1 km pixel dimension of the grid is slightly less than the radial resolution of the beamformed data.

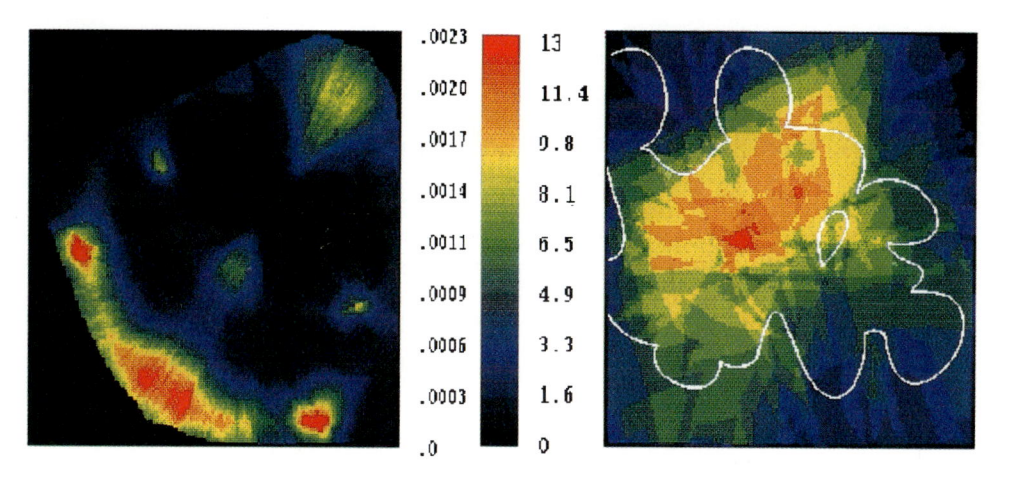

Figure 2. Left image, synthetic reverberation data, $Q_S(r_0,\theta_0)$, for a single observation s of region A.
Figure 3. Right image, number of redundant observations of each pixel in region A. White curve indicates boundary containing best resolution data.

The equivalent beamwidth of the array for steering angles from broadside, $\theta = \pi/2$, to angles approaching endfire is

$$\beta(\theta) = K\frac{\lambda}{L}\frac{1}{\sin\theta} \qquad (1)$$

where $K = 1.36$ for a Hamming window. We do not use data near endfire for reasons to be discussed later. The reverberation measured for a point a distance r from the array at angle θ from endfire consists of reverberation integrated over an annular sector of area $r\beta(\theta)\Delta r$ centered about that point plus the reverberation of a symmetric annular sector centered at r, $-\theta$, due to the array's right-left ambiguity.

Simulated array data is computed in the following way

$$Q_s(r_0,\theta_0) = \sum_{r=r_0-\Delta r}^{r_0+\Delta r} \sum_{\theta=\theta_0-\beta(\theta_0)}^{\theta_0+\beta(\theta_0)} \left\{\frac{\sigma(r,\theta)}{2}f_s(r,\theta) + \frac{\sigma(r,-\theta)}{2}f_s(r,-\theta)\right\} \qquad (2)$$

where $Q_s(r_0,\theta_0)$ is the reverberation measured at pixel $x_0=r_0\cos(\theta_0+\varphi_s)+x_s$, $y_0=r_0\sin(\theta_0+\varphi_s)+y_s$ for observation s and array axis orientation φ_s. The scattering coefficient $\sigma(r,\theta)$ for pixel $x=r\cos(\theta+\varphi_s)+x_s$, $y=r\sin(\theta+\varphi_s)+y_s$ has scattering strength $\Sigma(r,\theta)=10\log(s(r,\theta))$. For a given observation, we do not compute reverberation for pixels whose integration regions extend beyond the image boundaries. We also do not compute reverberation for pixels whose integration regions exceed a threshold number of pixels. In general, for the inversion to work, this threshold must be less than or equal to the number of observations S. For the present simulation we have chosen the threshold to be 7 without investigating whether the choice is optimal.

As is stated in equation (2), both the right-left ambiguity and the assembly of individual scattering pixels within an annular sector are incorporated by incoherent summation. The function $f_S(r,\theta)=I^2(r,\theta)/I_0$ where $I(r,\theta)$ is the acoustic intensity on the ocean bottom at horizontal position r,θ and I_0 is the source power. The function $f_S(r,\theta)$ reflects two way propagation from source to scatterer, employing reciprocity on the return trip. For inversion with actual data, the intensity on the ocean bottom $I(r,\theta)$ is computed using a range dependent propagation model such as the WRAP adiabatic mode formulation [5] or the wide angle parabolic equation. However for this simulation we have assumed zero transmission loss and unity source power, ie $f_S=1$. Weighting the data with a complicated propagation model would be lost on our simulation since the weights would simply be removed in the preprocessing stages before the actual inversion.

Due to the right-left ambiguity of the beamformed array data we expect that $Q_S(r_0,\theta_0)=Q_S(r_0,-\theta_0)$. However, for both simulated data, as shown in Figure 2, and real data [1] this is only approximately true in general. This is because the axis of the array is not always symmetric with respect to the Cartesian grid used to digitally define the image. The approximation becomes better as the pixel size decreases with respect to the radial resolution of the array, Δr. However, it is impractical to choose a grid size much smaller than Δr. As the grid becomes finer, the redundant information decreases and the inversion time prohibitively increases.

3. GLOBAL INVERSION

To perform a global inversion we initialize our scattering coefficient estimate $\sigma_g(x,y)$ at some value within an allowable range $\sigma_{min} < \sigma_g(x,y) < \sigma_{max}$ for all grid points $x=1$, nx and $y=1$, ny. We construct replica data $R_S(r_0,\theta_0)$ with our estimated scattering coefficient as in equation (2), replacing $Q_S(r_0,\theta_0)$ with $R_S(r_0,\theta_0)$ and $\sigma(x,y)$ with $\sigma_g(x,y)$. We then compute a cost function which sums the difference squared between replica data and true data for each observation and grid point

$$E(\sigma_g) = \sum_{s=1}^{S}\sum_{y=1}^{ny}\sum_{x=1}^{nx}(R_s(x,y)-Q_s(x,y))^2 \qquad (3)$$

We perturb our estimated scattering coefficient at a single grid point, and recompute the cost function. If the cost function E' including the perturbation is less than the original cost function E we accept the perturbation. Otherwise we retain the original value. We continue this process for each grid point. That comprises the first global iteration. We then iterate globally until the cost function reaches a minimum value, E_{min}, with respect to the σ_g. This method is known as the gradient method since the cost function can only decrease or remain constant. It is possible that E_{min} is a local rather than global minimum. In this case a simulated annealing approach is taken where the cost function comparison is based upon a Boltzmann distribution which allows E to increase enough to escape local minima

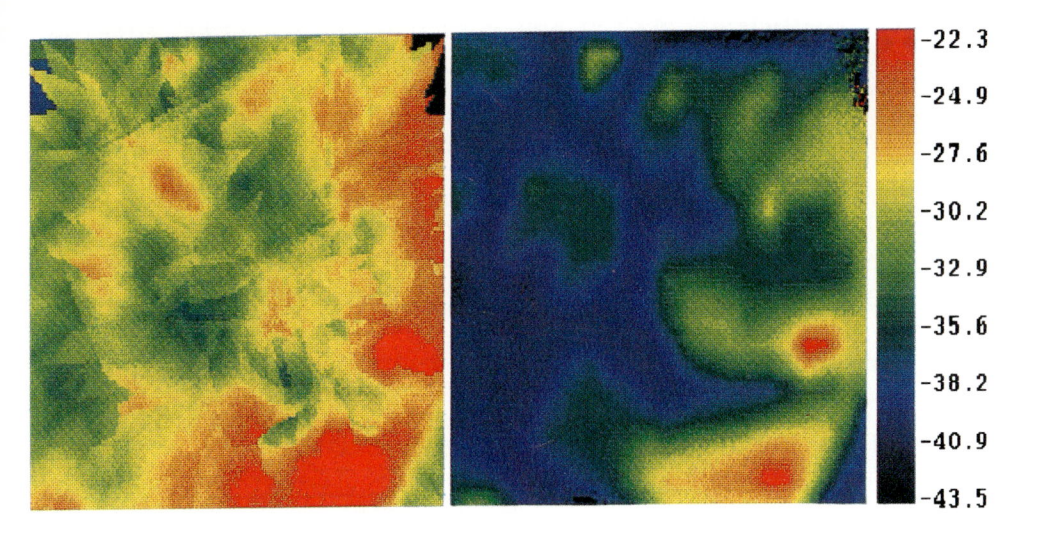

Figure 4. Left image, scattering strength estimate of region A via stacking, $\Sigma_s(x,y)=10\log(\sigma_s(x,y))$.

Figure 5. Right image, scattering strength estimate of region A via global inversion, $\Sigma_g(x,y)=10\log(\sigma_g(x,y))$.

[6]. However, for the present simulation this was not necessary. The gradient method provides a solution with final cost function E_{min} =7.5x10^{-5}, which is 4 orders of magnitude smaller than the sum square of the data, E_0 =.628. (Obtain E_0 by setting the replica data $R_s(r_0,\theta_0)$ to zero in equation (2) .)

In comparing E to E' it is inefficient and prohibitively time consuming to compute each cost function as shown in equation (3). Since only a single test grid point has been changed there is no difference between many corresponding terms in each summation. Therefore, instead of summing over all nx and ny grid points for each perturbation, only a smaller influence region about the test grid point need be considered. In general the influence region is approximately equal to the local integration region for the given observation s, or the integration threshold, whichever is less. For observation points related by the array's right-left ambiguity, the influence region must be large enough to account for the lack of perfect reciprocity about the array's axis. As noted earlier, this occurs for array orientations which are not symmetric with respect to the Cartesian grid.

The number of observations for each grid point for the 15 observations previously described are shown in figure 3. The white line defines the boundary within which the best resolution data, which has an integration region of no more than one pixel, is located.

4. RESULTS

We first show the image resulting from a stacking procedure. We add all observations of a particular pixel and divide by the number of observations of that pixel. The resulting scattering coefficients $\sigma_s(x,y)$ are converted to scattering strengths, $\Sigma_s(x,y)=10\log(\sigma_s(x,y))$, which are plotted in figure 4. This image bears a poor resemblance to the actual scattering strength image in figure 1. The magnitude and shape of estimated scattering features are significantly different from the actual ones. Artificial features of high scattering strength are also evident in the stacking estimate.

The scattering strength estimate resulting from the global inversion, $\Sigma_g(x,y)=10\log(\sigma_g(x,y))$, is plotted in figure 5. As expected from the dramatic decrease in the cost function, the global inversion image appears to be identical to the original image of figure 2. The right-left ambiguity of the data has been removed and the image is resolved to one pixel in regions where observations of such high resolution do not exist.

Even if there are no synthetic data observations at a particular grid point, due to thresholding or nearness to a border, the scattering coefficient can still be determined by global inversion if the integration region of adjacent observation points leaks over the unobserved point enough times. Comparing the regions observed, figure 3, to the regions imaged in figure 5, we note that this is the case for a variety of pixels near the image edges and corners. It is also clear that the leakage is insufficient to properly estimate scattering strength for a few infrequently observed and unobserved pixels in the upper right corner.

5. Discussion

We have developed a method for imaging the ocean basin by optimal use of reverberation data. By a global inversion, this method removes the right-left ambiguity associated with beamformed towed-array data and resolves the data to as fine a resolution as possible. In its present formulation, the global inversion technique can work well for traditional long-range monostatic reverberation experiments, such as [3], where propagation angles only vary over a shallow vertical width and the horizontal orientation of the scatterers do not vary from observation to observation. It can also work well for spatially distributed observations of omni-directional scattering "hotspots", as shown in the simulation. In order to image more complicated data sets, such as the ARC, our method is being extended to search for both the horizontal and vertical dependence of regional scattering strength. We note that the method presented is not restricted to estimating scattering strength, but may be used to invert for any physical parameters of the ocean bottom necessary in a realistic reverberation model.

Finally, we contend that it is possible to determine the optimal set of observations needed to image the ocean bottom with methods related to those presented. That is, assuming some or no knowledge of regional bottom scattering, and given constraints upon our resolution, ship maneuverability, time, site locations, etc, we could invert for the optimal course of the research vessel. We call this the Travelling Acoustician Problem. Qualitatively, we have begun this process by analyzing the number and location of redundant observations before proceeding to the global inversion, see figure 3.

References

[1] J.M. Berkson, N.C. Makris, R. Menis, T.L. Krout and G.L. Gibian, 'Long-range measurements of sea-floor reverberation in the Mid-Atlantic Ridge area,'This volume (1992).

[2] F.T. Erskine, G.M. Bernstein, S.M. Brylow, W.T. Newbold, R.C. Gauss, E.R. Franchi and B.B. Adams, 'Bathymetric Hazard Survey Test (BHST Report No. 3), Scientific Results and FY 1982-1984 Processing,' NRL Report 9048 (1987).

[3] J.R. Preston, T. Akal and J.M. Berkson, 'Analysis of backscattering data in the Tyrrhenian Sea,' J. Acoust. Soc. Am. *87*, 119-134 (1990).

[4] R.A. Wagstaff, 'Iterative technique for ambient-noise horizontal-directionality estimation from towed line-array data,' J. Acoust. Soc. Am. *63*, 863-869 (1978).

[5] W.A. Kuperman, M.B. Porter and J.S. Perkins, 'Rapid computation of acoustic fields in three-dimensional ocean environments,' J. Acoust. Soc. Am. *89*, 125-133 (1991).

[6] W.A. Kuperman, M.D. Collins, J.S. Perkins and N.R. Davis, 'Optimal time-domain beamforming with simulated annealing including application of *a priori* information,' J. Acoust. Soc. Am. *88*, 1802-1810 (1990).

ARSRP RECONN RESULTS AND BISSM MODELING OF DIRECT PATH BACKSCATTER

J.W. Caruthers and E.J. Yoerger

Naval Research Laboratory,
Stennis Space Center, MS 39529

ABSTRACT Modeling support designed to aid in mapping, interpreting, and analyzing direct-path bottom reverberation was developed during the ARSRP Reconnaissance Cruise of July and August, 1991. The Bistatic Scattering Strength Model (BISSM) provided scattering strength estimates and a simple ray model provided propagation loss and other parameters required to calculate reverberation. This model successfully predicts scattering from bottom features known deterministically to 0.5nm and statistically to 0.1nm. The BISSM model and its ability to explain the reconn data are discussed. (This work was supported by ONR and CNOC.)

1. INTRODUCTION

The Bistatic Scattering Strength Model (BISSM) (Caruthers et al. (1990)) has the form and notation given in Brekhovskikh and Lysanov (1982). It contains a backscatter term based on Lambert's law and a forward scatter term based on the high-frequency evaluation (reflections from a distribution of randomly oriented plane facets) of the Helmholtz/Kirchhoff integral as first given by Eckart (1953). Ellis and Crowe (1991) used a simplified version of this algorithm to model bistatic reverberation measurements.

BISSM has two modifications introduced by Caruthers et al. (1990): First, to provide for database inputs for most of its required parameters (vice determining these parameters empirically), it has been coupled with a high-resolution (0.5nm) database of bathymetry, mean slope and orientation of 0.5nm plane facets, and rms roughness and slopes in two perpendicular directions. Second, microroughness on the facets are allowed by the low-frequency evaluation of the Helmholtz/Kirchhoff integral also given by Eckart (1953). The first modification provides for the use of local angles of incidence and scattering at the facets. The second factor allows for additional scattering and coherence loss due to small scale roughness.

To support the ARSRP reconn cruise, BISSM was coupled with a simple ray calculation and vertical source beam pattern calculation to give direct-path and surface-reflected energy on the bottom at geographic grid points under the ship with a mesh size of 0.5nm. The real effects of highly variable bathymetry in the region on the propagation parameters (TL, travel time, angles, etc.) were taken into full account. Moreover, the local slopes were used to calculate local grazing and scattering angles. In other words, a flat-bottom assumption was not made for either the propagation or scattering calculation. We show here that this has a significant effect on scattering, even for Lambert's law backscatter, which is the primary effect discussed here. We have adopted a simple model for Lambert's coefficient (i.e., $\mu = 0.005\ h^2 + 0.0007$) where h is the height residual from the 0.5 nm planes. (The standard -27dB assumption, we feel is not workable for all small regions in the ARSRP reconn area.)

2. APPLICATION OF BISSM TO DIRECT-PATH ARSRP DATA

A reconnaissance experiment was conducted by the Acoustic Reverberation Special Research Program (ARSRP) in the region of the Mid-Atlantic Ridge in the summer of 1991. Several waveforms centered at 250Hz were transmitted from a 10-element, 33-meter vertical line array. The source-receiver geometry was assumed to be monostatic. Results presented here are confined to 2-sec, 70-hz bandwidth HFM sweeps. The source level was 233 dB re 1μPa @ 1m. The receiver was a horizontal towed array with 126 hydrophone groups forming 126 beams with a nominal width of 1-deg width at broadside. The resolvable limits of the system might be set at about 40m in range and 350m cross-range (at 20km).

The area of the reconn is characterized by highly variable bathymetry in the depth range of 2 to 5 km. There is a significant amount of exposed basalt and other hard rock materials. At most, only a thin veneer of sediments cover the higher sloped areas with deeper sedimentation in low-lying areas. In the larger valleys there is a significant amount of sediment ponding to thicknesses of several hundred meters. The conjugate depth for the source-receiver depth was about 4000m.

Reverberation in the area is characterized by strong discrete, often spike-like returns. Large-scale ridges several miles long and about a mile wide are visible and highly repeatable in numerous reverberation records. In many cases features seen at the resolvable limit repeat between successive pings. Hence, any modeling of these data should be capable of identifying these more salient features. Figures 1-4 show the results of modeling two areas utilizing the BISSM model. In all cases, agreement between model results and data was quite good.

Figure 1 represents data modeled from the time 0948Z of Julian day 218 of the above reconn data. Figure 1a is a BISSM reverberation level polar plot of the modeled area (32nm square). The modeled reverberation accounts for source beam pattern and propagation and scattering effects produced by actual bathymetry, 0.5nm plane mean slopes and orientation, and their statistics. It should be noted specifically that bathymetry enters the problem through its effects on transmission loss, travel time, and horizontal grazing angle at the actual depth (not the assumed flat-bottom depth). On the other hand, bathymetry determines scattering strength only if it has sufficient resolution to define local facet slope and azimuths. The track of the ship and heading of the towed array is about 310°.

To compare these modeled data to the reconn data, a transformation from the geographic display (Fig. 1a) to the beam/time standard display format of reconn data (Figs. 1b-c) is necessary. An advantage of this technique is that it can resolve the left/right ambiguity in the predicted data (Figs. 1b and 1c, respectively). Of course, this distinction is not possible in the reconn data. Therefore, for comparison we show their combination in Fig. 1d. Figure 1d also shows an overlay of significant part of the reverberation data as line graphs. It can be seen that trends in the reconn data match well with the model results. One might note the ridge structure in the near-forward beams (beams 0 to 60) for both the modeled (mapped from the lower left quadrant of Fig. 1a) and actual data. However, it should be noted that the reconn data shows more detailed

205

features than that produced by the model. This is due to the lack of sufficiently resolved bathymetry.

Figure 2 shows similar plots for time 1233Z (location comparison between Fig. 1a and 2a can be made from the locations of the same ping positions on the two displays). Again, there is good agreement of reconn data trends with the modeled data. Note that the same ridge structure mentioned above

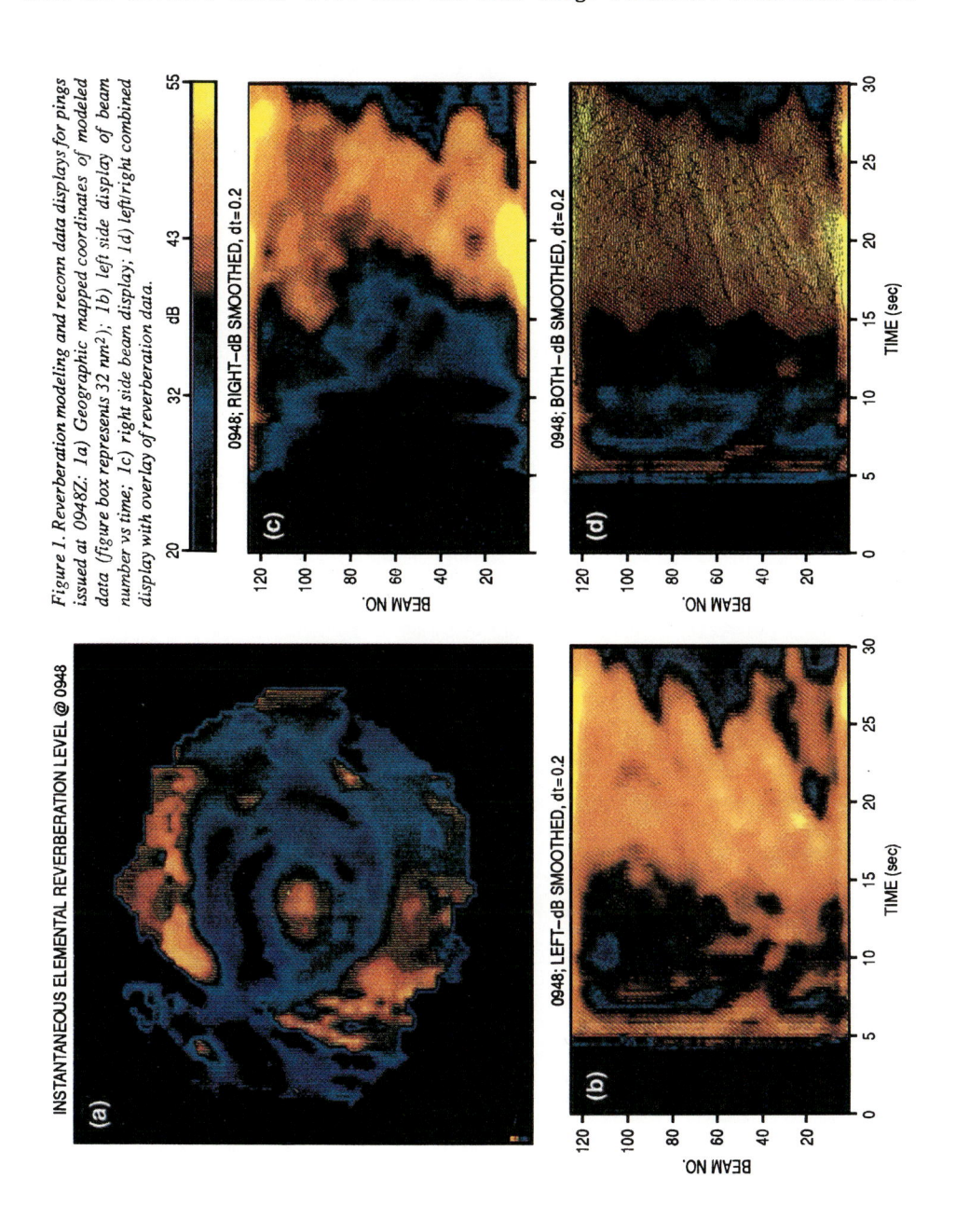

Figure 1. Reverberation modeling and reconn data displays for pings issued at 0948Z: 1a) Geographic mapped coordinates of modeled data (figure box represents 32 nm²); 1b) left side display of beam number vs time; 1c) right side beam display; 1d) left/right combined display with overlay of reverberation data.

is visible at the lower right in Fig. 2a and is visible on the modeled left aft beams (Fig. 2b). Problems of left/right ambiguity obscure the effects in the combined plots and the actual data (Fig. 2d).

To verify details at a finer scale, the 0948Z data are plotted at an expanded scale in Fig. 3. Shown are beams 6-78 for times 5 to 30 seconds. A beam/time plot of the corresponding reconn data is overlaid and again the

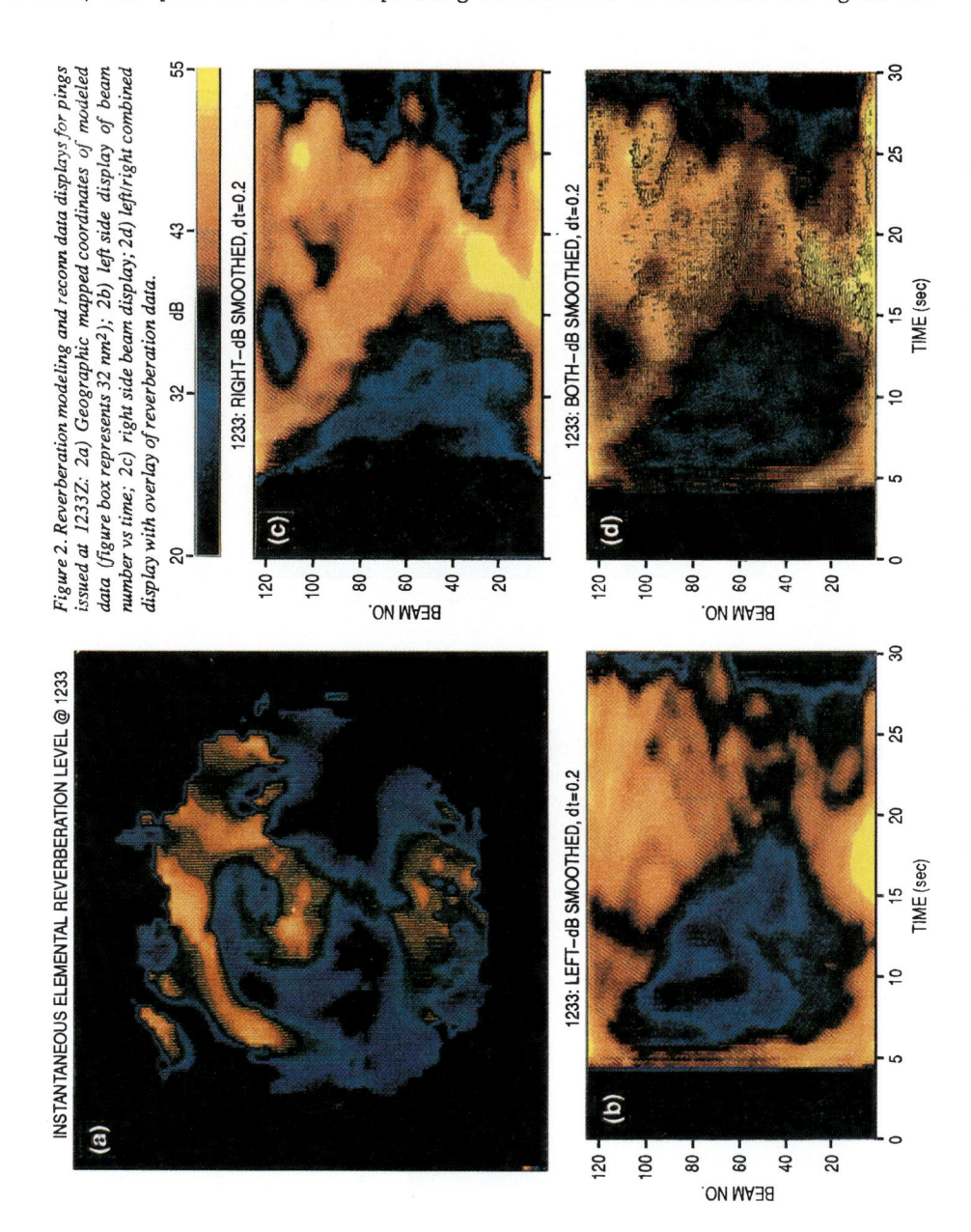

Figure 2. Reverberation modeling and reconn data displays for pings issued at 1233Z. 2a) Geographic mapped coordinates of modeled data (figure box represents 32 nm^2); 2b) left side display of beam number vs time; 2c) right side beam display; 2d) left/right combined display with overlay of reverberation data.

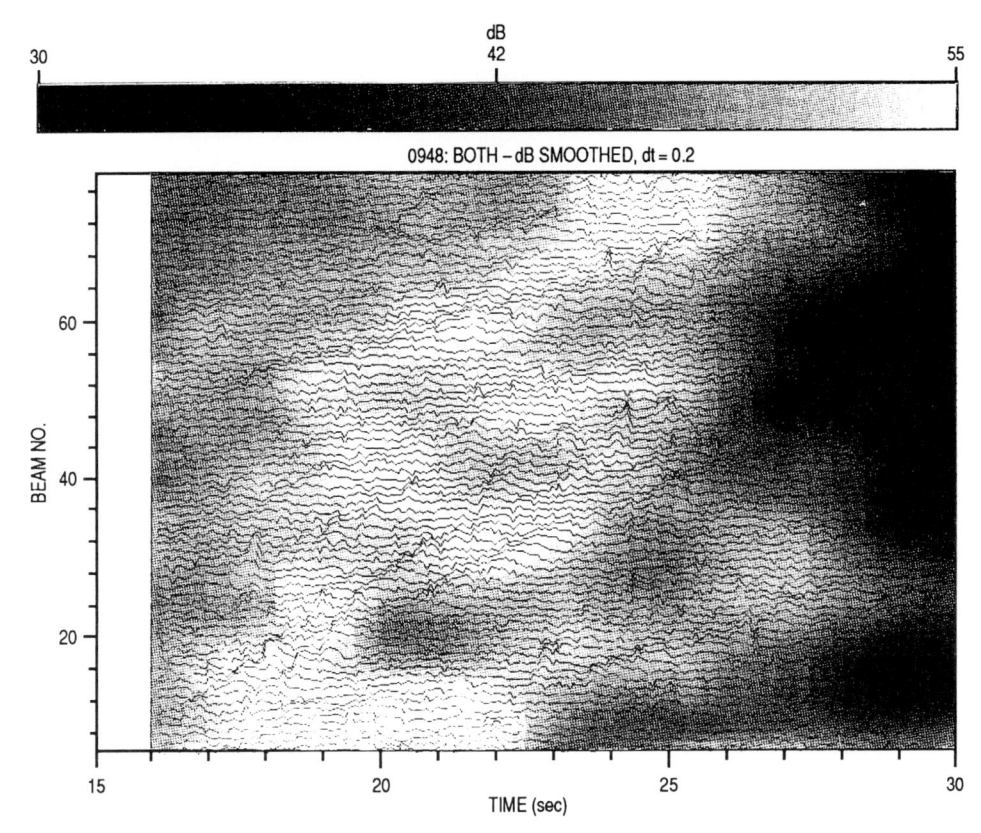

Figure 3. Reverberation modeling and reconn data displays for pings issued at 0948Z. Same as figure 1d with expanded scale.

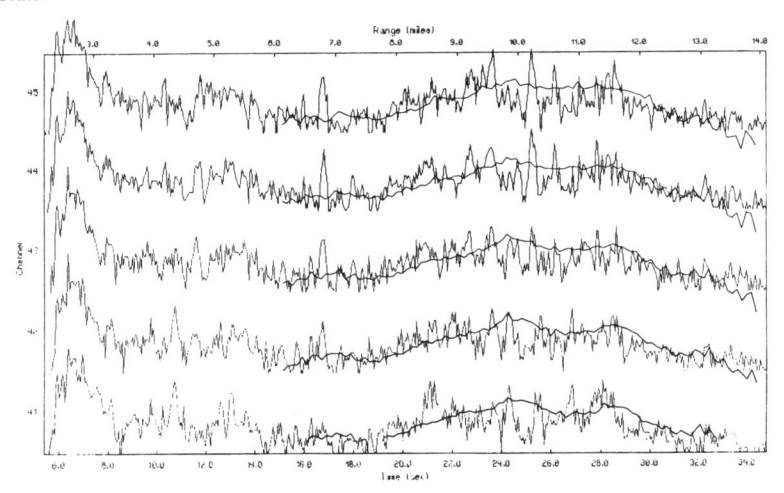

Figure 4. Reverberation modeling and reconn data displays for pings issued at 0948Z. Same as figure 3 with further expanded scale and line-graph representation of modeled data.

agreement is good with further indication of the lack of resolution in the modeled data. Figure 4 shows the reconn data at an even finer resolution with only 5 beams shown (41-45) and overlaid with a line-graph plot of the modeled data. It can be seen that the model predicts the actual data very well. It should be noted, however, that the magnitudes of the modeled data are only relative at this point. Further work will attempt to develop the relation for Lambert's coefficient given above through empirical fits with a broader range of reconn data.

3. CONCLUSIONS

The two direct-path bottom reverberation cases shown above is very typical of the data taken for other pings in the ARSRP reconn experiment. Backscattering is strong and distinct, and can be modeled very well with Lambert's law if actual bathymetry and local grazing angles are used. The flat-bottom assumption that is usually made in determining both propagation and backscattering strength would produce concentric rings of reverberation in geographic plots like Figs. 1a and 2a and vertical banding in the beam/time plots. Clearly those plots and the reconn data refute such an assumption.

It is important to note clearly that bathymetry affects scattering only through its effects on propagation, while local slopes and slope and roughness statistics affect scattering strength. The often-sought correlation between scattering and bathymetry is not a properly posed issue. Using the term "geomorphology" to mean bottom shape as well as depth, a better posed issue is to seek a correlation between geomorphology and scattering. Since, model results track well with the data, it appears BISSM can correctly model the more important features of the backscattering phenomena to the resolution available in the bathymetry.

4. ACKNOWLEDGEMENTS

The authors express their appreciation to James Showalter and Sergio Derada for assistance in developing the BISSM algorithm and coupling it to a propagation model. We are also grateful to Dr. Marshall Orr of the Office of Naval Research and Mr. Chester Wilcox of the Naval Oceanography Command for funding aspects of this project.

REFERENCES

Caruthers, J.W., Sandy, R.J., and Novarini, J.C. (1990) 'Modified Bistatic Scattering Strength Model (BISSM2)', NOARL SP023:200:90.

Eckart, C. (1953) 'The Scattering of Sound from the Sea', J. Acoust. Soc. Am., 25, 566-570.

Brekhovskikh, L. (1952) J. Exptl. Theor. Phys. USSR, 23:275-289.

Brekhovskikh, L. and Lysanov, Yu. (1982) FUNDAMENTALS OF OCEAN ACOUSTICS, Springer-Verlag, Berlin Heidlberg.

Ellis D.D. and Crowe, D.V. (1991) 'Bistatic Reverberation using a Three-Dimensional Scattering Function', J. Acoust. Soc. Am., 89, 2207-2214.

DIRECTIONAL PROCESSING OF THE SIMULATED DIRECT PATH, BROADBAND REVERBERATION DATA AND ITS INTERPRETATION

Ashok K. Kalra and James K. Fulford
Naval Research Laboratory,
Stennis Space Center, MS 39529-5004

ABSTRACT

Broadband ocean bottom/subbottom reverberation data are simulated for the Acoustic Reverberation Special Research Program's (ARSRP's) fine scale experiment to be conducted in FY 93. ARSRP objectives are to study the direct path bottom/subbottom reverberation processes at low frequencies (50 - 500 Hz) and low grazing angles (±15°). The data are simulated, over the Goff and Jordan sediment pond, for a horizontal array towed at an elevation of 25 m (from the sea floor) in order to generate the bottom/subbottom direct path, backscattered reverberations at low grazing angles (±15°). These data consist of the reverberation and/or specular reflection events from all angles, and we studied the application of the intercept time-ray parameter (τ-p) technique to filter out the undesired reverberations/reflections at high grazing angles. The low grazing angle (0 to 15°) data are then interpreted to identify the (location of) bottom and/or subbottom scattering patches or elements. These are further analyzed to study the reverberation hypotheses.

INTRODUCTION

The Acoustic Reverberation Special Research Program (ARSRP) of the Office of the Naval Research (ONR) is planning to acquire direct path, bottom/subbottom reverberation data, over a 5x5 km patch, in a fine-scale experiment to be conducted in FY 93. In the context of this project, direct path defines propagation paths that preclude interaction with the sea surface and minimize the effect of bulk variability within the water column. The objectives of this program are to understand the reverberation processes at low grazing angles (±15°) and at low frequencies (50 to 500 Hz). The data will be acquired in the ARSRP Atlantic Natural Laboratory located within the Abyssal plane in the Atlantic ocean.

A part of the reverberation data are to be acquired with the acoustic array of the Naval Research Laboratory's (NRL's) Deep Towed Acoustic/Geophysics System (DTAGS). DTAGS is identical to the horizontally towed seismic streamers but has two collinear receiver arrays (acoustic and geophysics), and in addition it can be towed close to the sea floor in water depths to 5000 m. Only the acoustic array data from DTAGS will be analyzed for the backscattered reverberation studies; the geophysics array data will be used (independently) to map the bottom/subbottom horizons and/or inhomogeneities that may produce the reverberations observed by the acoustic array.

As the ARSRP fine-scale experiment is not scheduled till FY 93, we have simulated the acoustic array data over a sediment pond for the acquisition parameters similar to that of DTAGS, and studied the intercept-time ray parameter (τ-p) technique to separate the low grazing angle data of interest to the ARSRP objectives (±15°) from the full spectrum of angles present in the data. The low grazing angle data are further interpreted to identify the bottom/subbottom scattering patches that may generate the backscattered reverberations. These data are further analyzed to study the reverberation processes.

FIELD ACQUISITION PARAMETERS FOR HORIZONTAL ARRAY

As mentioned earlier, the DTAGS acquires data with two collinear horizontally towed (acoustic and geophysics) arrays. The geophysics array is towed behind the acoustic which has a near offset distance of 72 m from the source. Each array consists of 24 channels and each channel is a group of 6 elements, 0.6 m apart. The channel separations in acoustic and geophysics arrays are 2.1 and 21 m respectively, and the acoustic and geophysics data are recorded simultaneously. The DTAGS source is an omnidirectional, 205 dB, 250 to 650 Hz, LFM resonator (marine vibrator).

In order to acquire the low grazing angle, backscattered reverberation data, the DTAGS will be towed at a distance of 300 m from the sea floor in 5000 m water depths. The acoustic array data will consist of the low grazing angle backscattered reverberations as well as the high angle backscattered and specularly reflected data. As a first step in processing, one needs to filter out the high angle data (backscattered and specularly reflected) from the full spectrum of angles present in the recorded data, and retain the low grazing angle data (of interest to the ARSRP objectives) for further processing and interpretation. We studied the application of time-intercept ray parameter (τ-p) technique on data that was simulated with acquisition parameters similar to that of DTAGS.

TIME-INTERCEPT RAY PARAMETER (τ-p) PROCESSING TECHNIQUE

The time-intercept ray parameter (τ-p) processing technique discriminates on the basis of the apparent velocities (v_a) of the plane wavefront recorded at the receiver arrays. Consequently, there is an inherent assumption that the data recording is done in the far field region.

Refering to Figure 1(a), the plane wavefronts, travelling at the medium velocity v, arrive at the recording array at a grazing angle γ. The apparent velocity of the waves along the recording array, v_a, is given by:

$$v_a = \frac{v}{\cos \gamma} \qquad (1)$$

The time-distance plot of the plane wavefront arrivals, is a straight line whose slope is the inverse of the apparent velocity as indicated in Figure 1(b). Consequently, we can write its equation as

$$t = \tau_0 + \frac{x}{v_a} \qquad (2)$$

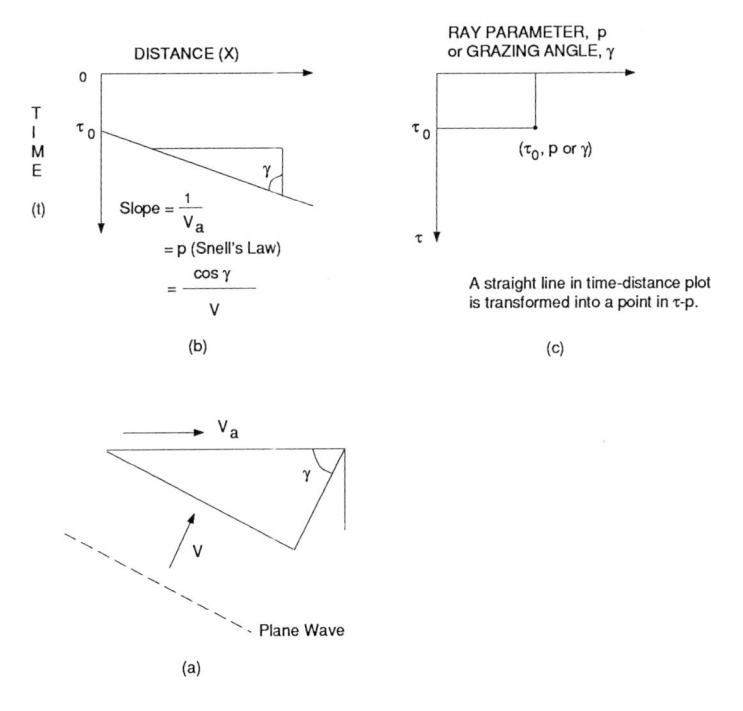

Figure 1

where τ_0 is the time-intercept and is the time taken to travel from the initiation of the source to the first element of the recording array. From Snell's law, we have that

$$\text{ray parameter } p = \frac{\cos \gamma}{v} = \frac{1}{v_a} \qquad (3)$$

Consequently, we can write the equation of the straight line in terms of the ray parameter p as,

$$t = \tau_0 + px \qquad (4)$$

where the ray parameter p is the slope of the straight line, and it represents the direction of arrival (DOA), γ, as given by (3) above.

A straight line is characterized by its slope (p) and the time-intercept (τ_0). In τ-p technique, one transforms a straight line into τ-p domain by plotting the sum of the responses along the line against its slope (p) and the time intercept (τ). As a result, a straight line in time-distance domain is transformed into a point in the τ-p domain as shown in Figure 1(c). The summation along the horizontal line (p=0) is the special case of a linear array response, where all the individual receiver responses are summed together with no delays and enhances the vertical arrival event (γ=90).

By this transformation, we decompose the time-distance data into planewaves and the process is correctly called 'planewave decomposition'. As mentioned above this transformation is gotten by summing or stacking the data along sloping or slanting straight lines, and so this process is also known as the slant stacking in exploration geophysics.

The slope p is the measure of the DOA, γ, according to (3), and consequently, one can also convert the p-axis into the DOA angle axis or grazing angle (γ) axis as indicated in Figure 1(c). The transformed γ-axis will, in general, represent all angles from -90 to +90°. We can retain the traces representing the angles of interest and reject those not of interest, for further processing. For example, for the ARSRP objectives, we will transform the specific traces representing the low grazing angles of interest ($\pm 15°$) to the frequency domain, in order to study the reverberation phenomena pertaining to the low grazing angles ($\pm 15°$). If we are interested to study the low grazing angle events in the time domain, we may want to inverse transform the selected traces corresponding to the angles of interest (such as $\pm 15°$) back into the time-distance domain.

The mathematical representations of the τ-p transform and its inverse are given by the following (Clarebout, 1985):

$$U_R(\tau,p) = \int u(x, \tau+px) \, dx$$

$$u(x,t) = \frac{1}{t^2} * \int U_R(p, t-px) \, dp$$

where * represents the convolution operation.

SIMULATED DATA, ITS PROCESSING AND INTERPRETATION

We have simulated the 24 channel acoustic array data for acquisition parameters similar to that of DTAGS. The data are simulated for single hydrophone receivers with the near offset distance of 72 m and hydrophone separations of 2 m over the 'Goff and Jordan' sediment pond model as shown in Figure 2. The source is an omnidirectional with frequency bandwidth of 65 Hz. The data are simulated for a tow elevation of 25 m (from sea bottom) in order to generate enough low grazing angle events of interest to the ARSRP objectives. The data are generated using the finite difference modeling algorithms.

Figure 3 is the simulated time distance plot for the 24 channel acoustic array. It indicates the higher angle specular reflections from the sea floor and the basement, along with the backscattered events from the front and back of the receiver array. The front and back events are evidenced by the positive and negative slopes of the linear events respectively as indicated in Figure 3. The strongest arrival is the direct blast, as expected. Figure 4 is the τ-γ transformed domain of the data of Figure 3, and represents the grazing angles from 0 to 90° as indicated.

Restricting our interests to the ARSRP objectives, we inverse transformed the grazing angles from 0 to 15° (in Figure 4) back to the time-distance domain and applied a non-linear gain to enhance the low energy events before plotting the result in Figure 5. In comparing Figure 5 with Figure 3, it is evident that the τ-p technique has succeeded in filtering out the high grazing angle events (backscattered and specularly reflected) from the full spectrum of high and low angle events present in Figure 3.

Figure 2

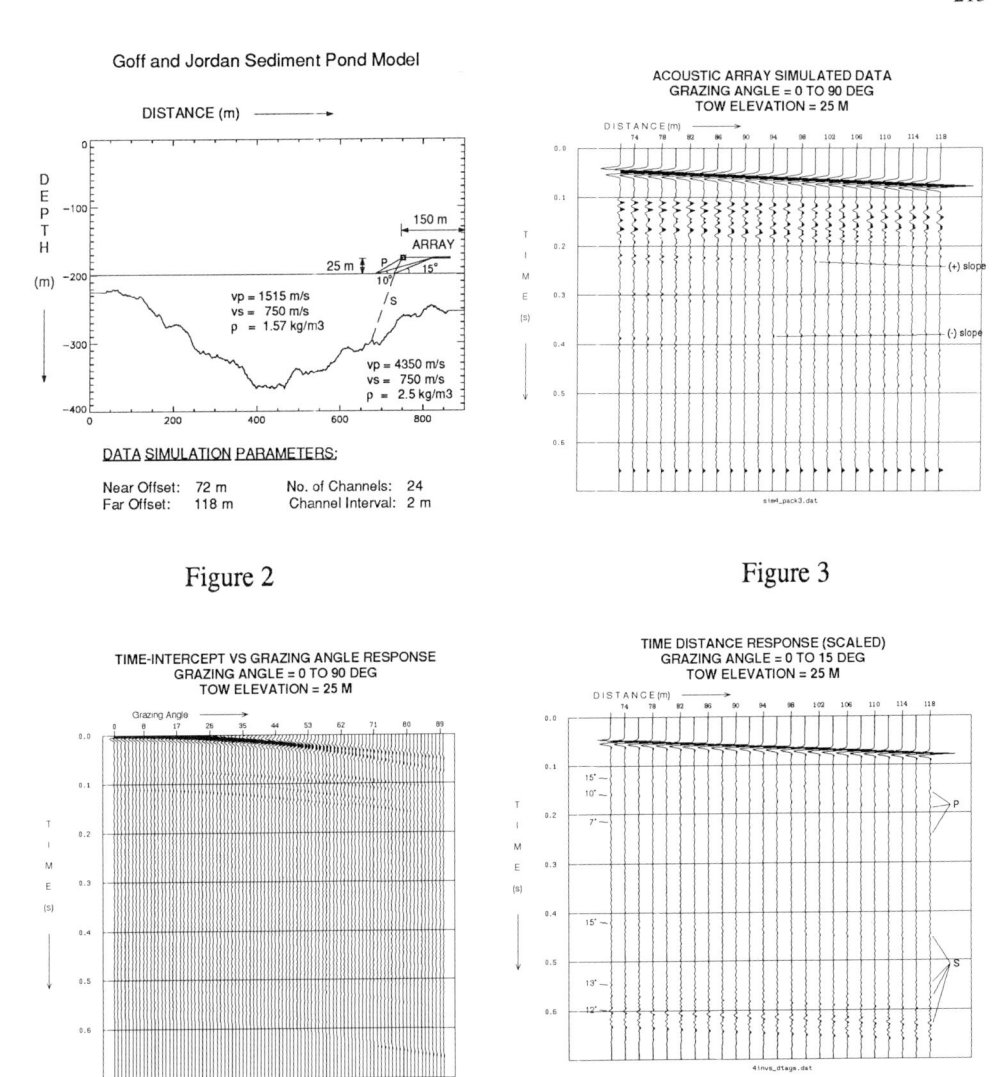

Figure 3

Figure 4

Figure 5

The direct blast event with the highest amplitude can be easily identified in Figure 5. We have also identified some of the later events as P- and S-wave events along with their arrival angles at the receiver array. The P-wave events are P-wave arrivals backscattered (at low grazing angles) from the sea bottom. The S-wave events are the ones converted to S wave at the sea bottom, backscattered from basement as S waves, backconverted to and received as P-wave events at the acoustic array. The exact arrival angles were determined from two way times by ray tracing the events. Some of the P and S events are ray traced on Figure 2 identifying the bottom and subbottom (basement) scattering areas generating the low grazing angle reverberation events of Figure 5.

REVERBERATION HYPOTHESIS

In order to study the reverberation processes, we transformed the low grazing angle (0 to 15°) backsacttered data in Figure 4 to frequency domain to determine the reverberation level response vs frequency and grazing angles. This result is shown in Figure 6. Comparing this response to that of the direct blast in Figure 7, which is considered to be identical to that of the incident signal, we determine that the backscattered reverberation is the result of a constructive and destructive interference of the incident signal with the subbottom

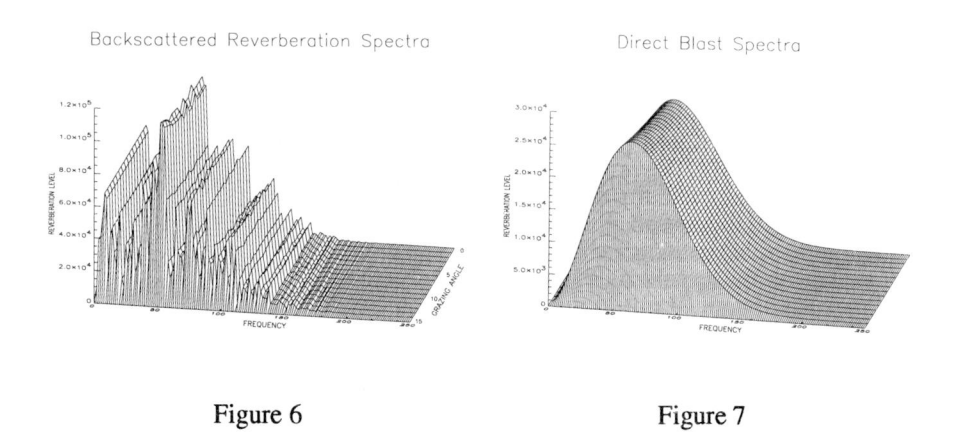

Figure 6

Figure 7

(basement) roughness. The rough basement acts as a diffraction grating, similar to Rayleigh scattering.

CONCLUSIONS

Based on the towed array simulated data, we have shown that the time intercept-ray parameter (τ-p) technique was very effective in filtering out the high angle reverberation and specularly reflected events and retaining the low grazing angle events of interest to the ARSRP objectives. The low grazing angle (0 to 15°) data were then interpreted to identify the bottom/subbottom scattering patches that produced the reverberations at these angles. These data were further analyzed to study the scattering processes.

ACKNOWLEDGEMENT

The data were simulated by Dr. Stuart Henrys at Rice University. This project was supported by the Office of Naval Research, Program Element No. 0601153N, under Program Manager Dr. Marshall Orr.

REFERENCES

Clarebout, Jon F. (1985) Imaging the Earth's Interior, Blackwell Scientific Publications.

NUMERICAL SCATTERING RESULTS FOR A ROUGH, UNSEDIMENTED SEAFLOOR

Alan Levander
Geology Department, Rice University
P. O. Box 1892, Houston, TX 77001

Alistair Harding and John Orcutt
Institute of Geophysics and Planetary Physics (0225)
Scripps Institution of Oceanography
La Jolla, CA 92093-0225

ABSTRACT

We have adopted a hybrid finite difference technique to investigate the role of elasticity in determining reverberation or backscatter properties of a realistic rough seafloor. Scattering from an elastic medium was found to exceed that from a model acoustic seafloor at shallow grazing angles by nearly 15 dB and the variability with wavenumber for the two seafloors was quite distinct.

1. INTRODUCTION

Acoustic reverberation or the forward- and back-scattering of acoustic energy by a rough seafloor has been studied for a number of years in the context of a Dirichlet boundary condition in which the displacement boundary condition is set to zero (e.g. Thorsos and Jackson, 1989). However, recent studies have shown that the seafloor is far from rigid; in fact the rigidity of materials such as basalt support shear velocities which are only 100's of m/s (e.g. Harding et al., 1989). Furthermore, the roughness of the boundaries is considerable and some authors conjecture that the seafloor topography is self-affine to very small scales (Goff and Jordan, 1988). In this paper we examine the effects of a flexible seafloor as well as topography which is fractal in character using finite difference methods.

Research sponsored by the Office of Naval Research through the Bottom/Subbottom component of the Acoustic Reverberation Special Research Program (ARSRP).

2. THE GENERATION OF SYNTHETIC REVERBERATION RESPONSES

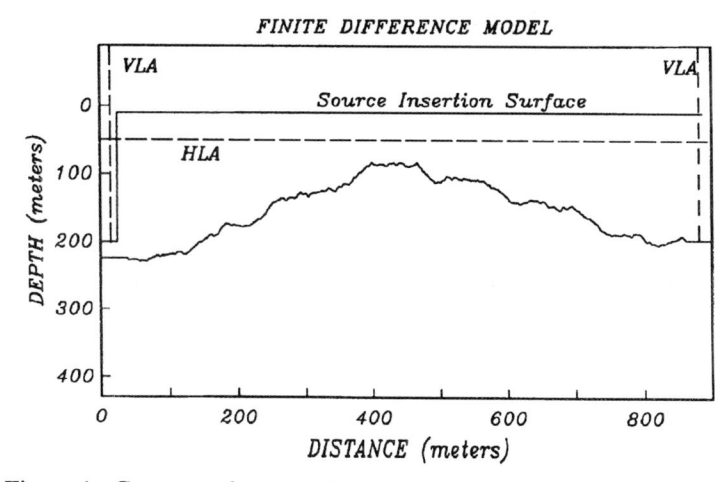

Figure 1. Geometry for numerical experiments used in this paper. The seafloor is 800 m in extent and over 400 m in depth. The topography is a realization of a fractal process (Goff and Jordan, 1988).

The second-order finite difference calculations were made in two dimensions with a semi-infinite planar beam at a low grazing angle of 5°. The frequency bandwidth extended from 0—150 Hz and the physical geometry is illustrated in Figure 1. The algorithm is, in fact, a hybrid technique which couples the propagation of acoustic energy in the water column to the finite difference grid and the extraction and propagation of acoustic energy to a distant receiver. The finite difference algorithm can be run with either an acoustic or elastic seafloor; the overlying ocean is always purely acoustic. The algorithms used for the synthetic computations presently run on a CRAY YMP 8/832 at the San Diego Supercomputer Center (SDSC).

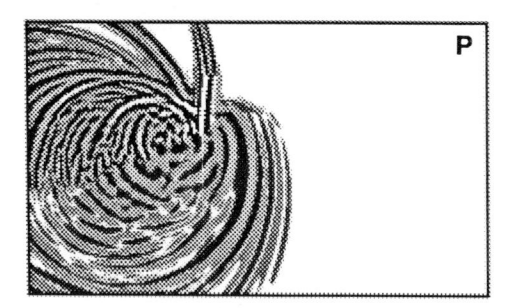

Figure 2. A snapshot of the compressional and shear wave propagation associated with an incident acoustic beam from the left.

The surface along which the source was continuously inserted is illustrated in Figure 1 and the positions of a horizontal line array (HLA) and two vertical line arrays (VLA) are indicated. The seafloor morphology is a realization of a Goff-Jordan (1988) model for Atlantic seafloor and is sampled at 1 meter intervals. This realization produced a modest (150 m) hill with considerable roughness extending to small scales (1 m). The compressional wave velocity in the seafloor was 2.6 km/s and the shear wave velocity was 1.45 km/s; these values are characteristic of those measured for the young seafloor.

3. SCATTERING OF ACOUSTIC ENERGY FROM THE MODEL SEAFLOOR

An example run of the fully elastic finite difference code is shown in Figures 2 and 3. In Figure 2 the beam has propagated part way up the abyssal hill. The divergence (P-waves) of the wave field (P) is illustrated in the top of the figure while the curl (S-waves) is shown in the bottom. The low-angle incident acoustic beam is very clear (propagating to the right) and considerable backscattering can be seen in the wavefronts propagating in an opposite direction (left). A head wave has been established in the water column ahead of the incident wave due to the higher velocities in the basement. The shear wave generated in the seafloor is very evident and wavefronts emanating from bathymetric features can be followed within this shallow crust.

Figure 3 follows the incident wavefront into the geometrical shadow created by the abyssal hill. A great deal of backscattered energy can be seen in the water column and some of this propagation is organized into clear wavefronts. The head wave in the water column is now well-established and the shear wave, even in the shadow, is a very strong and coherent feature. Note, in both cases that the shear wave follows the acoustic incident wave quite closely because of the proximity of the two velocities.

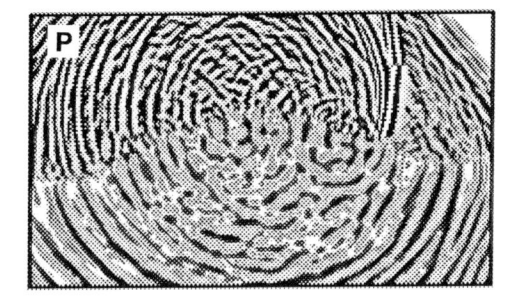

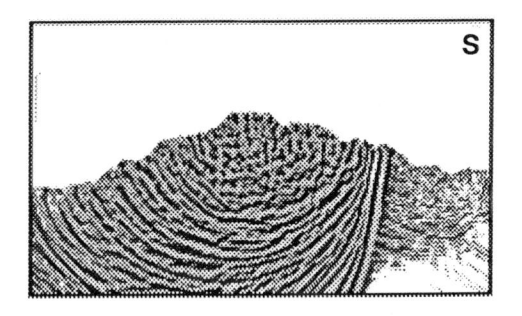

Figure 3. Same as Figure 2 except at a later time. The acoustic beam is now in the geometrical shadow of the abyssal hill.

The output from the horizontal line array records the passing wavefronts continuously and provides further intuition into the propagation and scattering processes. Figure 4 is a plot of the time series at each of the elements of the HLA as a function of time. The strong arrival along a straight line is simply the propagation of the acoustic beam along the line array. The energy which follows later is scattered back into the array by the underlying abyssal hill. Compressional head waves propagating through the high speed seafloor arrive before the incident wave at ranges in excess of approximately 400m. Those features in the long wavetrain which slope down and to the right are waves which are propagating in the forward direction while those waves which slope down and to the left have been scattered in the backward direction. Horizontal features are propagating vertically.

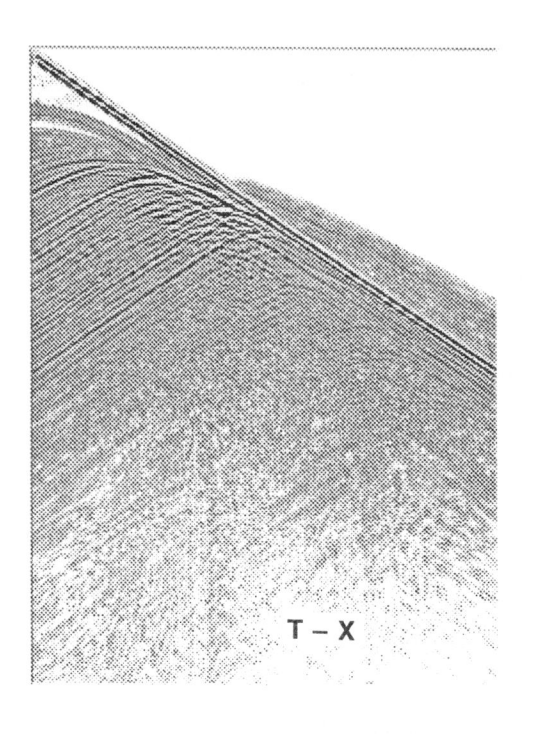

Figure 4. Time (vertical) vs. distance (horizontal) plot of energy received at the HLA.

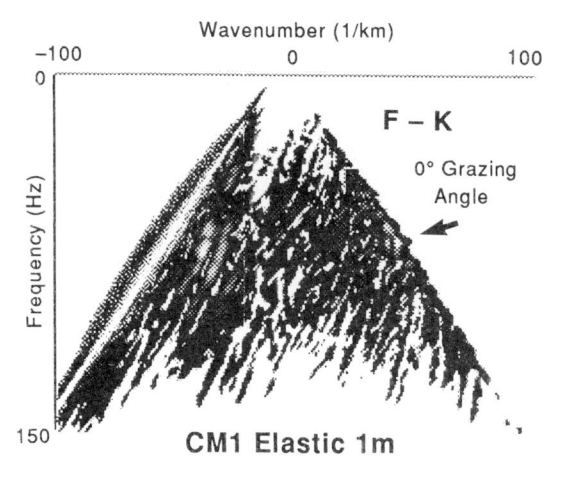

Figure 5. F-K transform of Figure 4.

The HLA information in travel time and distance can be Fourier transformed twice into the frequency-wavenumber domain. The results of this transform for this experimental setup are plotted in Figure 5. Those wavenumbers which are less than zero represent waves which have been scattered in the forward direction. The incident wave can be seen as a bright line sloping down and to the left. The gradient of this feature is the incident phase velocity. Most importantly, backscattering can be seen at wavenumbers greater than zero. The arrow in the figure points to those wave components which are backscattered toward the incident source (to the left in Figures 2 and 3).

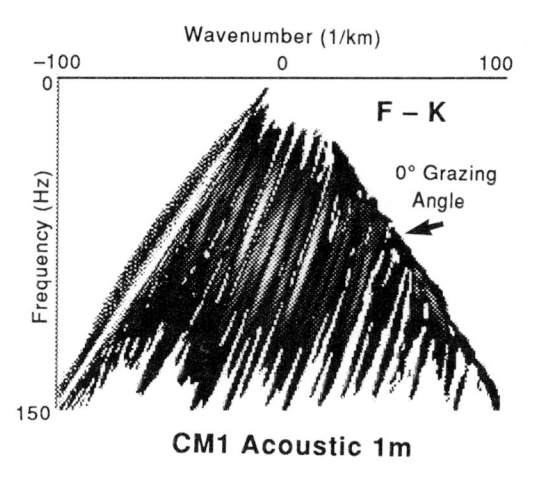

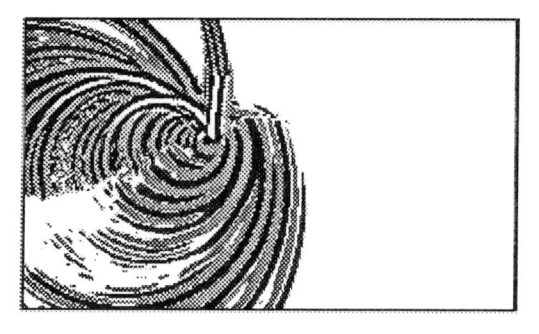

The acoustic case provides an interesting contrast to the elastic seafloor. Figure 6 shows the F-K spectra for the identical geometry and seafloor roughness except that the rigidity in the seafloor has been set to zero. In this case, there continues to be high levels of backscattering, but the distribution of the scattered energy across the wavenumber band is much more even. There does not appear to be a prefer-

Figure 6. F-K transform for acoustic seafloor.

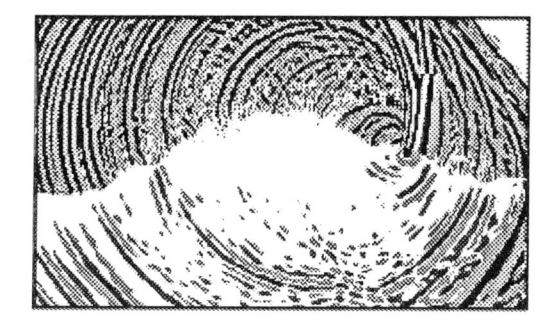

Figure 7: Two snapshots of propagation corresponding to the times illustrated in Figures 2 and 3 for a seafloor with no rigidity. Note the absence of a shear wave and the apparently better-organized backscattered energy in the overlying ocean. The shear propagation in the elastic case served to fill gaps in the wavefield and to lengthen the wavetrains scattered from the seafloor.

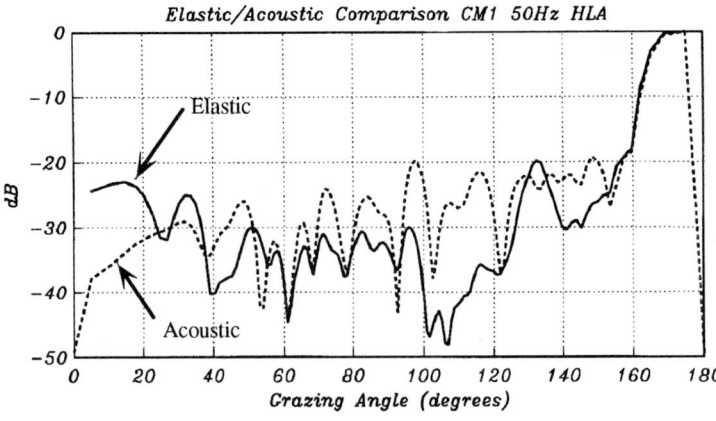

Figure 8. Scattering as a function of angle for the acoustic and elastic cases at a frequency of 50 Hz. Note the significant differences in the curves' shapes.

ence for scattering in the reverse direction as in the elastic case. The wave propagation snapshots are shown in Figure 7. Note that the energy levels in the seafloor have been reduced and that, of course, the shear wave in the basement has disappeared. Figure 8 plots scattering as a function of angle for the two cases. At low grazing angles, the elastic case exceeds the acoustic case by as much as 15 dB.

4. CONCLUSIONS

The effects of scattering in an ocean overlying an elastic seafloor can be simulated at high frequencies by the use of a hybrid finite difference algorithm. Backscattering is particularly important for an elastic medium in that the levels at some frequencies for 180° backscattering are quite high. In contrast, scattering from a rough, acoustic seafloor model is more even as a function of wavenumber. We speculate that the shear wave velocities in the seafloor, which are very close to the acoustic velocities in the overlying ocean, lead to significant backscatter, perhaps through the excitation of Rayleigh modes in the elastic seafloor.

REFERENCES

Goff, J.A. and Jordan, T.H. (1988) 'Stochastic modeling of seafloor morphology: Inversion of Sea Beam data for second-order statistics', *J. Geophys. Res., 93*, 13,589-13,608.

Harding, A.J., Orcutt, J.A., Kappus, M.E.., Vera, E.E., Mutter, J.C., Buhl, P., Detrick, R.S., and Brocher, T.M. (1989) 'Structure of young oceanic crust at 13°N on the East Pacific Rise from expanding spread profiles', *J. Geophys. Res., 94*, 12,163-12,196.

Thorsos, E.I. and Jackson, D.R. (1989) 'The validity of the perturbation approximation for rough surface scattering using a Gaussian roughness spectrum', *J. Acoustic. Soc. Am., 86*, 261-277.

THE EFFECTS OF SEAFLOOR ROUGHNESS ON REVERBERATION: FINITE DIFFERENCE AND KIRCHHOFF SIMULATIONS

John Orcutt and Alistair Harding
Institute of Geophysics and Planetary Physics (0225)
Scripps Institution of Oceanography, La Jolla, CA 92093-0225

Alan Levander
Geology Department, Rice University, P. O. Box 1892, Houston, TX 77001

ABSTRACT

Hybrid finite difference and Kirchhoff methods are used to examine the effects of varying seafloor roughness on energy scattered from the seafloor from an impinging acoustic wave at a very low angle (5°). Scattering is found to be highly dependent upon scales of roughness smaller than an acoustic wavelength. For similar seafloors, Kirchhoff and acoustic finite difference schemes are shown to predict scattering behaviors which are quite similar, but which are very different than scattering from an elastic seafloor.

1. INTRODUCTION

Goff and Jordan (1988) have proposed that seafloor roughness can be characterized as a fractal in that the topography, on small scales, is self-similar. Many previous studies of acoustic reverberation have examined the effects of regular variations in seafloor topography at a variety of scales. We show that seafloor roughness on small scales can have a profound effect on reverberation. The importance of small scales (less than a wavelength) have important implications for seafloor characterization studies which are conducted in support of reverberation measurements. Finally, we show that the primary effect of shear wave conversion in the seafloor is to fill gaps in the reverberation wavetrain and to lengthen the reverberative wavetrain significantly.

Research sponsored by the Office of Naval Research through the Bottom/Subbottom component of the Acoustic Reverberation Special Research Program (ARSRP).

2. THE EFFECTS OF SEAFLOOR ROUGHNESS

Ideally, bathymetric surveys of the world's oceans at very fine scales should be conducted to permit the quantitative prediction of seafloor reverberation. Levander et al. (this volume, hereinafter referred to as Paper 1) demonstrated that a rough, self-affine seafloor sampled at one meter intervals, could generate significant scattering in both the forward and reverse directions. The sub-wavelength sampling is obviously difficult to achieve; it is important to establish the scales necessary for predicting reverberation quantitatively.

The finite difference numerical methods used were introduced in Paper 1 and the geometry was illustrated in Figure 1 of that paper. We have smoothed the seafloor arbitrarily by sampling the bathymetry every 30 m and then using a smooth spline interpolation to resample back to a 1 m grid in order to examine the effects of roughness. Figure 1 shows a snapshot of the compressional and shear wavefields at a time identical to that in Figure 2, Paper 1. In this case there is more coherent compressional energy in the seafloor and the backscattered energy in the water column appears to be better organized.

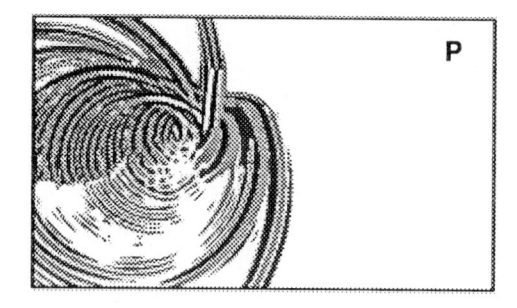

Figure 1: A snapshot of P and S wave propagation for a planar beam entering the left at a grazing angle of 5°. The seafloor is sampled every 30 m and is rather smooth.

Figure 2 is a plot of a snapshot at a later time (identical to Figure 3, Paper 1). Again the compressional energy levels in the upper crust are subdued and the scattered waves are better organized than in the case for the finer topography.

The frequency-wavenumber transform of the horizontal line array (HLA; see Paper 1) is shown in Figure 3. This transform is radically different from similar transforms for the finer topography (Paper 1, Figure 5). In this case only a minor amount of scattering has occurred in the reverse direction. Even though the shortest acoustic wavelengths in the calculation are 10m, the characterization of the seafloor to wavelength scales and even smaller is essential in predicting reverberation levels.

Obviously, the characterization of the ocean basins at these scales is impossible. However, we propose that reverberation levels can be predicted with reasonable certainty if deterministic topography is available on 100 m scales and a reliable means of extrapolating the statistics of the topography to much smaller scales is available. It is very likely that the relationships used for this extrapolation will have a profound effect on the scattered signal. Additional research to confirm proposed scaling laws for seafloor morphology is essential.

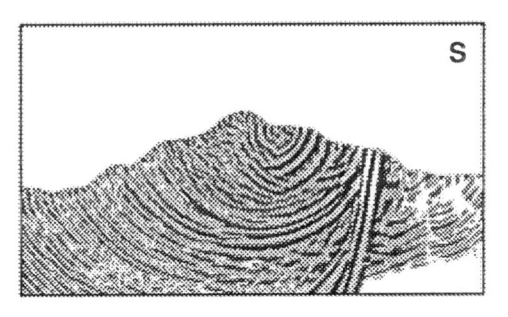

Figure 2. Same as Figure 1 except later in time.

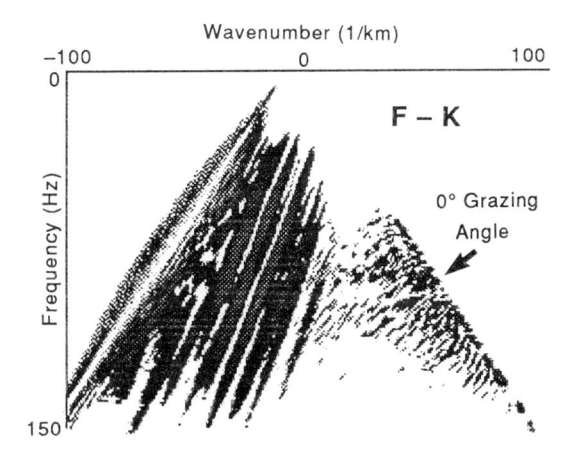

Figure 3. F-K transform of data recorded on the HLA.

3. KIRCHHOFF SCATTERING FROM A ROUGH SEAFLOOR

While finite difference or finite element methods permit the examination of seafloor scattering on small scales such as the 900 m of seafloor used in this study, the prediction of reverberation on ocean basin scales with these tools is impossible at the present time and is likely to remain so for the foreseeable future. Inexpensive alternatives to these finite difference studies are variations on the Kirchhoff integral method. We have adopted such a technique, which assumes an angle-dependent (akin to Lambert's Law) seafloor reflection coefficient, for comparison with the more time consuming finite difference method used here and in Paper 1.

Figure 4 plots the time-distance observations collected along the HLA for insonification of the rough seafloor using the Kirchhoff technique. As in Paper 1, the large arrivals which fall along a straight line which dips downward from left to right are the direct arrivals along the HLA. The later arrivals are reverberations from the underlying seafloor. Those patterns which move upward from left to right represent energy which is propagating in a direction opposite to the incoming beam. In this case, unlike Figure 4 (Paper 1), there are no crustal arrivals traveling ahead of the direct wave at large distances since energy does not return from below the surface of integration (the seafloor). In addition, the reverberation cuts off in the shadow of the abyssal hill unlike the beam computed using finite differences.

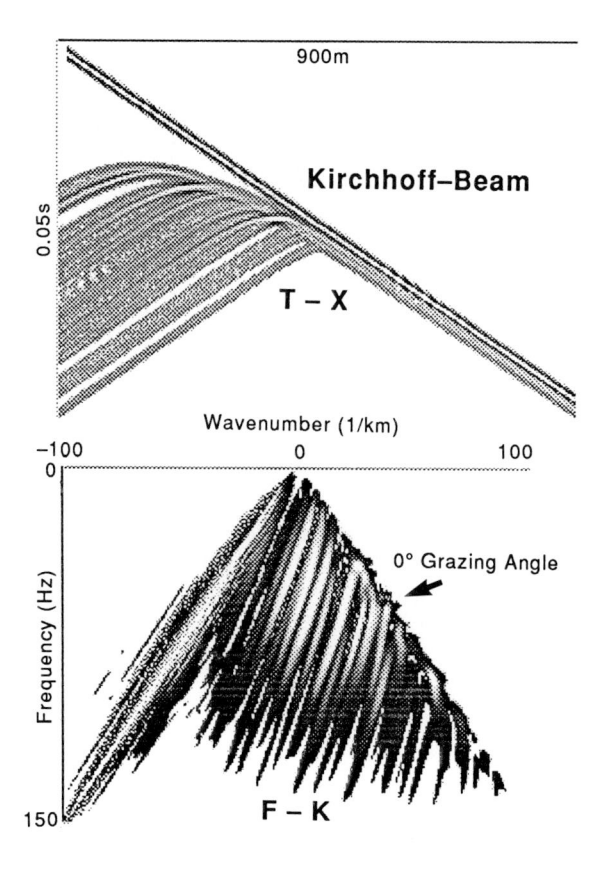

Figure 4. A time-distance plot of a recording along the HLA for the Kirchhoff case (top) and the F-K transform (bottom). Note that the F-K transform is similar to the acoustic finite difference case (Paper 1, Figure 6).

The F-K transform of the scattered energy is plotted at the bottom of Figure 4. Perhaps surprisingly, the F-K transform is very similar to the computation for the acoustic case (Figure 6, Paper 1). The scattered energy is broadly distributed over wavenumber and is generally not concentrated in the backward direction as was the case for an elastic seafloor (Figure 5, Paper 1).

A comparison of time histories is plotted in Figure 5 for a hydrophone along the HLA at a range of 250 m. The top *hydrogram* was collected from the finite difference run which employed an elastic seafloor. The second trace is from the finite difference algorithm, but with an acoustic seafloor. Finally, the bottom hydrogram was computed using the Kirchhoff

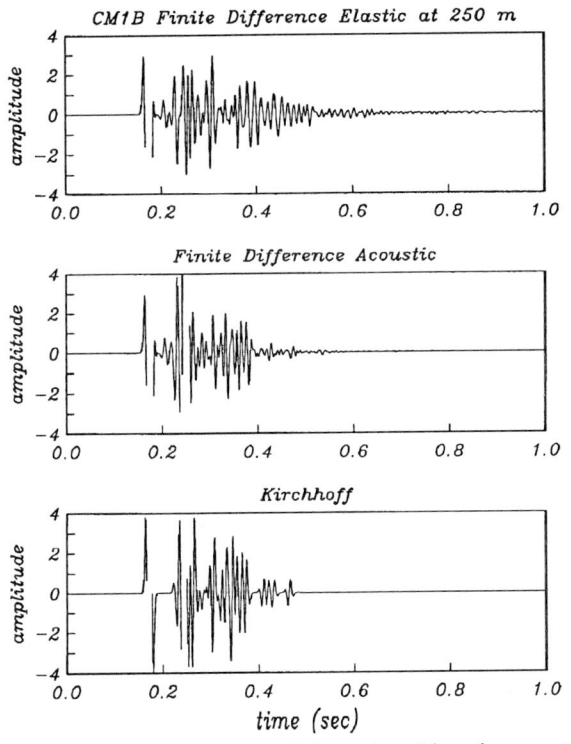

Figure 5. A comparison of three time histories recorded on an element of the HLA at a range of 250 m. Note that the Kirchhoff and acoustic finite difference results are very similar.

technique. As might have been predicted from a comparison of the F-K spectra, the Kirchhoff hydrogram is quite similar to the acoustic finite difference trace and differs from the elastic case.

The amplitudes of the traces are similar and the time behavior of the acoustic and Kirchhoff computations are nearly identical. Generally, the isolated impulsive arrivals are associated with small topographic features. These same topographic effects can be seen in the elastic finite difference calculation and, in fact, dominate the character of that hydrogram. The primary differences between these traces arises from the increased length of the elastic reverberation and the infilling between distinct arrivals.

4. CONCLUSIONS

The characterization of the statistics of seafloor behavior at small scales is very important for the quantitative prediction of seafloor reverberation and particularly for backscattering. It is essential that confidence in the reliability of topographic models be developed in order to facilitate these predictions. While Kirchhoff techniques fall short in the prediction of the entire wavetrains associated with seafloor reverberation, it is likely that the method will allow the prediction of the important features in the observed seafloor reverberation. Given that the use of finite difference methods is limited to small areas of the seafloor, the utility of Kirchhoff techniques is very important.

REFERENCES

Goff, J.A., and T.H. Jordan, Stochastic modeling of seafloor morphology: Inversion of Sea Beam data for second-order statistics, *J. Geophys. Res., 93*, 13,589-13,608, 1988.

A NUMERICAL SCATTERING CHAMBER FOR STUDYING REVERBERATION IN THE SEAFLOOR

R.A. STEPHEN
Woods Hole Oceanographic Institution
Woods Hole, MA 02543
USA

ABSTRACT. A numerical scattering chamber, based on the finite difference solution to the two-way elastic (or anelastic) wave equation in the time domain, is a powerful and convenient approach to studying the physics of surface and volume reverberation at the seafloor. Scattering from both surface roughness and volume heterogeneities at scale lengths comparable to wavelengths can be treated. The method includes all shear wave and interface wave effects and all multiple interactions between scatterers. Bottom parameters varying from soft sediments (with shear wave velocities much less than water velocity) to hard basalts (with shear velocities higher than water velocity) are studied. We use a Gaussian pulse-beam as the incident field and we compute the resultant scattered field (in compressional and shear wave energy density) on arrays of receivers surrounding the scattering region. Backscatter coefficients, defined as the ratio of the energy in the scattered beam at a given angle to the energy in the incident beam, can be computed. For example, for an incident beam at fifteen degrees grazing angle, the coefficient for direct backscattering from a very rough, basaltic seafloor is -17dB.

1. Introduction

Many problems in bottom interacting ocean acoustics require a knowledge of the scattered field from a rough, laterally heterogeneous seafloor. Recent studies [1,2] have shown that the physics of seafloor scattering can be quite complex with energy converting from compressional waves in the water to shear and interface waves at the seafloor and then reradiating as compressional waves back into the water. In order to investigate the physical mechanisms of scattering further and to quantify the magnitude, time spread and angle spread of the scattered field we have developed a numerical scattering chamber based on the finite difference solution to the two-way elastic (or anelastic) wave equation.

The finite difference method provides the capability to study full wave effects at the seafloor in range-dependent environments [3,4]. The approach is particularly useful for pulse sources, for strong backscattering, and for studies of the response at and below the seafloor. Calculations are carried out in the time domain and solutions for a given source pulse are obtained directly. (The method can also obtain a continuous wave (CW) solution simply by using a CW source and running the computations until steady state is reached [4].) Solutions are obtained for both forward scattering and backscattering including conversions to shear and interface waves. Multiple interactions between scatterers are also completely included. Since the finite difference method treats the whole model as a discrete grid, vertical and horizontal displacements (or velocities or accelerations) at the seafloor and within the bottom are obtained at the same time as the pressure field in the water. Insight into multi-pathing and bottom and sub-bottom scattering is obtained directly.

In this paper we describe the characteristics of the numerical scattering chamber and present an example of strong, low-angle backscattering from a rough, hard seafloor.

2. The Numerical Scattering Chamber

We consider a numerical scattering chamber which is completely surrounded by an absorbing region (Figure 1). (In the case described here we assume a two dimensional Cartesian co-ordinate system, however the technique can be extended to three dimensions [5].) The structure inside the chamber can be completely arbitrary and is specified on a grid with typically 10-50 grid points per wavelength both vertically and horizontally. At each grid point we define the compressional and shear velocities, the compressional and shear attenuations, and the density. For the seafloor scattering problem, we place in the chamber realizations of seafloor structure which contain either surface roughness and/or volume heterogeneities. The upper part of the chamber is water and the lower part represents sea bottom material such as sediments or basalts. The interface between the water and the seafloor continues into the absorbing regions on either side.

The scattering region is insonified using a beam generator which creates a beam of a given width at a given angle of incidence. We typically use a pulse beam so that we can track the energy propagation and multipathing in the scattering region.

Completely surrounding the scattering region we place a box of receivers. In the water column these measure the pressure field of the incident and scattered waves. In the bottom we process the displacement response at the receivers to give time series of compressional and shear energy density. When the shear modulus vanishes the compressional energy density equals the pressure.

Inside the numerical scattering chamber we compute solutions to the elastic wave equation by the finite difference method [1,3,4,6]. The scheme is based on second order, centered finite differences of the elastodynamic equations expressed in particle displacement. The initial conditions are that particle displacement and particle velocity are zero. The boundary conditions are the Sommerfeld radiation condition on all sides. This is implemented by solving the telegraph equation in a region surrounding the chamber. The parameters for the telegraph equation are selected to minimize artificial reflections. We treat attenuation within the chamber using the Padé approximant method [7].

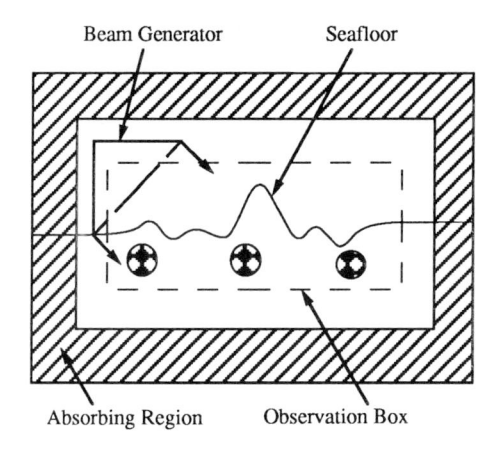

Figure 1: The numerical scattering chamber in two dimensional Cartesian co-ordinates is totally surrounded by an anechoic (absorbing) region. Inside the chamber one can 'place' arbitrary depth and range-dependent structure including surface and volume scatterers. These are defined by specifying compressional and shear velocity, compressional and shear attenuation and density on a grid with from ten to fifty grid points per wavelength. The structure is insonified by a beam from a beam generator which is transparent to the scattered energy.

3. The Implementation of Gaussian Beams

Many treatments of scattering are based on the infinite plane wave as the incident field. However in a time-space formulation as used here the infinite plane wave is very awkward. Since infinite plane waves at non-normal incidence to the seafloor interact with the seafloor at all times, the start-up field (implemented as a time dependent boundary condition) would have to consist of semi-infinite reflected and transmitted plane waves as well as a semi-infinite incident wave. (At sub-critical grazing incidence the transmitted plane wave would become an evanescent wave in the seafloor.) To avoid this problem we can truncate the lower edge of the incident plane wave so that it doesn't hit the seafloor until after the intial time for the computation. By truncating the top edge of the plane wave as well we can take advantage of a great deal of analysis that has been carried out for beams.

Point sources are easily implemented in time domain finite difference schemes but it is awkward obtaining a narrow angle, low grazing angle incident field from a point source without interference from head waves and diving waves. Arrays of point sources generate realistic beams but they always have side lobes which will affect the observed scattered field. In a time-space formulation attenuating the side lobe energy will distort the main lobe.

For the Gaussian beam the infinite plane wave is weighted along the wavefront with a Gaussian profile. However all beams spread because of energy leaking out the side of the beam. For a given propagation distance there is specified initial beamwidth which will give the narrowest beam over the whole path [8]. As path length increases the mimimum beam width increases. In bottom interaction studies the footprint on the seafloor increases with smaller grazing angle and the distance over which a finite beam must propagate to cover this footprint also increases. So the narrowest possible beam width increases with decreasing grazing angle.

For example, we consider a Gaussian beam propagating in a straight, constant width channel in homogeneous media. The channel corresponds to an idealized, finite-width 'ray' and we can imagine the channel reflecting and refracting at a sharp interface according to Snell's law. (On refraction at a sharp interface the channel width would change.) At fifteen degrees grazing incidence we take a channel width of 18.6λ (where λ is the wavelength in water, with a velocity of 1.5k/s, at the peak frequency of the source in pressure) and its footprint on the horizontal plane is 72λ. The narrowest Gaussian beam that can propagate across the footprint starts with a half-width of 4.7λ and is down 35dB on the edges of the channel. After propagating across the footprint the half-width is 6.6λ and the beam is down only 20dB on the edges of the channel.

For a low grazing angle problem there can be very many wavelength size scatterers within the footprint of the narrowest acceptable beam. Note that, although this issue is clearly evident in the time-space domain finite difference solutions, the physical constraints are placed by the wave equation and the physics of sound propagation.

4. A Rough, Basaltic Seafloor

As an example of the numerical scattering chamber we consider a rough, basaltic seafloor. The relief is about 10λ over a range of 70λ. For a peak frequency of the source of 250Hz this corresponds to 60m of relief over 420m. In this example the bottom is completely homogeneous. Figure 2 shows the compressional and shear wave fields in the scattering chamber 40 periods (0.16sec) after the beam initiated contact with the seafloor. In the absence of the seafloor this frame would show a nearly vertical planar wavefront, 3λ wide, at about 35λ range.

The weak energy in the compressional wave field between 42 and 60λ includes the head wave and energy that was forward scattered as compressional waves in the bottom. In the shear wave frame there is strong energy between 25 and 42λ. The right-most energy here is converted shear waves but the circular wavefronts are scattered shear waves propagating downward and backward (to the left). Packets of high energy near the interface in both compressional and shear frames are interface waves generated by secondary scattering (Huygen's Principle) from roughness elements on the interface. The backscattered shear and interface waves convert back into compressional waves in the water (seen in the compressional frame to the left of 30λ) and contribute to the low angle backscattered field.

At the same time that the shapshots in Figure 2 were generated we acquired time series of the pressure field on the left and upper edges of the scattering chamber. These were processed to show the incident beam (from the left at fifteen degrees) and the directly backscattered beam (to the left at fifteen degrees). The backscattered beam is much less coherent and is spread out much more in time than the incident

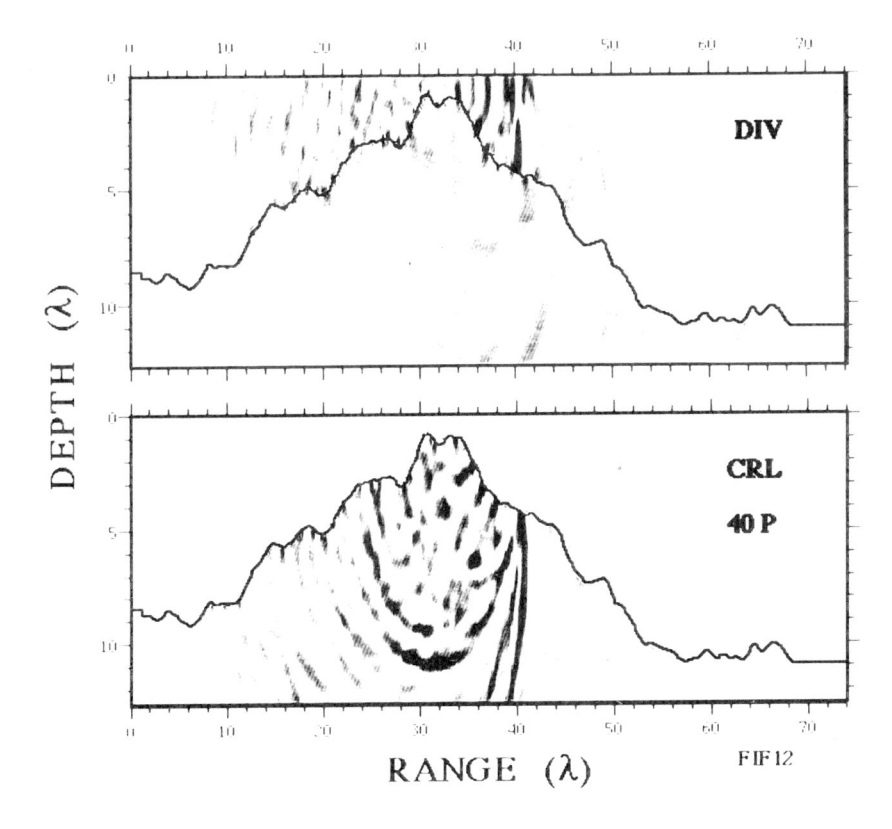

Figure 2: Compressional (DIV) and shear (CRL) wavefronts are shown 40 periods after a pulse-beam at fifteen degree grazing incidence hit a very rough, basaltic seafloor from the left. If the seafloor were not present the beam would have propagated cleanly to about 35λ. Energy to the left of 35λ is caused by backscatter and reverberation from the rough seafloor. Strong backscattered shear and interface waves can be seen in the shear wave frame. These convert back into compressional waves in the water and are a significant mechanism for strong, low angle backscatter. In this model the water has a compressional wave speed of 1.5k/s and a density of 1.0gm/cc. The bottom has compressional and shear wave speeds of 3.0 and 1.73k/s respectively and a density of 1.7gm/cc. In this particular model there is no intrinsic attenuation. Studies have shown that compressional and shear Q's as low as 20 reduce the backscatter coefficient less than 1dB.

beam. However a backscatter coefficient can be computed based on the ratio of the root-mean-square energy of the two beams. For the model shown here the backscatter coefficient is -17dB. This value is typical of backscatter coefficients for low grazing angles at actual seafloors.

5. Summary and Conclusions

The numerical scattering chamber based on the finite difference synthetic seismogram technique can be used for studying the backscatter of beams at small grazing angles from rough and laterally heterogeneous seafloors. Based on physical constraints the smallest footprint on the seafloor that can be considered in a time-space formulation for a beam at fifteen degrees grazing incidence is seventy-two wavelengths across. Many wavelength-size scatterers can fit within this footprint.

At a very rough, basaltic seafloor conversion of energy from compressional waves in the water to shear and interface waves at the seafloor is an important physical mechanism for generating strong backscatter. Specular reflection from facets is not necessary to create strong backscatter. An unsedimented rough basalt can have a backscatter coefficient at fifteen degrees grazing incidence of -17dB.

Acknowledgements

The work described in this paper was supported by the Office of Naval Research under Contract Number N00014-90-J-1493. Woods Hole Oceanographic Institution Contribution Number 8052.

References

[1] Dougherty, M.E. and Stephen, R.A. (1988) "Seismic energy partitioning and scattering in laterally heterogeneous ocean crust", J. Pure and Applied Geophysics **128**, 195-229.
[2] Dougherty, M.E. and Stephen, R.A. (1991) "Seismo/acoustic propagation through rough seafloors", J. Acoust. Soc. Am. **90**, 2637-2651.
[3] Stephen, R.A. (1988) "A review of finite difference methods for seismo-acoustic problems at the sea floor", Reviews of Geophysics **26**, 445-458.
[4] Stephen, R.A. (1990) "Solutions to range-dependent benchmark problems by the finite difference method", J. Acoust. Soc. Am. **87**, 1527-1534.
[5] Burns, D.R. and Stephen, R.A. (1990) "Three dimensional numerical modelling of geoacoustic scattering from seafloor topography", J. Acoust. Soc. Am. **88**, 2338-2345.
[6] Virieux, J. (1986) "P-SV wave propagation in heterogeneous media: Velocity-stress finite difference method", Geophysics **51**, 889-901.
[7] Day, S.M. and Minster, J.B. (1984) "Numerical simulation of attenuated wavefields using the Padé approximant method", Geophys. J.R. astr. Soc., **78**, 105-118.
[8] Červený, V., Popov, M.M. and Pšenčík, I. (1982) "Computation of wave fields in inhomogeneous media - Gaussian beam approach", Geophys. J.R. astr. Soc., **70**, 109-128.

Section 5

Low Frequency Measurements

LOW-FREQUENCY DIRECT-PATH SURFACE AND VOLUME SCATTERING MEASURED USING NARROWBAND AND BROADBAND PULSES

Roger C. Gauss, Joesph M. Fialkowski[†], and Raymond J. Soukup

U.S. Naval Research Laboratory
Washington, DC 20375-5000

[†]*Planning Systems Incorporated*
McLean, VA 22102

ABSTRACT. The spectral and temporal characters of surface and volume reverberation were interrogated during a series of recent deep-water, direct-path experiments using both narrowband (CW) and broadband (HFM,PRN) transmissions. Measures of low-frequency backscattering were obtained over a range of wind speeds (12 to 27 kts) for very low grazing-angle data: Doppler-sensitive waveforms (CW,PRN) were used to examine the spectral character (frequency shift(s), spread) and high-range resolution waveforms (HFM,PRN) were used to examine the spatial (range extent) and temporal (event lifetimes) characters. The major conclusion is that subsurface bubbles are inferred to be primarily responsible for the observed elevation of the empirical Ogden-Erskine/Chapman-Harris surface scattering strengths over predictions based on air-sea-interface scattering theory.

INTRODUCTION

In April 1990 in the Gulf of Alaska, during the 4th experiment of the Critical Sea Test (CST) Program, 12 1-hr direct-path measurements used short-duration (1-2 s), narrowband and broadband waveforms to assess the Doppler, spatial and temporal characters of very low-grazing-angle surface and volume scatter as functions of frequency (200-1000 Hz) and environmental conditions. The motivation for these tests was to help determine the source(s) of the enhanced backscattering strength (over theoretical predictions) vs. sea state seen in broadband SUS data [2] and, consequently, to help choose between the various hypotheses advanced to explain the anomalous scattering.

MEASUREMENT and PROCESSING DESCRIPTIONS

The experimental configuration consisted of a vertical array (deployed at midships through a center well) of 10 low-frequency (LF) and 10 high-frequency (HF) high-level sources and a high-resolution, horizontal receiving array cut for 1000 Hz towed 1 km aft of the ship. Both were towed at a nominal depth of 150 m at 3 kts. For these data, LF was an 80-Hz band centered at 250 Hz and HF was 170-Hz band centered at 915 Hz. This mildly bistatic run geometry limited both the duration of the waveforms (1.25 to 2.4 s) and the range of grazing angles (1.5 to 3.5 deg). (The upward-refracting sound-speed profile permitted very low grazing-angle backscatter.) Transmissions used CW, HFM (Hyperbolic Frequency Modulation) and PRN (Pseudo-Random Noise) waveforms. Signal durations were designed to be capable of resolving mean zero-Doppler components (bubbles, fish) from Bragg components (air-sea interface scattering). In general, waveforms were transmitted 10 times with the pulse-repetition intervals (PRI) of 70-90 s at LF and 45 s at HF. Table 1 summarizes the data collected and the environmental conditions during each run. The runs are ordered by a "modified wind speed," included to give a more accurate measure of the sea-state conditions than the instantaneous wind speed.

TABLE 1. Run-by-run summary of environmental conditions.

	Run	\<Wind Speed\> True (kts)	Std dev	Exp-wtd	\<Significant Waveheight\> TSK (m)	W-Buoy	Peak Wave Pd (s)	Ampl (m-sq/Hz)	Peak Orb Vel (m/s)	Rel Wind Dir (deg R)	COMEX	Dur (hr)
LF	51F	28.7	3.2	26.9	6.3	-	-	-	2.0	30	108 0100 Z	1.75
	8A	26.2	2.7	25.8	5.2	5.1	10	30	1.6	0	95 1000	1.5
	8AR	16.8	1.6	19.1	4.2	3.7	10	20	1.3	90-160	96 1840	1.67
	3A	15.7	1.4	18.6	3.2	3.0	8.4	10	1.2	340	97 0100	2.0
	50B	18.5	1.2	14.9	2.3	2.1	9.1/5	4/1	0.8	245	107 1035	1.0
	48B	12.1	0.9	11.8	3.5	2.6	8.4	1	1.3	235	106 2210	2.0
HF	51C	22.4	2.4	24.2	5.3	3.5	9.1	20	1.8	25	107 2120	1.25
	53B	20.6	2.2	24.1	3.9	-	-	-	1.2	30	108 1200	1.25
	50F	24.5	2.4	22.9	3.2	3.6	14/7	6/15	0.7/1.4	220	107 1615	1.25
	6B	20.9	2.4	21.8	3.5	4.2	10	20	1.1	280	96 1210	0.67
	60B	18.7	1.9	19.1	3.6	3.4	10	30	1.1	3	110 2200	1.2
	36B	19.2	1.9	16.9	1.6	2.1	8.4	5	0.6	0	103 1425	1.0

Subsets of hydrophone data (to avoid near-field effects) were filtered and then frequency-domain beamformed: at HF, 50 phones (at 0.75-m spacing) were used to form 51 beams spanning 180 deg; at LF, 34 phones (at 2.25-m spacing) were used to form 35 beams. These data were subsequently processed using either spectrum analyzers (with 90-% overlap) for pulsed-CW waveforms or matched filters for broadband signals. The Doppler-sensitive data were (bistatically) corrected for ownship motion.

MEASUREMENT RESULTS

Overall, strong evidence of volume scattering was observed at LF and strong evidence of surface scattering at HF.

1. Reverberation Level

Figure 1 shows representative LF and HF HFM histograms, where the unnormalized correlator-squared power values have been accumulated over all pings from all beams and appropriate time samples. The left figure displays the highest (Run 51F; 27 kts) and lowest (Run 48B; 12 kts) sea-state LF cases as well as the peak volume-reverberation case (Run 3A; 19 kts). The right figure shows the corresponding cases for HF. Sea-state dependence was manifested by a translation of the full

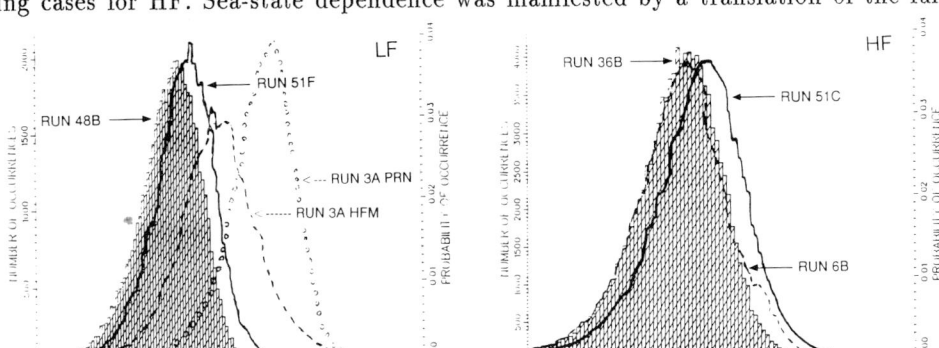

FIGURE 1. Histograms of LF HFM (210-245 Hz) and Run-3A PRN (255-290 Hz) [left] and HF HFM (870-905 Hz) [right] unnormalized correlator output.

range of power values, as opposed to an increase in just the tails of the distributions, implying that the measured backscatter is generally not due to scattering off a few, dominant bubble clouds, but to the aggregate of ensonified bubble clouds. Volume scatterers were temporally patchy throughout CST 4: e.g., the differences in mean levels of the Run 48B and 51F distributions were much smaller than predicted by Ogden-Erskine [2], evidence of a "volume-reverberation floor" in Run 48B; while, in Run 3A, the volume scatterers were locally abundant and are clearly seen by not only the strong increase in the HFM mean level but by the relative levels of the (simultaneously transmitted) PRN to that of the HFM, consistent with the PRN frequencies being closer to the observed volume-scatter resonance frequencies [1]. At HF, the range of wind speeds (17-24 kts) was smaller but, as with the scattering strengths derived from the CW data (not shown), the relative levels of the Run-36B and -51C HFM data were consistent with Ogden-Erskine values.

2. Spectral Character

Typical observed LF and HF mean spectra are shown in Figure 2, where we have averaged over selected beams and ten transmissions. With all runs, the spectral peak frequency had a mean equal to the transmitted frequency. Bragg peaks were rarely observable in the data (only at 210 and 220 Hz) even at the lowest wind speed (12 kts), presumably obscured by the spreading from the always dominant zero-Doppler component (due to an unknown combination of microbubble and biologic sources). As the primary surface mechanisms believed to control spectral character (spread and shift) are proportional to frequency (orbital velocity) or the square root of the frequency (Bragg scattering), one expects both the spread and any spectral-peak shifting (from zero Doppler) to increase with frequency. The orbital velocity is also proportional to the significant waveheight, so that at a given frequency, the spread should also increase with sea state. Table 2 summarizes statistics of both spectral spreads

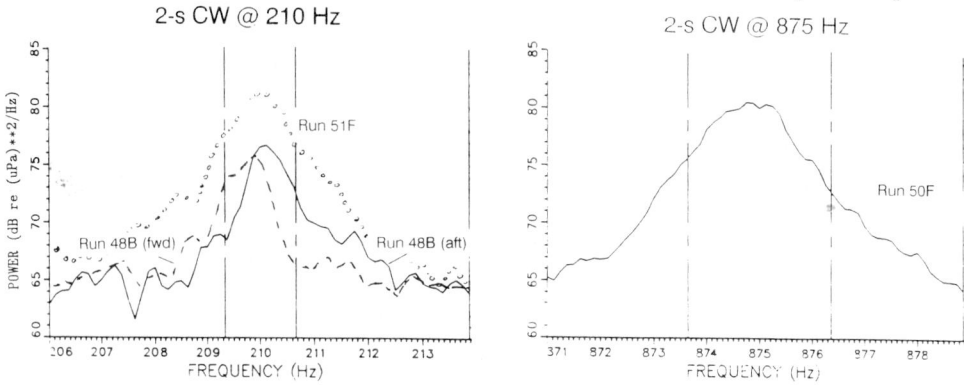

FIGURE 2. Typical LF [left] and HF [right] CW spectra. (Vertical lines correspond to up/down Bragg frequencies.)

and peak Doppler shifts for each run using 1.5-s CWs at LF and 1.25-s CWs at HF. (Spreading loss $\equiv$ 6-dB-down area over that of an ideal spectrum.) No sea-state or azimuthal dependencies of either the spectral spread or peak-Doppler shift statistics were seen at LF or HF, though the presence of biologics makes quantifying the influence of the surface/bubbles difficult at LF. (The systematic azimuthal dependence of the HF spectral-peak are believed due to the imprecise ship speed values used in correcting for ownship motion.) At HF, where we have some confidence that we are seeing surface scattering, the spreading shows little wind-speed dependence compared to values predicted from orbital-velocity calculations (Table 2).

TABLE 2. Statistics of spectral character computed vs. beam set: A = aft, B = broadside, F = forward.

```
                      MEAN SPREAD (Hz)                    MEAN SPREADING    SPECTRAL PEAK (Hz)
        PREDICTED   3-dB           6-dB          10-dB      LOSS (dB)        MEAN         STD DEV
    RUN  SPREAD    A   B   F     A   B   F    A   B   F    A   B   F       A   B   F    A   B   F
    ---  ||------|----------- |------------ |------------||------------||--------------|------------||
LF  51F ||  1.4  |0.9 0.9 0.8 |1.4 1.3 1.3  |2.4 1.9 1.6 ||1.9 1.5 1.2 || 0.1  0.0 -0.1 |0.3 0.3 0.3||
    8A  ||  1.0  |0.8 0.8 0.8 |1.1 1.0 1.1  |1.6 1.6 1.6 ||1.3 1.0 1.2 || 0.1 -0.1 -0.1 |0.3 0.3 0.3||
    8AR ||  0.8  |0.8 0.8 0.6 |1.0 1.1 1.0  |1.5 1.5 1.4 ||1.0 1.1 0.7 ||-0.1  0.0 -0.1 |0.3 0.3 0.3||
    3A  ||  0.8  |0.6 0.8  -  |1.0 1.1  -   |1.5 1.6  -  ||0.8 1.0  -  || 0.0 -0.1  -   |0.3 0.3  - ||
    50B ||  0.6  |0.8 0.9 0.8 |1.1 1.3 1.1  |1.6 1.8 1.8 ||1.1 1.8 1.3 || 0.0 -0.1 -0.1 |0.3 0.3 0.3||
    48B ||  0.8  |0.8 0.8 0.8 |1.4 1.0 1.1  |1.5 1.4 1.5 ||0.8 0.8 0.9 || 0.1  0.0 -0.1 |0.3 0.3 0.3||
    IDEAL SIGNAL  |    0.6     |    0.8      |    1.0     ||    0       ||      0        |    0      ||

HF  51C ||  4.4  |1.0 1.0 0.9 |1.4 1.4 1.4  |2.3 2.3 1.9 ||1.5 1.3 1.1 || 0.3  0.1 -0.1 |0.4 0.5 0.5||
    53B ||  3.0  |1.0 1.0 1.0 |1.5 1.5 1.4  |2.3 2.5 2.1 ||1.5 1.5 1.4 || 0.5  0.1 -0.2 |0.5 0.5 0.5||
    50F ||  3.4  |0.8 0.9 0.9 |1.3 1.4 1.3  |2.0 2.4 2.1 ||0.8 1.2 0.8 || 0.4 -0.1 -0.2 |0.4 0.4 0.5||
    6B  ||  2.8  |0.9 0.9 0.9 |1.3 1.4 1.4  |2.0 2.1 2.0 ||1.1 1.1 1.1 || 0.1 -0.1 -0.1 |0.4 0.5 0.4||
    60B ||  2.8  |0.9 0.9 0.9 |1.4 1.4 1.4  |2.3 2.1 2.0 ||1.2 1.2 1.2 || 0.2  0.0 -0.5 |0.5 0.5 0.5||
    36B ||  1.4  |0.9 0.9 0.9 |1.4 1.3 1.3  |2.1 2.0 1.9 ||1.3 1.1 1.0 || 0.4 -0.1 -0.3 |0.4 0.4 0.4||
    IDEAL SIGNAL  |    0.7     |    1.0      |    1.2     ||    0       ||      0        |    0      ||
```

3. Spatial and Temporal Characters

In CST 4, (energy-)normalized match-filtering of reverberation following broadband transmissions often revealed isolated (< 10 m) point scatterers, the significant sources of which appeared to be volume scatterers (salmon) at LF and surface scatterers (presumably bubbles) at HF as will be argued in the next section. Figures 3a and c respectively show some representative normalized matched-filter output at LF and HF during the highest sea states - seen are a few prominent returns atop the generally noise-like background - and from LF Run 3A (Figure 3b), when volume scatterers were hypothesized to be dominant - seen are many strong returns.

An algorithm to compute the lifetimes of the strong features seen in Figures 3a-c was developed and applied (using a 0.2 threshold) to successive broadband pings (Figures 3d-f, respectively). The height of the bar gives the value of the correlation

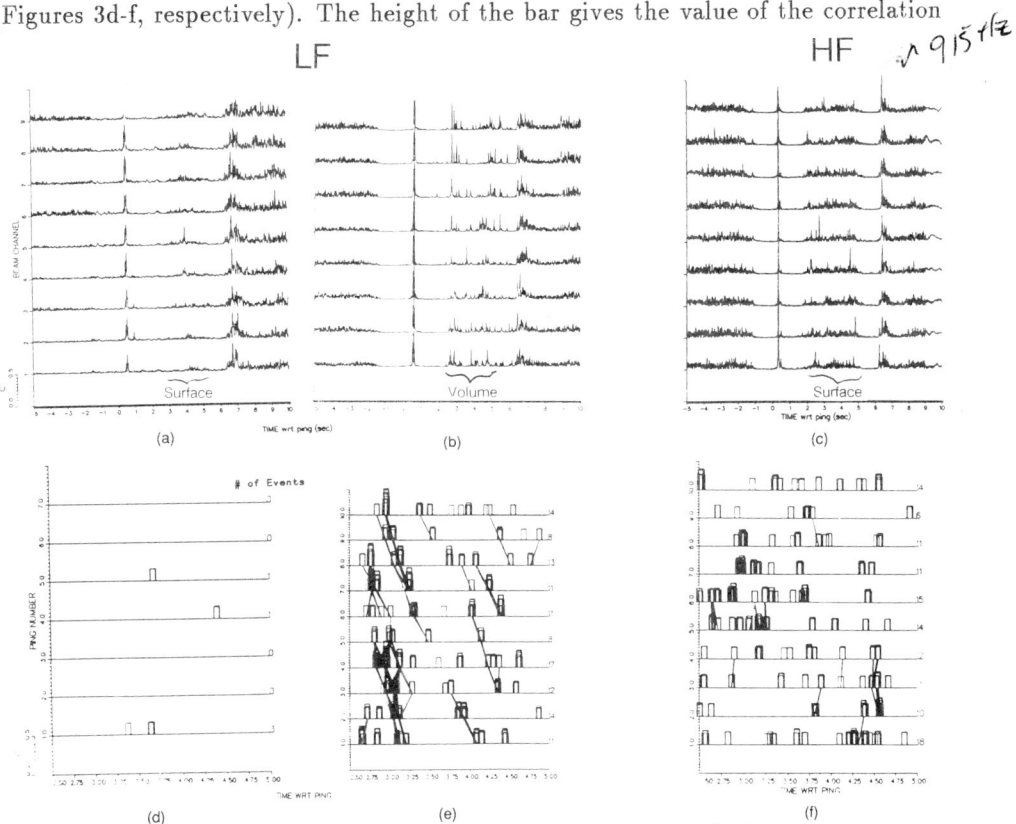

FIGURE 3. (a)-(c) Normalized HFM correlator output (single ping).
(d)-(f) Feature lifetimes for data from Runs displayed in (a)-(c).

coefficient. The results from the specified beams - here, all beams not looking at the ship - are combined. The connections between bars represent recurrences of the same phenomena for successive transmissions. For two bars to be connected, they must have been obtained from the *same* beam and a time window that accounts for ship motion and scatter motions (up to 1.6 m/s). In the high sea-state cases, we see a relatively large number of exceedences for the HF signals compared to the corresponding LF signals, consistent with bubbles as driving surface backscatter (when whitecaps are present), since bubbles become increasingly significant scatterers with increasing frequency. The relatively small number of connections in the HF data suggest that most clouds that produce prominent returns have lifetimes less than 45 s (the time between pulses). At LF, we have hypothesized that volume scatterers were sometimes dominant, such as in Run 3A (Figure 3), where HFM data exhibited long lifetimes. Their parallel slopes are also consistent with a typical salmon cruising speed (1.5 m/s) and direction. Quantification of the probability of survival between pings versus run is provided in Table 3. The number of features in this table also is consistent with an earlier hypothesis that the PRN saw more volume scatterers than the HFM at LF and, moreover, more surface scatterers were seen by both signals at HF than at LF (consistent with bubbles as the mechanism). Examination of the number of features or probability of connection at HF showed little dependence on sea state (though the range of sea conditions was not great). (Because of spatial resolution limitations, these probabilities are upper bounds.)

TABLE 3. Normalized survival-function statistics for simultaneous HFM/PRN transmissions. P(1) = probability of survival to next ping; P(2) = probability of 3-ping survival.

	RUN	HFM Ave # of features per ping	P(1)	P(2)	Zero-Doppler Ave # of features per ping	P(1)	P(2)	PRN Estimated speed (m/s)	Noise Ave # of features per ping	P(1)	P(2)	Ping rate (s)
LF	51F	0.2	0		8	.10		-	<1	0	0	70 s
	8AR	36	.30		84	.40		1.0	<1	0	0	90
	3A	31	.25		96	.40		0.8	<1	0	0	90
	50B	4	.05		15	.15		-	<1	0	0	70
	48B	7	.15		27	.25		0.9	<1	0	0	70
HF	51C	21	.10	.05	4	.05	0	-	1.2	<.20	0	45
	53B	12	.05	<.05	1	0	0	-	1.7	0	0	45
	50F	24	.25	.05	5	.15	0	-	1.8	0	0	45
	60B	26	.20	.10	10	.15	<.05	1.0	2.7	<.10	0	45
	36B	19	.10	<.05	4	0	0	-	1.3	0	0	45

CONCLUSIONS

Our analysis revealed the following: <u>Spectral character.</u> Spectral maxima consistently at zero Doppler. Spectral spreading showed no wind-speed dependence, contrary to orbital velocity calculations. <u>Reverberation level.</u> Sea-state dependence was seen by a translation of the full range of power values, implying not just a few bubble clouds but the aggregrate contribute to the observed backscatter. <u>Spatial character.</u> Normalized matched-filter data revealed strong, isolated (< 10 m) features. <u>Temporal character.</u> Surface scatterer lifetimes were generally < 45 s. Volume scatterers were very patchy in intensity from run to run. <u>Azimuthal dependence.</u> No systematic dependence on the direction of the seas for *any* phenomena observed.

<u>ACKNOWLEDGEMENTS.</u> This work was supported by SPAWAR, PMW 183-3 and by the Office of Naval Technology, Code 234.

REFERENCES

1. J. Monti & C. Hayek, "Low-Frequency Volume Scatter and Distant Reverberation: 200 - 1500 Hz," SACLANT Ocean Reverberation Symposium, May 1992.

2. P. Ogden & F. Erskine, "Low-Frequency Surface and Bottom Scattering Strengths Measured Using SUS Charges," Ocean Reverb. Sym., May 1992.

LOW-FREQUENCY SURFACE AND BOTTOM SCATTERING STRENGTHS MEASURED USING SUS CHARGES

Peter M. Ogden and Fred T. Erskine

Acoustics Division
U.S. Naval Research Laboratory
Washington, DC 20375-5000

ABSTRACT During a series of at-sea experiments, we have measured low-frequency (70–1000 Hz) surface and bottom backscattering strengths for a variety of sea states and bottom types using broadband SUS charges as sources. The measured surface scattering strengths cover a range of grazing angles from 30 deg down to about 5 deg and a range of wind speeds from 1.5 to 14 m/s. We discuss the dependence of the surface scattering strengths on frequency, grazing angle, and some environmental parameters. Bottom scattering strengths are presented for grazing angles from 50 deg down to about 27 deg and for a variety of bottom types. We discuss the dependence of the bottom scattering results on frequency, grazing angle, and various geophysical parameters.

1. INTRODUCTION AND DESCRIPTION OF EXPERIMENTS

Scattering strengths from ocean boundaries are required inputs for models that predict acoustic reverberation for given sets of acoustic and environmental parameters, as well as for simpler, sonar-equation calculations. At present, most models that predict reverberation at 1 kHz and below make use of long-established empirical scattering strength formulas such as Chapman-Harris for surface scattering (Chapman and Harris 1962) and Mackenzie for bottom scattering (Mackenzie 1961). The conditions under which these formulas are appropriate, however, are not well known.

One of the goals of the Critical Sea Test (CST) series of at-sea experiments, sponsored by the U.S. Space and Naval Warfare Systems Command, has been to measure boundary scattering strengths at frequencies of 1 kHz and below and investigate their dependence on various acoustical and environmental properties. In this paper we discuss surface and bottom scattering strength results from the first five CST tests: CST-1 (August 1988) in the Norwegian Sea over the Norway Basin and the Aegir Ridge, CST-2 (April 1989) in the Icelandic Basin and Hatton Bank areas, CST-3 (August 1989) in the western Atlantic over the Bermuda Rise and Hatteras Abyssal Plain, CST-4 (April 1990) in the Gulf of Alaska over the Aleutian Abyssal Plain, and CST-5 (June 1991) in the eastern Mediterranean over the Ionian Abyssal Plain and the Malta Ridge.

During these first five CST tests, we have conducted a total of 40 direct-path scattering strength measurements using SUS charges as broadband sources, with each measurement using 10 to 20 SUS charges. For the surface scattering data sets, wind speeds ranged from 1.5 to 14 m/s, with sea states ranging from 0 to 4.5. Most of the bottom scattering data sets have been collected from flat-bottomed, thickly-sedimented areas, though some data have been obtained from regions with significant bathymetric features.

In each scattering test run, SUS charges with nominal detonation depths of 457 m were deployed from the aft end of a research ship with drag plates attached to slow their descent. By the time the charges reached detonation depth, they were directly beneath a horizontal line array towed by the same ship (normally at a depth between 100 and 200 m). Individual shots were processed into 4 Hz bands and then beamformed. The shots were combined into average reverberation curves at analysis frequencies that were harmonics of the bubble frequency. A raytrace program was used to calculate the geometric aspects of the run such as ensonified areas and grazing angles. Finally, the average reverberation curves were combined with source levels and geometric parameters to solve the sonar equation for scattering strength as a function of frequency and grazing angle.

2. SURFACE SCATTER RESULTS

Figures 1 and 2 give surface backscattering strength results for a portion of the SUS data that covers the range of sea conditions encountered during the CST measurements. These figures show measured scattering strengths plotted against mean grazing angle, with the average wind speed over the course of the run (corrected to a height of 19.5 m) shown in the legend. Figure 1 shows scattering strengths averaged over a 20-Hz band centered at 70 Hz, while Fig. 2 shows results for the same bandwidth centered at 930 Hz. Each figure also presents three reference curves. Two are Chapman-Harris (CH) curves for the lowest and highest wind speeds represented in the data set, while the third shows the prediction of first-order perturbation theory for the case of a simple k^{-4} wave spectrum (where k is the wavenumber). Perturbation theory is used to predict the levels to be expected from scattering from a rough air-sea interface (Thorsos 1990). In general, perturbation theory predicts only a small dependence of scattering strength on frequency and wind speed, in contrast to the empirical Chapman-Harris formula.

Figure 1 shows that at 70 Hz, the surface scattering strengths are clustered at or below the perturbation theory level, with little wind speed dependence. The measured values are in disagreement with Chapman-Harris, even at the highest wind speeds. At this frequency, perturbation theory offers a better description of the scattering strength data. At 930 Hz, however, a wind speed dependence similar to Chapman-Harris is seen, particularly for the higher wind speeds. Note that even at 930 Hz, the lower wind speed cases fall close to the levels predicted by perturbation theory.

Examination of such plots at a number of frequencies below 1 kHz has led to several conclusions about surface scattering. The most important of these are summarized in Fig. 3. We have identified three regimes in the parameter space defined by frequency and wind speed. For relatively calm seas at all frequencies, and for rougher seas at lower frequencies, the scattering strengths are reasonably well characterized by perturbation theory, indicating that air-sea interface scattering is the dominant mechanism. At higher wind speeds and higher frequencies, the Chapman-Harris empirical formula gives reasonable results. Since the frequency and wind speed dependence for Chapman-Harris is entirely different from that predicted by perturbation theory, it is a strong probability that another mechanism is responsible for scattering in this regime. This mechanism is most likely scattering from subsurface bubble clouds. In between the two regimes just noted, there is a transition region in which the two mechanisms are presumably competing; the result is a region in which the scattering levels are difficult to predict with any accuracy. Another conclusion is that the boundary between the air-sea interface scattering regime and the onset of dominance by bubble scattering is best described not

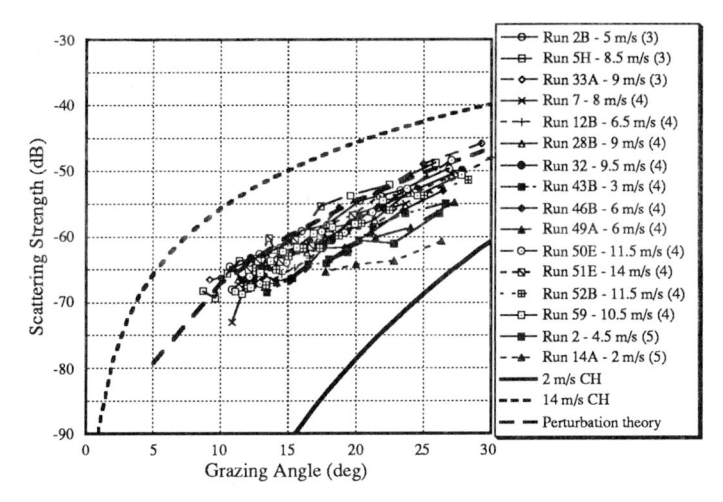

Figure 1. Surface scattering strength vs grazing angle at 70 Hz

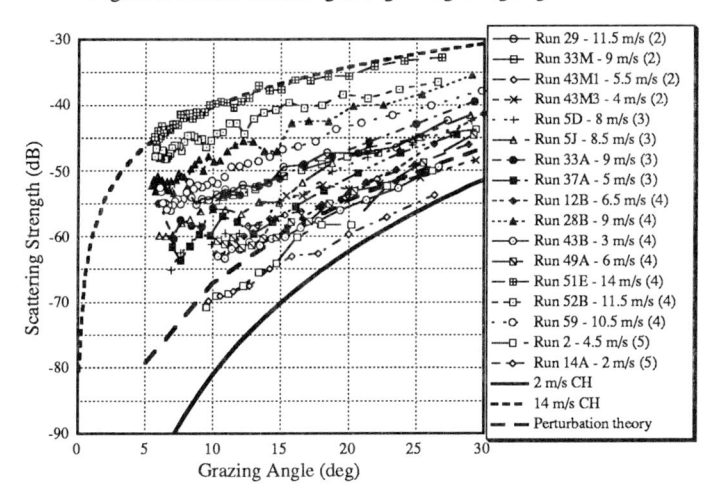

Figure 2. Surface scattering strength vs grazing angle at 930 Hz

by wind speed but by the transition from sea state 2 (few whitecaps) to sea state 3 (whitecaps common). This is part of a broader observation that instantaneous wind speed is a poor predictor of scattering levels; parameters such as sea state that incorporate wind history appear to do a better job of prediction.

3. BOTTOM SCATTER RESULTS

The same data that were used to produce the surface scattering results were also analyzed to get bottom backscattering strengths. The results reported here were obtained by using

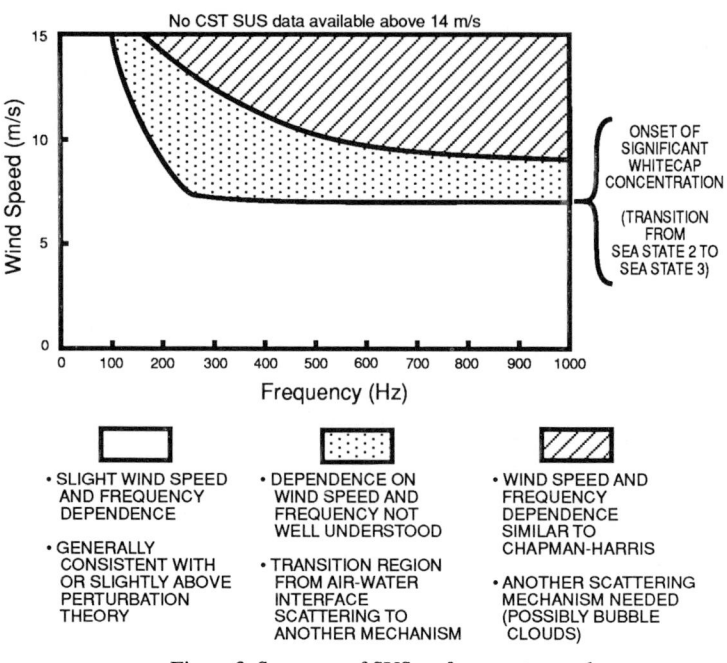

Figure 3. Summary of SUS surface scatter results

the reverberation from shortly after the first bottom fathometer return to the beginning of the second fathometer return. This translates into a lower grazing angle cutoff of about 25 to 27 deg. All results shown were obtained on the broadside beams.

Figure 4 shows the frequency dependence of the bottom scatter results at 30 deg grazing angle for typical runs in the various test sites. The Mackenzie scattering level at 30 deg is included in this figure as a reference. Virtually all the sites surveyed show scattering strengths that either decrease or stay relatively flat with increasing frequency. The Bermuda Rise and the Norwegian Basin results show the most frequency dependence, while the Aleutian and Ionian Abyssal Plain results show a decrease between 70 and 500 Hz, then are relatively flat. The exceptions occurred in the Hatteras Abyssal Plain. One site showed the only case we have found so far of an increase in scattering strength with increasing frequency, while another site showed a U-shaped frequency dependence. (One site in the Aleutian Abyssal Plain also showed the U-shaped dependence.) Since the Hatteras Abyssal Plain results represent single sites, most of the sites surveyed to date show a decrease in scattering strength with increasing frequency.

Figures 5 and 6 give examples of scattering strength results vs grazing angle at 130 and 940 Hz, respectively, for the deep-water, thickly-sedimented sites surveyed. At the highest grazing angles plotted, virtually all frequencies and sites show a rapid increase in scattering strength with increasing grazing angle, though the effect is more pronounced at low frequencies. This increase is probably caused by normal-incidence energy that penetrates into the subbottom and is backscattered. Since the grazing angles are

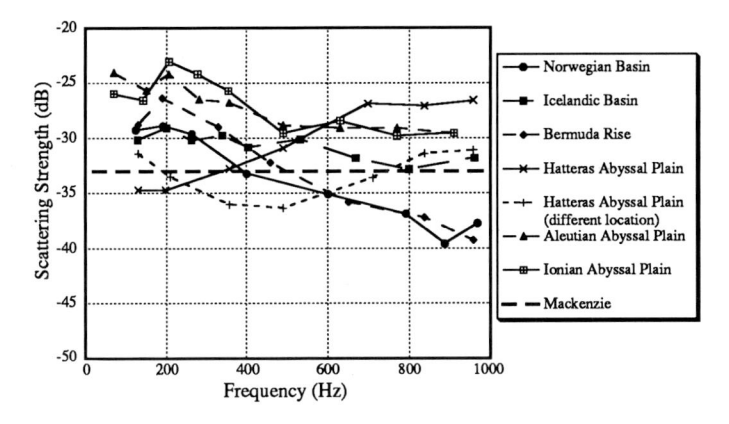

Figure 4. Frequency dependence of bottom backscatter at 30 deg grazing angle

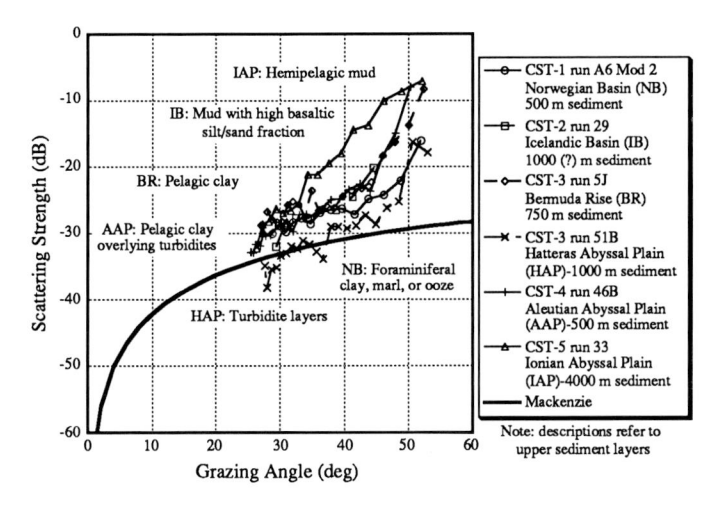

Figure 5. Bottom scattering strength vs grazing angle at 130 Hz

calculated assuming the scattering takes place at the water-sediment interface, the time-delayed subbottom returns are assigned to a lower grazing angle than they should be. The net effect is to enhance the higher grazing-angle returns above the true diffuse backscatter levels. However, examination of multiple beams and frequencies suggests that for grazing angles of 30 to 35 deg and below, almost all the subbottom contamination is gone and true diffuse backscatter is seen. Comparing Figs. 5 and 6 shows that at 130 Hz, the diffuse backscatter levels from all the sites surveyed are similar (with the exception of the single Hatteras Abyssal Plain site), while at 940 Hz there is a wide variety of backscattering strengths seen at 30 deg grazing angle. Thus the 940 Hz backscatter seems to be sensitive to the composition of the upper sediment layers, while the 130 Hz backscatter is mostly invariant with sediment type.

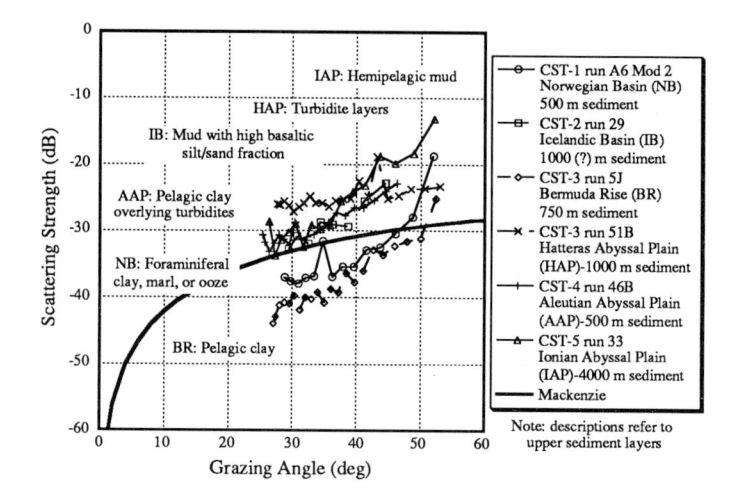

Figure 6. Bottom scattering strength vs grazing angle at 940 Hz

4. SUMMARY AND CONCLUSIONS

Direct-path measurements of surface and bottom scattering strengths using SUS charges as sources have been carried out in a variety of surface conditions and locations. For the surface, we have found evidence for two different scattering mechanisms—air-sea interface scattering and scattering from something like subsurface bubble clouds—that depend differently on frequency and wind speed. A frequency-wind speed diagram has been constructed that shows the areas of dominance for the mechanisms. For the bottom, we have observed that most of the sites surveyed exhibited scattering strengths which decreased with increasing frequency at 30 deg grazing angle. In addition, the scattering strengths near 1 kHz appear to be much more dependent on sediment type than are those close to 100 Hz, which show little sensitivity to sediment type.

5. REFERENCES

Chapman, R. P. and Harris, J. H. (1962) "Surface Backscattering Strengths Measured Using Explosive Sound Sources," *J. Acoust. Soc. Am.* **34**(10), 1592–1597.

Mackenzie, K. V. (1961) "Bottom Reverberation for 530- and 1030-cps Sound in Deep Water," *J. Acoust. Soc. Am.* **33**, 1498–1504.

Thorsos, E. I. (1990) "Acoustic scattering from a 'Pierson-Moskowitz' sea surface," *J. Acoust. Soc. Am.* **88**, 335–349.

UPSLOPE PROPAGATION DATA VERSUS TWO-WAY PE

Garry J. Heard, D.J. Thomson and G.H. Brooke
Defence Research Establishment Pacific
FMO Esquimalt, B.C., Canada V0S 1B0

ABSTRACT. Transmission losses measured in a range-dependent ocean environment are compared to two-way parabolic equation (PE) predictions. The field experiment employed SUS charges as the acoustic sources and a towed line array as the receiver in a region where the bottom shoals almost uniformly from 1500 to 500 m along a 14.6-km track. Simulations of upslope propagation indicate that two-way PE codes are more accurate than one-way PE codes. In addition, the two-way PE implementation has the capability of modelling backscatter.

1. INTRODUCTION

Two-way parabolic equation (PE) codes are currently being developed to model reverberation in range-dependent environments. This development was preceded by the realization that one-way PE codes do not conserve energy in range-dependent oceans. As a result, even the outgoing field given by one-way PE models can be inaccurate. In this paper, two-way PE predictions are compared with measured data obtained during an upslope propagation experiment off the west coast of Canada.

2. THE EXPERIMENT

In May 1987, a propagation experiment was carried out near the continental shelf off the west coast of Vancouver Island in a region where the water depth decreased from 1500 to 500 m. The site of the experiment, in situ sound speed profile and bathymetry along the propagation track are shown Fig. 1.

Shot waveforms were recorded onboard CFAV ENDEAVOUR using the Canadian Ocean Acoustic Measurement System (COAMS) [1] array towed at a nominal speed of 4 kt. A 1200-m sub-array with inter-element spacing of 38.1 m was used for the data collection. Operating conditions caused an array tilt of approximately 1° and changes in tow-speed caused the mean array depth to vary during the experiment. Figure 2(a) shows this depth variation as a function of time, while Fig. 2(b) shows the sub-array's shape mapped to range and depth co-ordinates at each shot reception.

The experiment was designed to give fine sampling of the transmission loss (TL) versus range. This was accomplished by towing the array radially away from the source ship, CSS PARIZEAU, which maintained a fixed position during the experiment and deployed 0.82 kg SUS charges at approximately 10-minute intervals. Each shot waveform allowed the determination of TL at 32 equispaced ranges. In this way, 352 measurements of TL were obtained along the 14.6-km track from just 11 shots.

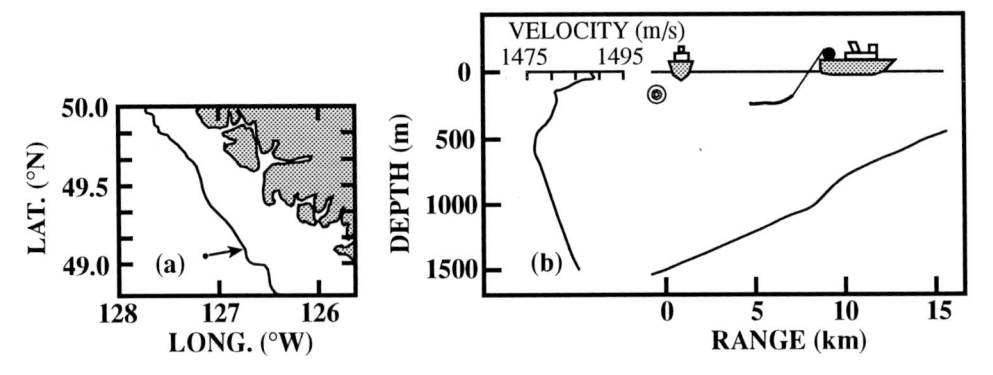

Figure 1: (a) The location of the experiment. (b) The bathymetry and sound speed profile along the receiver track.

3. TWO-WAY PE

One-way PE models have been used to interpret low-frequency acoustic propagation near coastal regions, where changing bathymetry provides a mechanism for mode-stripping [2] or enhancement [3] effects. Recently, Porter *et al.* [4] showed that one-way models do not conserve energy in such environments and that systematic errors in transmission loss can accumulate in the calculation of the outgoing field. At a recent workshop [5], two-way PE codes were introduced that make use of a single-scatter approximation to overcome the inherent limitation of one-way models. In addition to improving the accuracy of the outgoing field, the two-way PE codes allow the backscattered field to be determined [6].

Simulations of upslope propagation in an idealized wedge-shaped environment [7] indicate that one-way PE errors can reach 3 dB over 4 km at 25 Hz. This descrepancy is shown in Fig. 3(a), where both one-way [8] and two-way PE [9] predictions are compared to two-way coupled-mode results. It is evident that the two-way PE code accurately models the outgoing component of the field in this case. Upslope propagation for the experimental

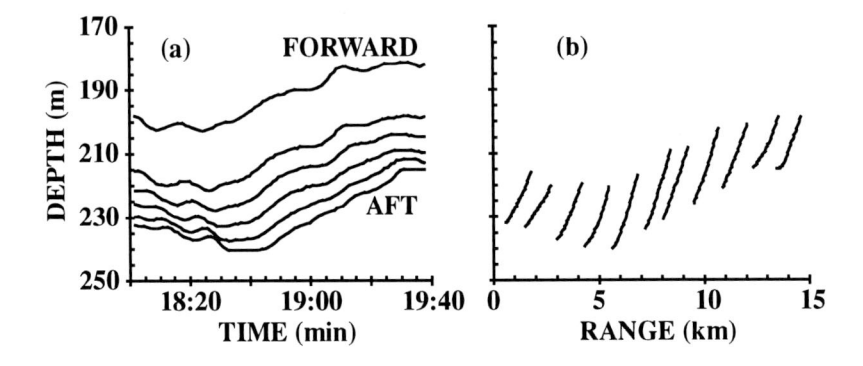

Figure 2: (a) Variation of array depth with time. (b) The array position at each shot time.

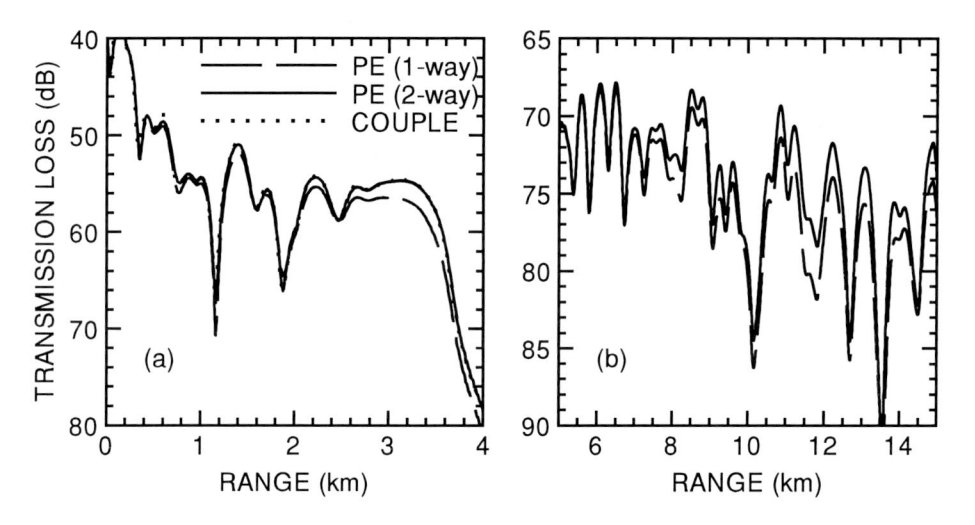

Figure 3: One-way and two-way TL comparisons at 25 Hz for (a) a benchmark lossy wedge [7] (b) the experimental configuration.

configuration is expected to be more complicated due to refraction by sound speed gradients in the medium [10].

A geoacoustic model for a location within 40 nmi of the upslope site was determined by Dosso and Chapman [3] by analyzing the refracted arrivals from deep shots. A different geoacoustic model for the same location was determined by Sen et al. [11] by analyzing the reflected signals from shallow shots. In this study, the features of both geoacoustic models were combined as shown in Fig. 4.

Based on these environmental inputs, the one-way and two-way PE predictions were carried out for the experimental track. Fig. 3(b) compares the predictions at 25 Hz for nominal source and receiver depths of 188 m and 221 m respectively. Although these TL curves exhibit more structure than those for the idealized wedge, it is apparent that one-way PE can overestimate the transmission losses by 3 dB over the bottom-interacting region of the slope.

4. FIELD DATA ANALYSIS

For each SUS charge, the time series received on each hydrophone group was processed to obtain the energy flux density in the 25-, 50- and 100-Hz 1/3-octave bands. For each hydrophone group, TL was calculated by subtracting the received energy flux density from the known source level [12]. TL determined this way can be assigned an RMS uncertainty of 2 dB based on system performance specifications and source level fluctuations.

A significant part of the data processing effort involved the reconstruction of the array location (depth and range) at the time of shot arrival. Relative ship positions were determined by interpolation between differential LORAN positions taken closest to these shot times. The combined shapes of the COAMS array and tow cable were used to estimate each source-to-hydrophone range offset. The uncertainties in these position estimates,

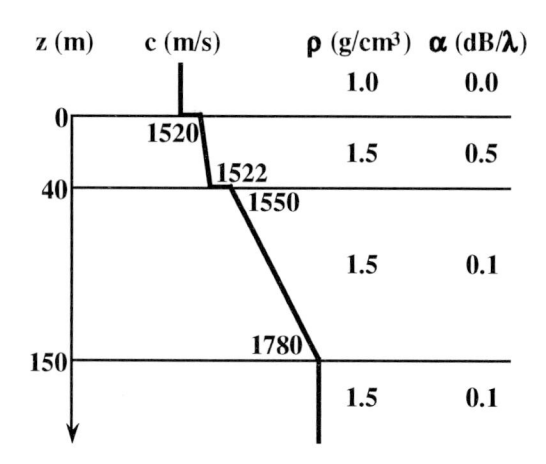

Figure 4: Composite geoacoustic model based on [3, 11].

including the inherent energy flux measurement times, lead to hydrophone group position uncertainties within 100 m in range and 1 m in depth.

5. COMPARISON OF DATA AND MODEL RESULTS

Simulations showed that variations in array depth along the propagation track can produce variations in TL of several dB at given ranges. To ameliorate this effect, the PE transmission losses were computed along specific sensor trajectories. Figure 5 shows measured TL compared to single-frequency PE predictions calculated along the estimated trajectories of the first (fore) and last (aft) hydrophone groups in the COAMS sub-array aperture. A 100-m range average has been applied to reduce the PE multipath structure. The jumps observed in the measured TL between adjacent shots arise from differences in depth between the fore and aft hydrophone groups due to array tilt (see Fig. 2). These jumps are also evident in the PE predictions. Overall, the comparison between theory and experiment is quite good and is consistent with the expected uncertainty in the empirical results. The predicted losses are systematically lower than the measured losses in the Lloyd Mirror region (0–4 km) for each band, possibly due to the cable losses in the array. The TL differences in the bottom-interacting region of the track (4–15 km) are probably due to our incomplete knowledge of the geoacoustic environment.

6. CONCLUSIONS

A fixed-source, moving-receiver propagation experiment was carried out in an upslope region off the west coast of Canada. Closely spaced estimates of TL along a 14.6-km track were determined by using a 1.2-km towed array to record broadband signals from 11 shots. The measured TL was modelled using a two-way PE code. Simulations indicate that two-way PE predictions are more accurate than one-way PE predictions in range-dependent, bottom-interacting environments. Good overall agreement between measured

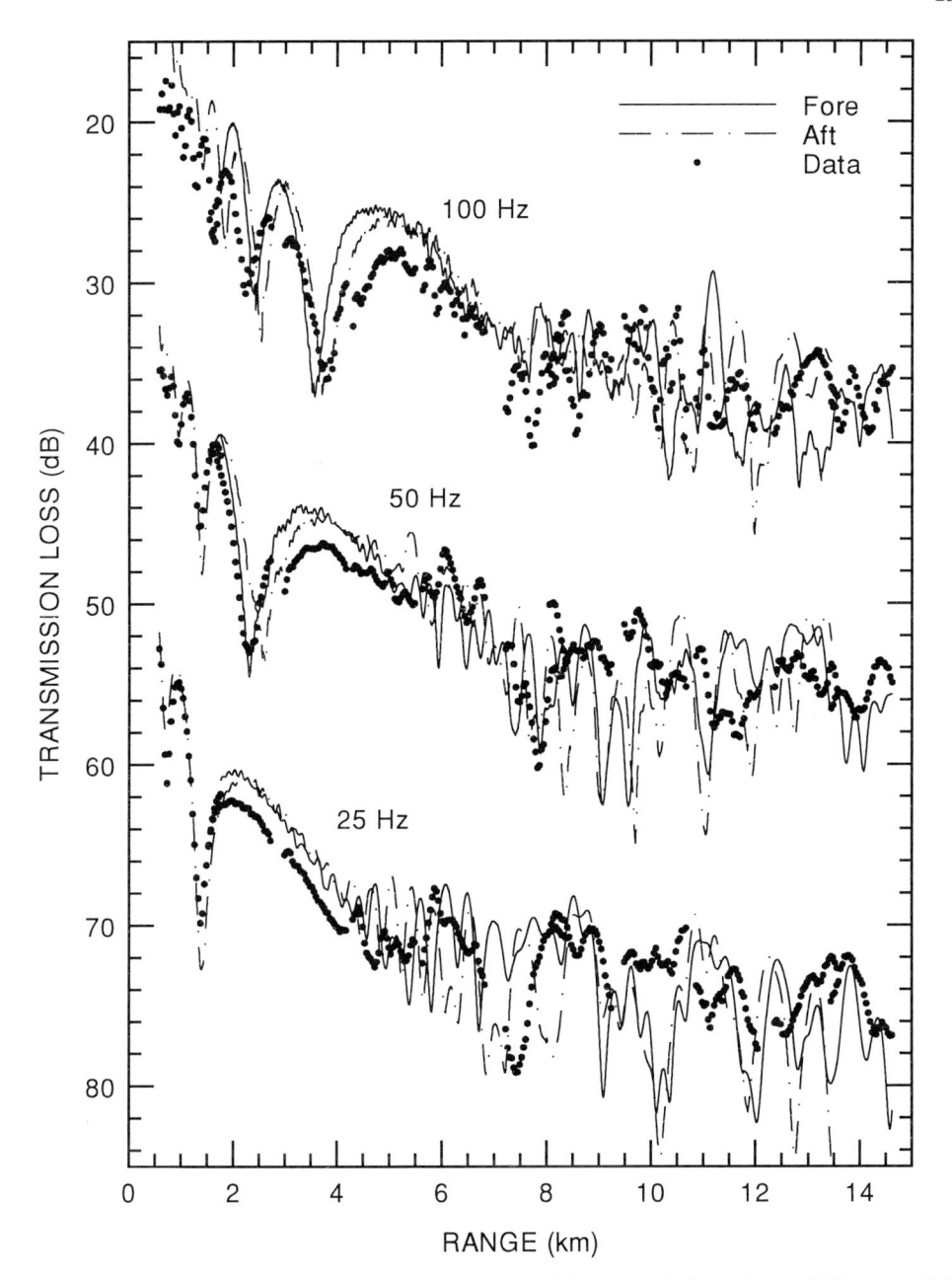

Figure 5: PE and measured TL values for 25, 50 (offset -20 dB), and 100 (offset -40 dB) Hz. PE curves were generated by following the trajectories of the fore and aft hydrophones in the COAMS sub-array.

and predicted TL in 1/3-octave bands centered at 25, 50 and 100 Hz were obtained using a composite geoacoustic model based on nearby estimates of the subbottom structure.

In future work, we plan to analyze the measured waveforms to obtain local estimates of the geoacoustic parameters. In addition to traditional analyses of bottom losses, the towed-array data can be processed using time-domain beamforming techniques. Further development of the two-way PE model will allow the measured and predicted backscattered fields to be compared.

REFERENCES

[1] Craig, D.W. and McKee, L. (1990). "High performance shipboard acoustics data acquisition, imaging," *Sea Technology*, **31**, 51–53.

[2] Jensen, F.B. and Kuperman, W.A. (1980). "Sound propagation in a wedge-shaped ocean with a penetrable bottom," *J. Acoust. Soc. Am.* **57**, 1564–1566.

[3] Dosso, S.E. and Chapman, N.R. (1987). "Measurement and modeling of downslope acoustic propagation loss over a continental slope," *J. Acoust. Soc. Am.* **81**, 258–268.

[4] Porter, M.B., Jensen, F.B. and Ferla, C.M. (1991). "The problem of energy conservation in one-way models," *J. Acoust. Soc. Am.*, **89**, 1058-1067.

[5] King, D.B., Chin-Bing, S.A., Davis, J.A. and Evans, R.B. (1992). "A review of the Parabolic Equation Workshop II," *J. Acoust. Soc. Am.*, **91**, 2389.

[6] Collins, M.D. and Evans, R.B. (1992). "A two-way parabolic equation for acoustic backscattering in the ocean," *J. Acoust. Soc. Am.*, **91**, 1357–1368.

[7] Jensen, F.B. and Ferla, C.M. (1990). "Numerical solutions of range-dependent benchmark problems in ocean acoustics," *J. Acoust. Soc. Am.*, **87**, 1499–1510.

[8] Thomson, D.J. (1990). "Wide-angle parabolic equation solutions to two range-dependent benchmark problems," *J. Acoust. Soc. Am.*, **87**, 1514–1520.

[9] Brooke, G.H. and Thomson, D.J. (1991). "A single-scatter formalism for improving PE calculations in range-dependent media," Presented at the PE Workshop II, 6–10 May, Slidell, LA (to be published in proceedings).

[10] Collins, M.D., Ali, H.B., Authement, M.J., Nagl, A., Überall, H., Miller, J.F., and Arvelo, J.I. (1988). "Low-frequency sound interaction with a sloping refracting ocean bottom," *IEEE J. Ocean. Eng.*, **13**, 235–244.

[11] Sen, M.K., Frazer, L.N., Mallick, S. and Chapman, N.R. (1988). "Analysis of multipath sound propagation in the ocean near 49° N, 128° W," *J. Acoust. Soc. Am.*, **83**, 588–597.

[12] Yee, W. (1991). "Source levels of small explosive charges," Jasco Research Ltd., *DREP Contractors Report* **91-18**.

Section 6

Volume Scattering

LOW-FREQUENCY VOLUME REVERBERATION MEASUREMENTS

Tuncay Akal[1], Robert K. Dullea[2], Guido Guidi[1], John H. Stockhausen[3]
[1] *SACLANT Undersea Research Centre, La Spezia, Italy*
[2] *NUSC, New London, CT 06320, USA*
[3] *DREA, Dartmouth, Nova Scotia, Canada*

An experimental technique has been developed to measure broadband low-frequency volume reverberation. Experiments were performed in the North Atlantic with broadband explosive sources in conjunction with the end-fire beam of a vertical line array to provide scattering strength contours from measurements sampled in both frequency ($\sim$ 12 Hz) and depth ($\sim$ 4 m). The average effective resolution in frequency ($\sim$ 100 Hz) and depth ($\sim$ 30 m) resulted from signal processing procedures and geometrical effects associated with the measurements. During the measurements a SIMRAD scientific echo sounder system was utilized. This system incorporates split-beam processing technology to provide the most accurate in-situ assessment, analysis and display of biological target strength and fish length distributions. The experimental and analysis techniques are described along with representative examples that compare high and low scattering levels observed during the experiment.

INTRODUCTION

It is generally acknowledged that volume reverberation for a wide range of frequencies is primarily attributable to biological activity. A technique has been developed to measure volume scattering strength from 100 Hz to 1500 Hz. The technique basically consists of using a broadband explosive source in conjunction with the end-fire beam of a vertical line array (VLA) to obtain scattering strength both as a function of depth and frequency.

In the North Atlantic, which is inhabited by large populations of commercial sized fish, the potential exists for encountering high levels of volume reverberation at low frequencies. Volume reverberation data obtained with this technique were sampled in both frequency ($\sim$ 12 Hz) and depth ($\sim$ 4 m). The average effective resolution in frequency ($\sim$ 100 Hz) and depth ($\sim$ 30 m) were obtained due to signal processing procedures and geometrical effects associated with the measurement technique.

In the following sections details of the experimental technique, data processing and scattering strength formulation are described, and VLA measurements characteristic of both low and high scattering levels are presented.

1. EXPERIMENTAL TECHNIQUE

The experiment and analysis techniques used in this experiment were developed from the experience SACLANTCEN gained in the 1970s [1,2].

For this experiment a longer, vertical line array was used in order to lower significantly the frequency range of the measurements (100–1500 Hz) compared to the earlier technique which covered a much higher frequency range (1.6–16 kHz). The experimental arrangement for the measurement of volume scattering strength as a function of depth is shown in Fig. 1. A line array of hydrophones was suspended vertically at a depth intended to be greater than that of most of the significant scatterers. An explosive charge was detonated below the array and scattered sound was received at the array from those scatterers within an upward-looking beam. The detonation depth was chosen to ensure, whenever possible, that the surface reflected arrival preceded the bottom reflection.

The beam was sufficiently narrow so that at any time the contributing scatterers lay within a limited depth interval; thus the dependence of scattering strength on depth could be determined. Figure 2 shows an typical signal from the upward-looking end-fire beam beginning with the direct arrival from the source, followed by sound scattered from those parts of the scattering layers above the array that lie within the beam. Reflections from the surface (or sometimes from the bottom) define the limit of the measurements.

Both the array and reference hydrophone (which was used for source level measurements) were decoupled from the ship with a floating tube, and the signals were fed by cable to the ship (Fig. 1). Simultaneously, in conjunction with the low-frequency VLA experiments, measurements were conducted with a high-frequency, scientific echo sounder (SIMRAD) to obtain estimates of fish size distributions.

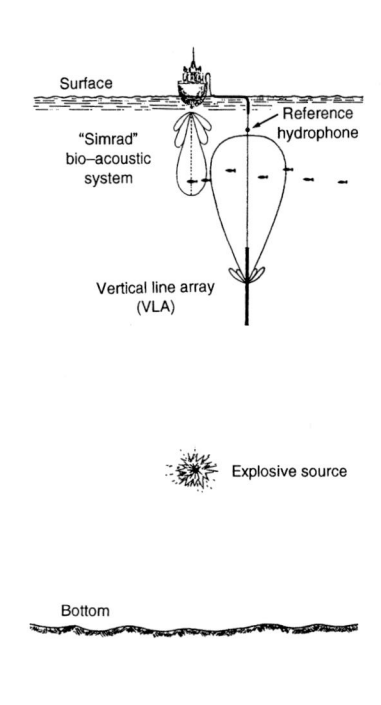

Figure 1 *Experimental geometry.*

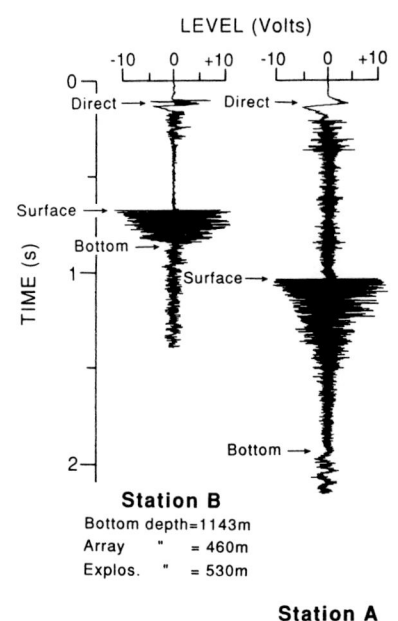

Figure 2 *An example signal from the upward-looking end-fire beam showing high (A) and low (B) reverberation data.*

2. DATA PROCESSING

A. Geometrical considerations

As shown in Fig. 3, the sound reaching the receiver has traveled from the source to the scatterers and hence to the receiver. Scatterers contributing at any time lie within an ellipsoidal shell defined by the total path length and by the source pulse duration. The contributions from scatterers within this shell are weighted according to the spherical spreading losses incurred on the two paths and by the directional sensitivity of the array beam.

It is clear from the figure that the effective resolution in depth depends on the curvature of the ellipsoidal shell within the beam; resolution is therefore better with narrower beams at earlier arrivals.

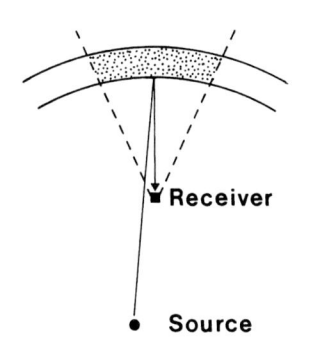

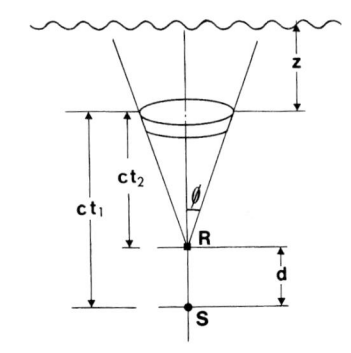

Figure 3   Geometrical considerations.

Figure 4   Parameters for the derivation of the scattering equation.

B. Derivation of the scattering equation

Referring to Fig. 4, it is assumed that the scattering volume is a disk and that all parts of the disk are equally distant from the source and from the receiver. Consider the effect at the receiver of an interval $\delta\tau$ in the pulse of intensity I_0 (at unit range) emitted by the source. Sound from this pulse interval, scattered from thickness $c\delta\tau/2$, arrives at the receiver at the same time, where c is the speed of sound.

The intensity I_R of scattered sound at distance r from a volume V of scattering strength m, when insonified with intensity I_i is

$$I_R = \frac{mI_i V}{r^2}.$$

Thus the intensity of scattered sound at the receiver of Fig. 4 is

$$\delta I_R = \frac{I_0 m(z) V}{(ct_1)^2 (ct_2)^2},$$

where

$$V = \pi(ct_2 \tan\phi)^2 c\delta\tau/2;$$

thus
$$\delta I_R = \frac{I_0 m(z)\pi \tan^2 \phi}{(ct_1)^2} \times c\delta\tau/2.$$

Since the total travel time t is $t = t_1 + t_2$ and $ct_1 = ct_2 + d$, then $ct_1 = \frac{1}{2}(ct + d)$ and

$$\delta I_R = \frac{I_0 m(z)\pi \tan^2 \phi}{\frac{1}{2}(ct + d)^2} \times c\delta\tau/2.$$

The total energy leaving the source E is given by

$$E = 4\pi \int I_0 \, d\tau,$$

so that the received intensity due to the complete source pulse is

$$I_R = E c m(z) \tan^2 \phi / 2(ct + d)^2.$$

The result of the full derivation [1] is

$$I_R = \frac{Em(z)}{2td} \ln \frac{1}{1 - [2ctd/(ct+d)](1 - \cos\phi)}.$$

If $\phi < 35°$, this expression is approximated within 5% by

$$I_R = \frac{Ecm(z)}{(ct+d)^2}(1 - \cos\phi). \tag{1}$$

This slightly more accurate approximation relative to the expression containing $\tan^2 \phi$ was used in the reduction of data in the present experiment.

C. Beamforming

The required upward-looking receiving beam was obtained with an interpolating time domain beamformer [4], which was applied separately to the three sub-arrays.

Before beamforming, sharp-cutoff digital low-pass filtering was performed on all of the hydrophone signals, followed by decimation to a sampling rate of 3000 Hz. The cut-off frequencies were 375 Hz for hydrophones used in the long array with 2 m spacing, 750 Hz for the 1 m array and 1500 Hz for the 0.5 m array.

With a sound speed of 1500 m/s the three arrays would be spatially aliased, with the generation of a complete backlobe nominally at 375, 750 and 1500 Hz, respectively. It would be prudent then to use each sub-array only up to some appropriate, slightly lower frequency. These limiting frequencies were further lowered by the sound speeds typically encountered in the experiment.

D. Spectral analysis

The beam time series were then processed with a sequence of fast Fourier transforms of short segments, so that a two-dimensional matrix of spectral power in time (depth) and frequency was obtained.

A measure of smoothing in the frequency domain was achieved by Hanning windowing and by zero-padding. Smoothing in time (depth) resulted from overlapping the segments. The parameters used are shown in Table 1.

Estimates were thus obtained at 4 m intervals, each of which was an average over an interval of ca. 16 m. For those estimates near the surface, the window length was reduced to avoid contamination by the strong surface reflection.

The last reflection-free estimate was assigned a depth between 2 and 6 m, depending on the length of the small unused interval remaining before the surface reflection.

Table 1 *Parameters for time/frequency transform*

	Points (no.)	Time (ms)	Depth interval (m)
Transform length	256	85 1/3	–
Hanning window length	128	42 2/3	~ 32
Effective window length (depth resolution)		21 1/3	~ 16 –
Step size	16		~ 4
Frequency interval ~ 12 Hz			
Frequency resolution ~ 34 Hz			

E. Noise correction

Since it is possible that some values in the time/frequency matrix of spectral density may have low signal-to-noise ratios, a noise correction was applied. A length of the beamformed time series prior to the direct arrival was subjected to the same analysis as the reverberation record and the resulting spectral estimates (as linear powers) were subtracted leaving an estimate of noise-free reverberation. Where the scattered signal was low, it sometimes happened that estimates of noise exceeded those of reverberation. Below 152.36 Hz, the data are frequently not reliable for this reason and are not presented. For the instances above 152 Hz when the reverberation level was less than the estimated noise level, the uncorrected scattered signal was used in the computation of scattering strength.

F. Computation of scattering strength

Equation (1) is inverted to give scattering strength in terms of known and measured quantities:

$$m(z, f) = \frac{I_R(t, f)(ct + d)^2}{E(f)c[1 - \cos \phi(f)]}. \tag{2}$$

This form indicates explicitly the dependences of quantities in the equation on time and frequency. The depth z at which the scattering at time t occurs is related to t and array depth z_a by

$$z = z_a - (ct - d)/2. \tag{3}$$

3. SCATTERING STRENGTH

A. The scattering strength matrix

When all of the foregoing observations and corrections were applied, the result was a matrix

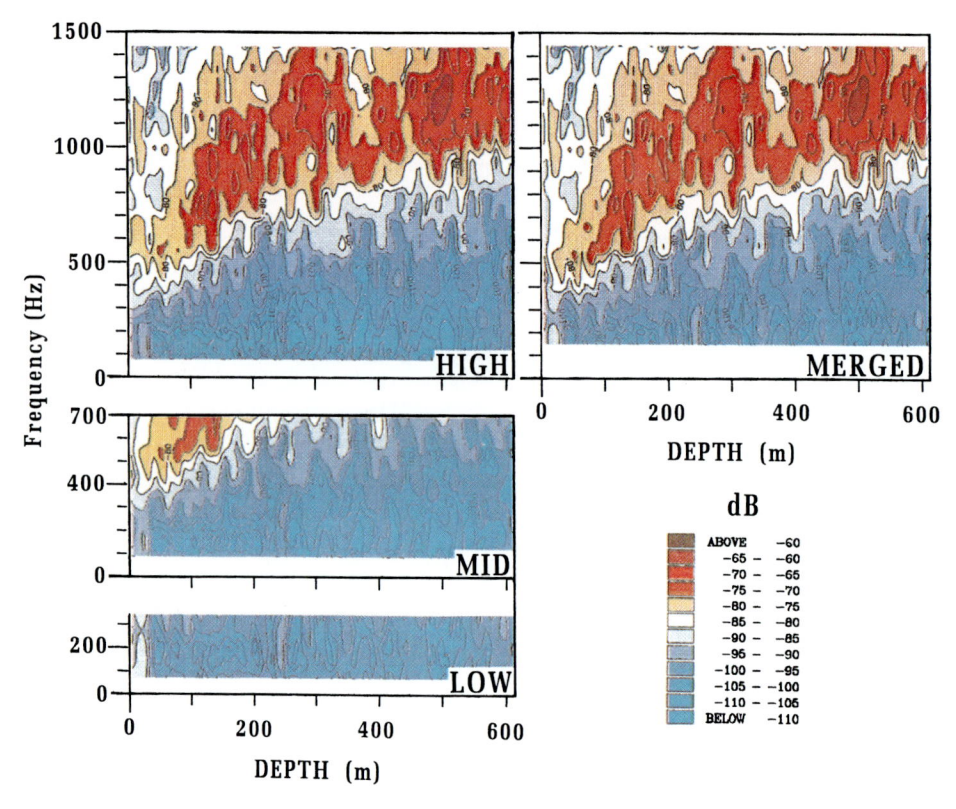

Figure 5 a) *Examples of scattering strength vs frequency and depth for three sub-arrays*
b) *Final result after merging.*

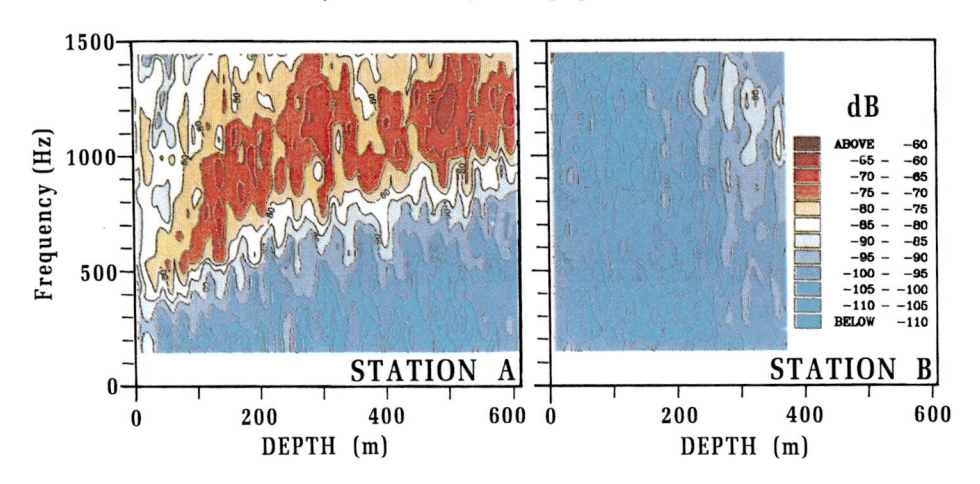

Figure 6 *Two examples of contour presentations for scattering strength vs frequency and depth.*

of volume scattering strength in depth and frequency. The frequency interval was 11.72 Hz, and the depth interval was nominally 4 m but varied slightly as sound speed at the array different from the nominal 1500 m/s, was taken into account.

As mentioned earlier, and shown in Fig. 2, when the record was terminated by the surface reflection, the last uncontaminated point varied in depth between 2 and 6 m. There was one exception: when the shot was so deep, or the bottom so shallow that the bottom reflection preceded the surface reflection at the array, part of the volume reverberation was masked by bottom reverberation and depths are calculated using Eq. (3).

B. Merging of matrices from sub-arrays

The data from the three sub-arrays result in three scattering strength matrices, each of which is valid and optimum in a relevant frequency range. The three matrices are combined into one by selecting the appropriate elements. Data from the low-frequency array are used from 152.4 to 339.8 Hz, the mid-frequency array from 351.6 to 691.32 Hz, and the high-frequency array from 703.2 to 1441.46 Hz (Figs. 5a and 5b). Within these ranges the depth resolution due to ellipsoidal shell curvature varies typically from 20 to 35 m near the surface, from 10 to 15 m at 300 m depth, and is nominally zero at the array, usually ca. 600 m deep. When the length of the FFT data windows is taken into account the resolution becomes ca. 35–50 m at the surface, 25 m at 300 m depth and 15 m at the array. The resolution in frequency was controlled by the smoothing applied to reduce spurious peaks at the frequencies of the source spectrum nulls, and with the windows length used it was ca. 100 Hz. The result is a representation of volume scattering strength over a frequency range greater than three octaves with rather uniform depth resolution throughout.

C. Display format

Contour

The two-dimensional matrix of scattering strength *vs* depth and frequency lends itself to contour presentation, especially as a standard colour contouring and printing package is available at the Centre, where the analysis was performed. This format was used by Stockhausen and Figoli and by Doutt [1,2].

Figure 6 shows two examples of this presentation, chosen to illustrate the range of scattering strength encountered in the experiment.

4. CONCLUSIONS

The combination of wideband scattering strength profiling with a vertical line array, target strength estimation with a scientific echo sounder, and bio-acoustic modelling can yield a detailed understanding of the low-frequency scattering environment in the survey area [5]. The volume reverberation experiments also demonstrated that the measurement technique with the vertical array is a valuable research tool for obtaining high resolution (spectral and spatial) measurements of low-frequency scattering strengths. Such measurements are required to obtain a greater understanding of the fundamental mechanisms of volume scattering processes.

REFERENCES

[1] Stockhausen, J.H. and Figoli, A. (1973). 'An upward-looking bistatic research system for measuring deep scattering layers in the ocean,' SACLANTCEN TR-225. La Spezia, Italy, SACLANT Undersea Research Centre. [AD 764 395]

[2] Doutt, J.A. (1977). 'Broadband measurements of volume scattering strength in the Mediterranean,' SACLANTCEN SR-17. La Spezia, Italy, SACLANT Undersea Research Centre.

[3] Barbagelata, A. and Guerrini, P. 'Specifications of SACLANTCEN vertical array,' In preparation.

[4] Wardale, J.P. (1979). General purpose beamformer hardware handbook, Plessey Marine Research Unit Technical Note 69/79/97.

[5] Akal,T., Dullea, K.D., Guidi, G. and Stockhausen, J.H. (1993). 'Low-frequency volume reverberation measurements.' Journal of the Acoustical Society of America, April 1993.

LOW FREQUENCY VOLUME SCATTER AND DISTANT REVERBERATION: 200-1500 Hz

Joseph M. Monti
Naval Undersea Warfare Center Detachment
New London, CT 06320
U.S.A.

C. Scott Hayek
Johns Hopkins University
Applied Physics Laboratory
Laurel, MD 20723, U.S.A

ABSTRACT: Low frequency volume scatter and distant reverberation measurements have been made in the Icelandic Basin, Gulf of Alaska and the Ionian Sea. Volume scattering data showing effects of fish swimbladder resonance along with spatial and temporal dependencies have been obtained as a function of depth to 600 m and frequency from 200-1500 Hz using explosives and a vertical line array receiver. Distant reverberation results, obtained using controlled waveforms from a vertical line array of projectors and a horizontal line array receiver (i.e., a near-monostatic configuration), also show effects of biological scattering. Both ray and mode based distant reverberation model predictions, utilizing *in situ* volume scatter data, agree well with measurements.

INTRODUCTION

Biologically induced reverberation from fish located at shallow depths (0-100 m) has affected sonar performance since the early days of convergence zone operations. It is well understood that fish swimbladder resonance is a major cause of this reverberation which can be highly variable over the operational bands of contemporary sonars. As part of the 1987 Tactical ASW Environmental Acoustic Workshop, held at the U.S. Naval Research Laboratory [1], a committee on biological reverberation stated that the cause of biologically induced low-frequency reverberation is similar to that for higher frequencies — *a resonant frequency phenomenon*. Recent findings of the U.S. Space and Naval Warfare Systems Command (SPAWAR) Critical Sea Test (CST) Program support this position and data support high levels and highly variable biological reverberation which extends to lower frequency bands, i.e., below 1500 Hz.

Estimates of the acoustic resonance frequency for swimbladdered fish, as a function of fish length and water depth, have been derived by Foote [2] from measurements of pollack and saithe based on a relationship between swimbladder size and length. Similarly, the resonance frequency formula of Andreeva [3] has been used, taking the real part of the complex shear modulus of fish tissue, $\mu_r = 10^5$ dynes/cm^2. Based on these studies, swimbladdered target strength as a function of

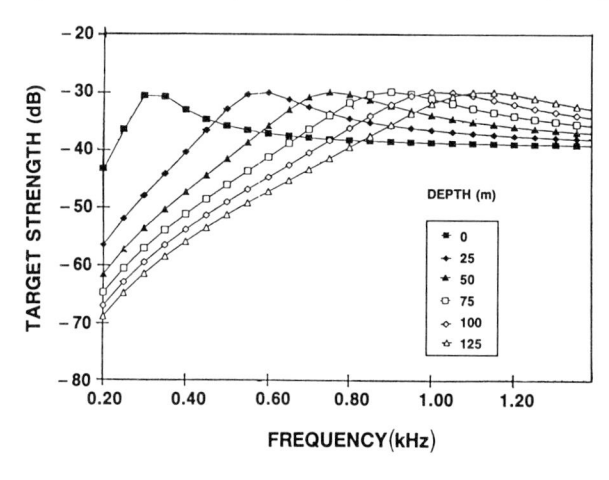

Fig. 1 Target strength of 30 cm swimbladdered fish with Q = 3 based on Andreeva's model simplified by Saenger [4].

water depth for 30 cm (length) fish with a Q = 3 have been estimated by Saenger [4]. The resulting family of curves, shown in Fig. 1, indicate that swimbladdered fish on the order of 30 cm in length can cause resonant scattering over the low-frequency band of interest.

VOLUME SCATTERING STRENGTH MEASUREMENTS

A technique developed by Stockhausen and Figoli [5], which uses a Vertical Line Array (VLA) receiver and deep explosive charges as an acoustic source to collect volume reverberation, was used to acquire data over the Icelandic Basin [6], in the Gulf of Alaska [7] and in the Ionian Sea [8]. The VLA used in these measurements contained 64 hydrophones configured as three 32-element nested apertures of half wavelength spacing with design frequencies of 400, 800, and 1600 Hz. The VLA was submerged to a depth of 600 m and hydrophones were phased to produce an upward looking endfire beam. MK59 Mod 1a SUS charges were detonated below the array at a depth of 900 m. Two desensitized hydrophones, at the top and bottom of the array, were used to receive undistorted waveforms of the high-level SUS signals required for data processing. The VLA also contained two tilt and depth sensor packages. Once the volume reverberation data were acquired, *volume scattering strength* [i.e., $S_V(f,z)$ in dB//m^{-3}] was determined for increments of 3 m in depth and 16 Hz in frequency. Depth-profile values of S_V were integrated over 100 m intervals (from 0-600 m) to provide *layer scattering strength* results [i.e., $S_L(f; z_1, z_2)$ in dB//m^{-2}] and over the entire ensonified water column to provide *column scattering strength* results [i.e., $S_C(f)$ in dB//m^{-2}].

Examples of Ionian Sea 250 and 1000 Hz nighttime volume scattering strength (S_V) profiles are shown in Fig. 2 to illustrate the similarity between slope water (continental rise) and shallow water sites. As can be observed, biological layering is evident at both sites, and volume scattering strength values are very similar as a function of frequency.

Fig. 2 Ionian Sea 250 and 1000 Hz S_V nighttime profiles

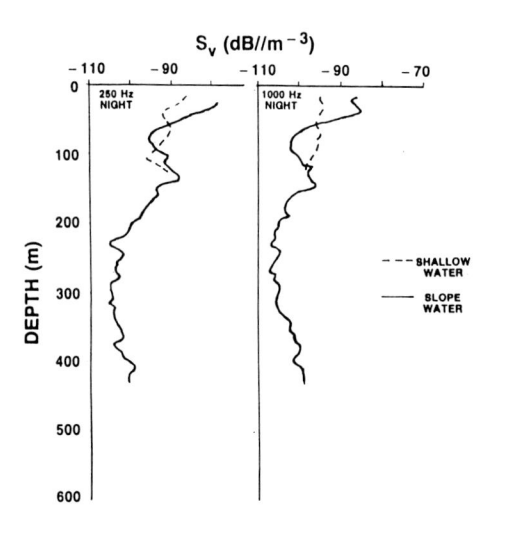

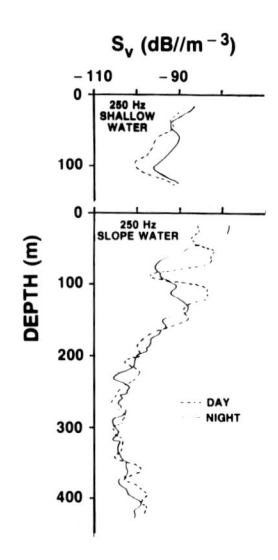

Fig. 3 Ionian Sea 250 Hz daytime and nighttime S_V for different water depths.

As illustrated by Fig. 3, there is very little day/night dependence for the two sites. Layering is still prevalent during daytime for the slope water site, but it is bimodal where some migration causes a secondary layer to form at around 100 m.

The column scattering strength (S_C) spectra, from 200-to-1500 Hz, derived from the Ionian Sea volume scattering strength (S_V) data are shown in Fig. 4. Swimbladder resonance effects are not evident in the slope and shallow water S_C spectra, although a hint of a resonance is seen at 350 Hz for the slope and shallow water areas. In addition, comparisons can be made in Fig. 4 with S_C results acquired by Stockhausen in 1971 [9] and Doutt in 1974 [10] for measurements taken in the Ionian Sea. Day and night 1600 Hz results are plotted for both data sets. The 1971 results show a moderate day/night dependence ($\approx$10 dB). The 1974 results exhibit a weak dependence (2-3 dB) which is similar to the recent S_C levels for the slope and shallow water sites.

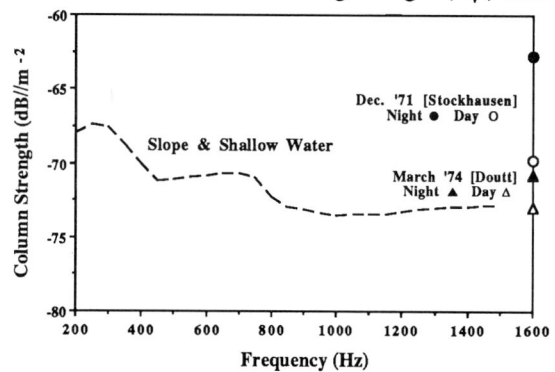

Fig. 4 Ionian Sea S_C spectra for different water depths and SACLANTCEN 1971 and 1974 1600 Hz day and night S_C values.

Representative 400 Hz S_V profiles obtained in the Gulf of Alaska and Ionian Sea are shown in Fig. 5. The Gulf of Alaska profile portrays a dense scattering layer in the upper 25 m while that for the Ionian Sea portrays a relatively uniform distribution of scattering over the entire water column. Column scattering strength (S_C) spectral shapes have been obtained for measurements conducted over these ocean regions and over the Icelandic Basin. Shown in Fig. 6 are S_C spectra which are representative of the geometric average of data acquired for selected runs at these ocean regions. All data were collected at

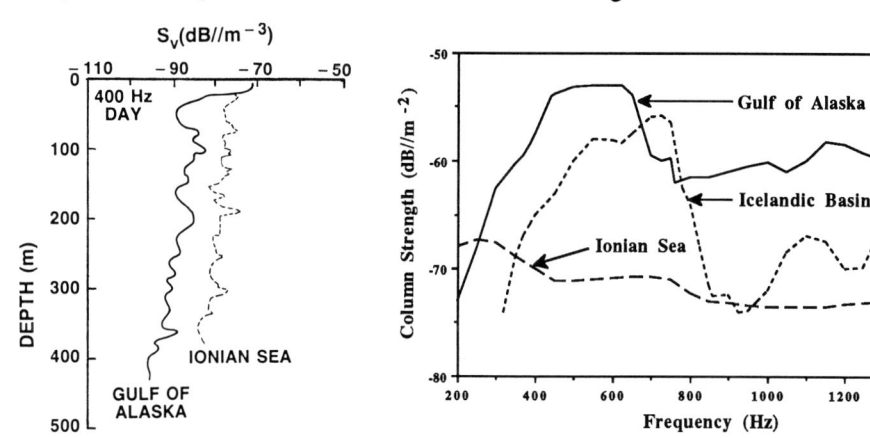

Fig. 5 Ionian Sea and Gulf of Alaska 400 Hz S_V spectra.

Fig. 6 Geographic comparison of S_C spectra.

relatively low windspeed (< 5 m/s). Strong swimbladder resonant effects are evident over the 400 - 800 Hz frequency regime, and above 1200 Hz, for the Gulf of Alaska and Icelandic Basin spectra.

For the Icelandic Basin, the cause of spectral peaks is believed to be swimbladder resonant scattering from blue whiting fish which are located in the upper 100 m of the water column and on the order of 20-40 cm in length. For the Gulf of Alaska, the cause of the spectral peaks is believed to be swimbladder resonant scattering from salmon located in the upper 25 m of the water column, which are on the order of 30-100 cm in length.

Fig. 7 shows spatial variability of volume scatter (S_C) for data collected in the Gulf of Alaska, as a function of latitude and longitude. For the results presented as a function of latitude, the windspeed varied from 9-to-12 m/s and there are large variations on the order of 5-20 dB in S_C across all frequencies. There does not appear to be a systematic trend in the fluctuations. By contrast, the results presented as a function of longitude are essentially the same over the entire spectrum for data gathered under different windspeed conditions, i.e., 4-to-11 m/s. An investigation into the oceanography for the Gulf of Alaska revealed that the test site was located along the Alaskan current which flows from west-to-east [11]. It is expected that relatively uniform salmon migration patterns should occur along the current flow and, because of this, S_C should be relatively uniform with longitude. North-south salmon migration across the current flow is expected to be more variable causing S_C to be more variable with latitude. The relative fluctuations observed in S_C for the latitude and longitude dependent stations seem to be qualitatively consistent with this hypothesis.

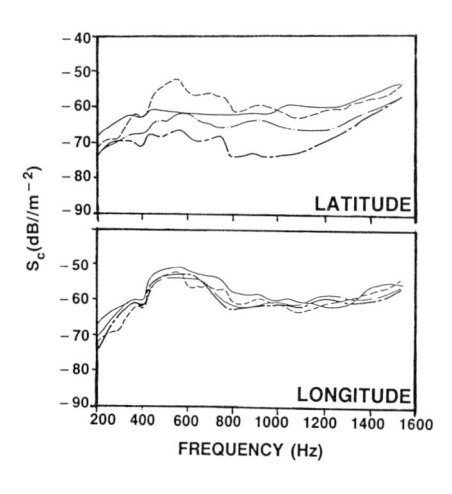

Fig. 7 Gulf of Alaska S_C spectra as a function of latitude and longitude.

The bounds of S_C spectra for the Icelandic Basin, Gulf of Alaska and Ionian Sea are shown in Fig. 8 along with results for the Norwegian Sea acquired by Love [12] using a downward looking conical receiver and surface vented explosive charges. They represent the average column strengths, not the standard deviation of the individual data sets. These bounds for S_C spectra are plotted against Chapman-Harris surface scattering strength empirical model predictions that were based on a grazing angle of 5°. For the time of year that these data sets were acquired and depending on frequency within the 200-1600 Hz band, it appears that for low-Doppler targets, active sonars can be volume reverberation limited at windspeeds less than about 6-10 m/s for the Icelandic Basin, about 8-15 m/s for the Gulf of Alaska and 4-8 m/s for the Ionian Sea. Estimates of the windspeeds at which the dominant scattering mechanism transitions from volume to surface as a function of octave bands from 200 to 1600 Hz, for the four ocean regions visited, is shown in Table I. This analysis assumes the transmission loss to surface and volume scatterers is the same and that bottom reverberation is lower than both the surface and volume reverberation.

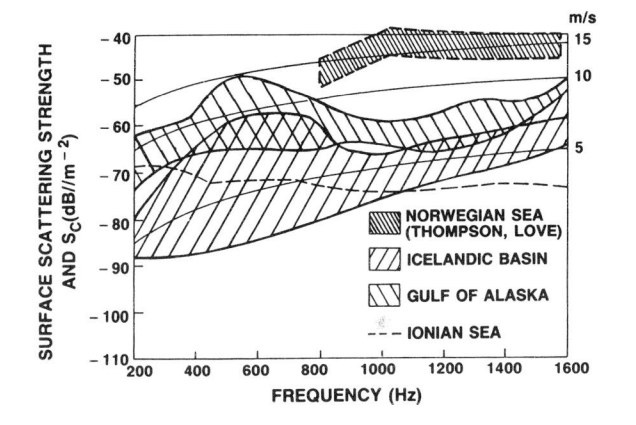

Fig. 8 Bounds of S_C spectra for the Norwegian Sea, Icelandic Basin, Gulf of Alaska and Ionian Sea.

TABLE I
Windspeeds (m/sec) at which dominant scattering transitions from volume to surface.

WINDSPEEDS (m/sec) AT WHICH DOMINANT SCATTERING TRANSITIONS FROM VOLUME TO SURFACE

Ocean Region	Frequency (Hz)		
	200 - 400	400 - 800	800 - 1600
Ionian Sea	8	7	4 - 5
Gulf of Alaska	12	13 - 15	7 - 9
Icelandic Basin	6 - 8	7 - 10	6 - 7
Norwegian Sea	—	—	16 - 20

DISTANT REVERBERATION MEASUREMENTS

Direct path volume scattering strength measurements were accompanied by distant reverberation measurements. A primary goal was to identify the origin of distant reverberation, e.g., sea surface, volume or bottom scattering. Since biological scatterers are known to exist close to the sea surface, the challenge was to separate sea surface scattering from sub-surface scattering. The measurements were conducted using a VLA of high-powered projectors (suspended to depths between 110 and 190 m) and a horizontal line array receiver (towed at a depth of 91 m). The VLA acoustic projectors were phased to produce a beam in the vertical plane with a nominal vertical beamwidth of 4° at 1000 Hz.

The ability to steer the source beam in the vertical plane is crucial in determining the cause of near-surface convergence zone (CZ) reverberation. For the CZ distant reverberation measurements made in the Norwegian Sea, advantage was taken of the sound speed profile by alternately steering the source at 10° and 4° downward. This resulted in respectively ensonifying or shielding the sea surface in the CZ. However, scattering from the first CZ was apparent for both the 4° and 10° downward steering transmissions, showing up at the expected arrival time of about 50 seconds. Since CZ reverberation from the 4° downward transmissions was as intense as that from the 10° downward transmissions, it was caused by subsurface scatterers. This conclusion is consistent with the fact that estimates of surface scattering strength, for the low windspeeds existing at the time, were -51 dB//m^{-2}, while measured volume scattering strengths were -43 dB//m^{-2} (see Fig. 8). The location of the main biological scattering layer was determined to be at a depth of 420 m for the time of day that the distant reverberation measurements were made. The earlier arrival and slightly higher initial level observed for the 4° downward steering reverberation relative to the 10° reverberation, as predicted by the ray-based transmission loss model CONGRATS, was additional evidence that a 420 m layer of blue whiting fish was the source of the reverberation.

Distant reverberation modeling was performed to better understand the propagation conditions and mechanisms responsible for the observed reverberation and to reconcile the levels of distant reverberation with the scattering strengths obtained using direct path measurement techniques. The most successful reconciliation between direct path volume scattering strength measurements and distant reverberation levels was attained with the Norwegian Sea data. This occurred because blue whiting fish, a source for high level low frequency swimbladder resonance, were omnipresent. Their scattering strength and their pervasiveness in space and time caused volume reverberation to dominate.

Presented in the upper panel of Fig. 9 is a waterfall plot showing the azimuthal and time structure of the envelope for Norwegian Sea reverberation resulting from 16-second, Hanning CW transmissions corrected to constant beamwidth. Reverberation levels for three transmissions were averaged over frequency bins selected according to transmitted frequency and own-ship Doppler correction. The plot shows volume reverberation which is fairly consistent across azimuth at 60 seconds, followed by bottom returns at certain azimuths. Presented in the lower panel of Fig. 9 is the data from an average of several transmissions on near-broadside beams plotted against (a) the ray theoretic model, RAYREV (an extension of CONGRATS), prediction for the case of a perfectly matched echo from a scattering layer centered at 420 m, and (b) the prediction for a return that has been split into an ensemble of echoes, each with its own Doppler shift. The data giving the Doppler "smear" of the echo ensemble were taken from the spectral widths measured in the corresponding Norwegian Sea distant reverberation run. Agreement between data and theory is excellent, reinforcing the postulate that the deep scattering layer observed in the distant reverberation measurement has a scattering strength close to that measured using the direct path volume scattering data acquisition techniques.

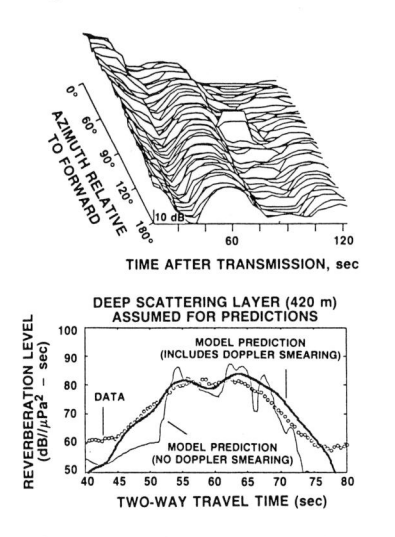

Fig. 9 Measured and predicted 900 Hz band distant reverberation (bottom) and waterfall plot (top) showing azimuthal and time variation.

For the Gulf of Alaska propagation conditions (half channel ducting in 5000 m water), it was necessary to separate bottom from surface and total water column reverberation. Bottom reverberation was segregated by steering the source VLA beam down 12° and analyzing reverberation corresponding to the time when rays cycle back to the surface. The reverberation in this "pseudo CZ" is dominated by the upper water column or sea surface scattering. Distinguishing between surface and volume reverberators was difficult and confounded by the fact that the salmon inhabited the upper 100 m of the water column. Under the half channel acoustic propagation conditions which existed, these shallow reverberators could not be isolated (range-resolved) from near-surface bubble clouds or long distance surface reverberation. To deduce the nature of the reverberators, determinations were made as to whether levels correlated with weather and analysis was conducted regarding the spatial distribution of the weather-correlated data relative to the distribution of the uncorrelated data. There were examples where reverberation clearly did not correlate with windspeed and was determined to be caused by biological scattering. Fig. 10 is presented to illustrate the azimuthal variability of distant reverberation, as seen through the zero-Doppler bandpass filter, for the transmission of 250 Hz, 12-second, Hanning weighted CW pulses. Uniform surface reverberation occurs under the higher 13.5 m/sec windspeed condition (left) and reverberation hot spots occur under the lower 5.5 m/sec windspeed condition (right). For the lower windspeed condition, there are strong reverberators which are distributed and those which are clustered into high target strength patches. This is an example of a reverberation return clearly not correlated with windspeed and believed to be due to volume reverberators.

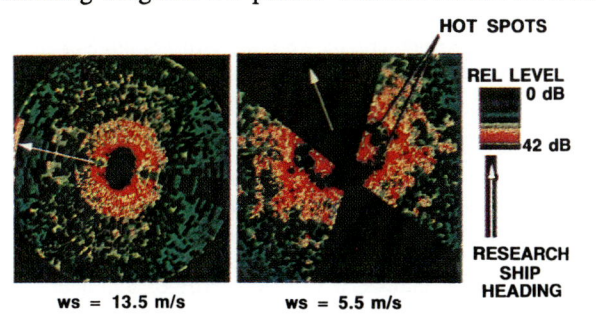

Fig. 10 Gulf of Alaska azimuthal variability showing 250 Hz volume reverberation hot spots for 5.5 m/sec windspeed (right) and uniform surface reverberation over 83 nmi for 13.5 m/sec windspeed (left).

The salmon in the Gulf of Alaska are not omnipresent in time and space throughout the test area, but rather appear in patches. This is confirmed by the direct-path volume scattering strength results which varied by more than 16 dB depending on location and time. Fig. 10 provides evidence that patches of high fish density caused distant reverberation levels to vary over a wide dynamic range. Unfortunately, the large fluctuation in possible target strength eliminated modeling as a discriminant for separating volume from surface reverberation. However, modeling did show that both an average maximum salmon scattering strength profile (acquired during the direct-path measurement), together with a bubble scattering strength profile, predicted reverberation comparable to that for observed high windspeeds.

SUMMARY AND CONCLUSIONS

Analyses of volume scattering strength acquired in the Norwegian Sea, Gulf of Alaska, Ionian Sea and over the Icelandic Basin have shown that populations of large swimbladdered fish, such as blue whiting and salmon, cause high volume reverberation levels at low frequencies (200-1500 Hz). Strong swimbladder resonance effects have been observed below 1500 Hz in measured layer and/or column strength spectra. These high reverberation levels are specific to the geographic areas where measurements were made and should not be taken as representative levels world-wide. However, it has become apparent that low densities of swimbladdered fish can cause significant low frequency volume reverberation levels due to fish swimbladder resonant scattering effects. The character of the volume reverberation ranged from relatively spatially homogeneous and temporally stationary in the Norwegian Sea during summer to extremely patchy and highly time variable in the Gulf of Alaska. Over the Icelandic Basin, during winter, the test site was on the outskirts of a population of blue whiting, the same type fish that caused high reverberation in the Norwegian Sea.

The principal result of the distant reverberation measurements is the confirmation that swimbladder scattering plays an important role at 250 Hz (salmon) and 1000 Hz (blue whiting). It appears that volume scattering should be an important contributor to distant reverberation at these frequencies in other ocean areas where relatively large swimbladdered fish exist. Integrated column strength was found to be inadequate for predicting distant reverberation under certain propagation conditions. It is necessary to use volume scattering strength as a function of depth when the transmission loss to different layer depths varies, such as under CZ propagation. Distant reverberation level is, therefore, closely tied to knowledge of the types and migration of fish inhabiting a test area at the time of an experiment. When the dominant scatterer is omnipresent, as were the Norwegian Sea blue whiting, distant reverberation can be modeled from direct path results to within 2 dB.

REFERENCES
1. K.W. Lackie, ONR, B.F. Cole, NUSC, and S.R. Santaniello, PSI, 1987. Identification of this document may be obtained form the authors of this report.
2. K.G. Foote, "Rather High Frequency Sound Scattering by Swimbladdered Fish", *J. Acoust. Soc. Am.*, Vol. 78 (2), pp. 688-700, August 1985.
3. I.B. Andreeva, "Scattering of Sound in Oceanic Deep Scattering Layers", in <u>Acoustics of the Ocean</u>, edited by L.M. Brekhovskikh (Nauka, USSR Academy of Sciences, Acoustics Institute, 1974).
4. R.A. Saenger, "Swimbladder Size Variability in Mesopelagic Fish and Bioacoustic Modeling", *J. Acoust. Soc. Am.*, Vol. 84 (3), pp. 1007-1017, September 1988.
5. J.H. Stockhausen and A. Figoli, "An Upward Looking Bistatic Research System for Measuring Deep Scattering Layers in the Ocean", SACLANTCEN Technical Report No. 225, 15 May 1973.
6. J.M. Monti, M.T. Sundvik, G.F. Sharman, D.M. Potter, and R.A. Saenger, NUWC, 1991. Identification of this document may be obtained form the authors of this report.
7. J.M. Monti, NUWC, C.S. Hayek, JHU/APL, C.H. Thompson, NOARL, 1992. Identification of this document may be obtained form the authors of this report.
8. J.M. Monti, et al., "CST-5 Volume Scattering Analysis Utilizing a Vertical Line Array Measurement System: Quick-Look Report", in publication.
9. J.H. Stockhausen, "Volume scattering strength vs. depth and frequency in the eastern Atlantic Ocean and the western Mediterranean Sea", SACLANTCEN Report SM-60.
10. J.A. Doutt, "Broadband measurements of volume scattering strength in the Mediterranean", SACLANTCEN Report SR-17, 1 June 1977.
11. D.G. Browning et al., "A Relationship Between Ocean Circulation and Volume Reverberation in the Subarctic Northeast Pacific Ocean Gulf of Alaska", NUSC Technical Document No. 7681, 26 June 1986.
12. R.H. Love, "Low Frequency Volume Scatterers in the Norwegian Sea", NOARL Report in publication.

GINRUNS - A 1991 VOLUME SCATTERING AND CONVERGENCE ZONE EXPERIMENT IN THE NORWEGIAN SEA

James A. Doutt
Woods Hole Oceanographic Institute, Woods Hole, MA 02543 USA

John R. Preston
SACLANT Undersea Research Centre, La Spezia, Italy

ABSTRACT Volume scattering experiments are discussed for two deep (> 1500 m) sites in the Norwegian Sea. Results are presented showing scattering strength as a function of depth and frequency below 1500 Hz. A day-night migration effect was visible at both sites. Four distinct clusters of scattering can be seen on these plots. The relation of stronger scattering regions overlaid by curves of resonant frequency as a function of fish bladder radii and depth is reasonably consistent with trawl data and Simrad fish echo data taken two weeks later. The dominant scatterers are two-year old blue whiting. Additionally, horizontal line array (HLA) measurements using SUS charges were made at a distance of a convergence zone while vertical volume scattering measurements were being taken. Results show strong evidence of volume scattering from one convergence zone on the HLA in the 900-1400 Hz frequency band.

1. INTRODUCTION AND EXPERIMENT DESCRIPTION

In the latter part of July, 1991, a volume scattering experiment was conducted in the Norwegian Sea. Instruments used included a vertical line array (VLA) for local volume scattering measurements at low frequencies, and a HLA for longer range data. Both VLA and HLA data were collected at several sites. A major objective of the experiment was to relate the measurements from the two arrays. The VLA results for two of the sites will be compared: site 6 which was the westernmost site in colder fresher water, and site 2 which was warmer and more saline. HLA results at one site will also be discussed and compared with GENERIC SONAR MODEL predictions. Blue whiting were expected to be the major scatterers in these deep waters of the Norwegian Sea and this has been confirmed both by the experimental data and by a fish survey done by the Faeroese two weeks after this experiment. That survey showed a preponderance of 2 year old blue whiting ranging in length from 22-35 cm.

1.1 EXPERIMENTAL SET-UP

The experimental set-up was designed by Stockhausen [1] in the early 1970's and was further refined by Stockhausen and Akal in the late 1980's and a nearly identical setup is described in detail in Akal et al [2]. It consists of a vertical array suspended at a depth of approximately 700 meters and steered to end-fire looking up towards the surface. An explosive SUS charge containing 4 lb. of cyclotrol, detonated at a depth of either 915 or 1220 meters provides a broadband source. Energy from this shot radiates outward and is received by the array if scatterers are present in the beam. The advantage of this technique over most others is that the scattering strength can be computed to within a few meters of the ocean surface (where it can be 20 dB higher than 20 meters deeper). The array actually consists of three nested arrays so as to cover the frequency range 150 to 1430 Hz. Each of these arrays is processed separately and the individual results are combined in the final stage of processing. Figure 1. shows a typical time series used to produce the color plots. This is for the high frequency sub-array including the direct arrival from the SUS, the signal overload region, and region of valid volume scattering.

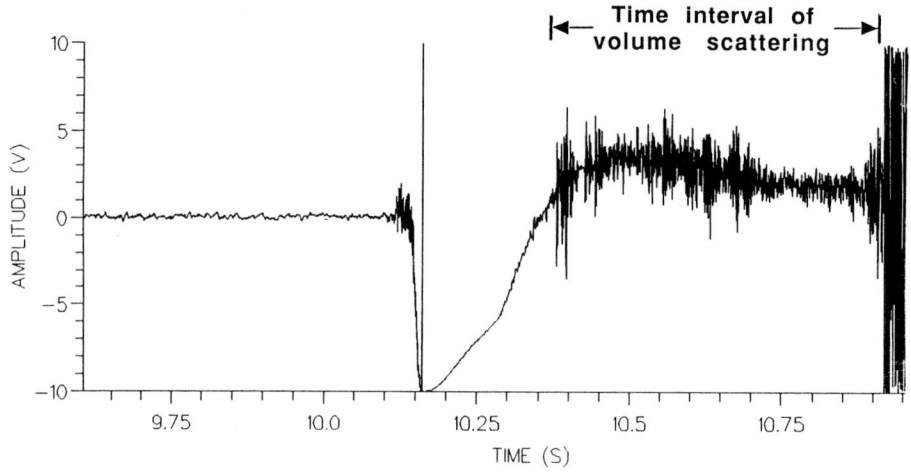

Figure. 1 Sample high frequency time series from upward endfire beam.

In addition to the array, a low-gain reference hydrophone deployed over the side at a depth of approximately 50 meters is used to monitor the source wave form. Thus the actual source spectrum of a shot is used for the data reduction of that shot. The detonation depth of the SUS and the depth of the reference phone are derived respectively, from the bubble period and the surface reflection visible in this signal.

1.2 ANALYSIS

The final output of the processing is a diagram of depth in meters on the vertical axis, frequency in Hz on the horizontal, and scattering strength in dB re m^{-1} represented by various colors with red corresponding to high levels and blue corresponding to low levels (Fig. 2). The depth resolution of this analysis was 16 m and the frequency resolution was 12 Hz but smoothing has been applied in both dimensions. Two artifacts of the processing are apparent in the diagrams. The vertical banding visible especially at low frequencies is due to inaccuracies in the data at the nulls in the source spectrum introduced by the bubble pulses. The horizontal region of low intensity at the bottom of the diagram arises because the electronics in the array have not yet fully recovered from the intense shock of the direct arrival from the SUS by this time.

1.3 INTERPRETATION OF DIAGRAMS

Andreeva [3] has shown that the resonant frequency F of a fish swim bladder (idealized as a sphere) is inversely related to its radius R and directly related to the absolute pressure P (P_0 + depth). Curves of resonant frequency vs. depth for a constant swim bladder radius have been superimposed on the scattering diagram in Fig. 3. These show that a swim bladder with a .75 cm radius at a depth of 110 meters will resonate at a frequency of 1500 Hz. As the fish rises, the resonant frequency of its swim bladder will follow the heavy curve until, at the surface, it will resonate at 590 Hz. Since the scattering

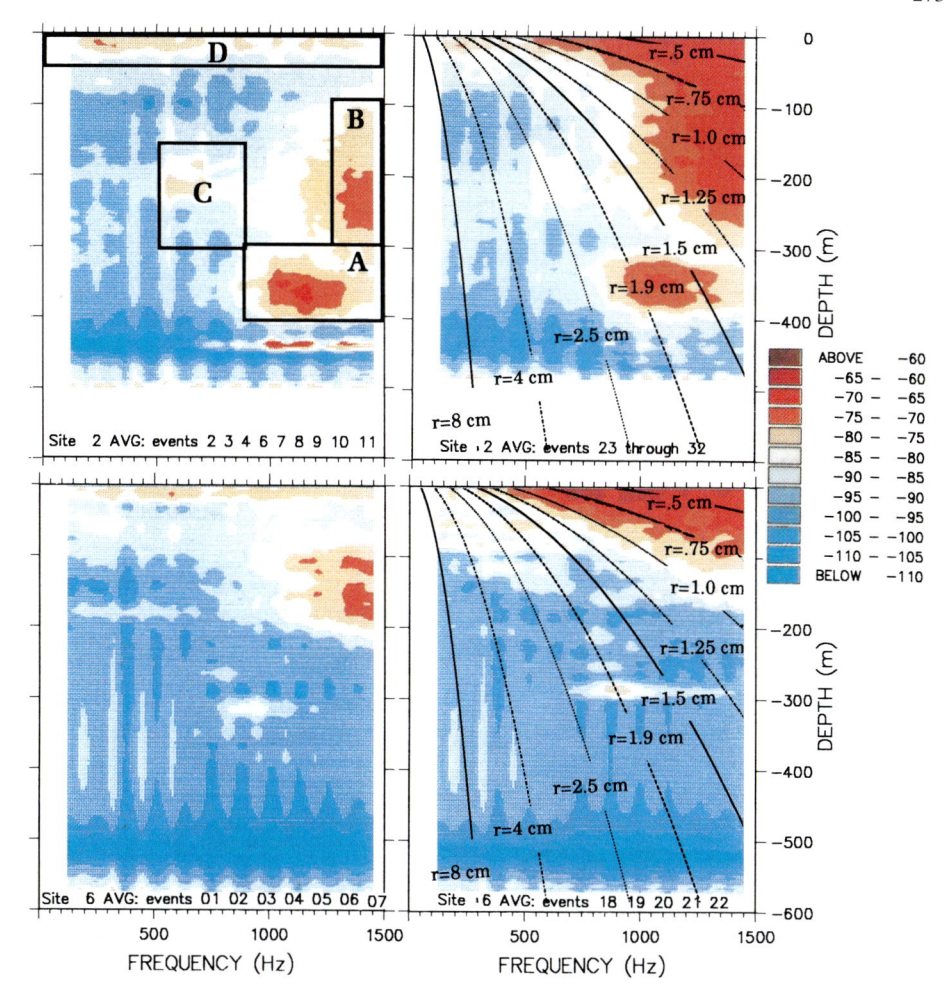

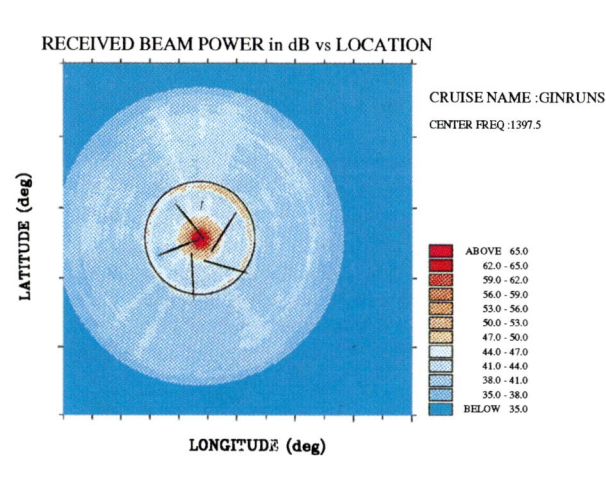

Figs. 2-3. (top) Volume scattering strength at site 2 day (L) and night (R) seen by the vertical line array (dB//m^{-1}).

Figs. 4-5. (middle) Volume scattering strength at site 6 day (L) and night (R) seen by the vertical line array (dB//m^{-1}).

Figure 6. (bottom) received levels from volume scattering at site 2 as seen by the horizontal line array (dB//1µPa/Hz).

is dominated by resonant scatterers, a useful although somewhat loose interpretation of peaks in the scattering diagram is: depth on the vertical axis, size of scatterer on the horizontal axis (with smaller scatterers corresponding to higher frequencies), and number of scatterers represented by the scattering strength. It must be remembered that the size-of-scatterer axis is different at each depth and follows the constant-size-bubble resonance curves.

2.0 DESCRIPTION OF DATA - GENERAL DAYTIME FEATURES

Figure 2, typical of daytime diagrams, is from site 2 and was produced by averaging (in dB) nine events over a 45 minute period. In a similar fashion to what was done in the Mediterranean by Doutt [4], five zones can be identified. Four are regions of high scattering strength and one is of low levels. These four high scatter zones have been defined specifically to differentiate between different populations of scatterers, and so the precise depth and frequency limits of the zones are somewhat arbitrary. The peak levels will vary from site to site (depending on the number of scatterers present). Also, at some sites, certain zones may be missing altogether.

Table 1.   Typical daytime parameters for the 4 high scatter zones

zone	depth range (meters)	frequency range (Hz)	bladder size (cm)	scattering strength (dB re m$^{-1)}$
A	280 - 440	85- >1430	1.4 - 1.9	-60 to -70
B	200 - 300*	1300 - >1430	.75 - 1.25	-70 to -80
C	150 - 300	500 - 900	1.9 - 3.0	-65 to -85
D	0 - 30	frequency Independent	N/A	-70 to -85

* Typically 200 to 300 meters, but shallower at site 6

The data shows that the upper frequency limit of zone B (and at some sites also zone A) extends to frequencies higher than the 1430 limit of this analysis. The results of Thompson [5] show that for zone B there is a broad peak from 1.5 to 2.7 kHz. In addition to these four zones of high scattering levels, low levels of less than -85 dB were found at all sites for depths greater than 30 meters and frequencies less than 500 Hz.

2.1 GENERAL NIGHTTIME FEATURES

Figure 3, also from site 2, is typical of nighttime diagrams. It was produced by averaging ten events over a 50 minute period. The most apparent feature is the wedge-shaped region of high scattering near the surface. This nighttime feature appears at all sites both in the present study and, at higher frequencies, in previous work in the Mediterranean [4]. Although zone D, the frequency-independent region near the surface, appears similar to the daytime regime, above 500 Hz this wedge of high scattering strength dominates the picture. Levels can dramatically increase: 10 to 20 dB above those seen in daytime. Many downward-looking experimental configurations are overloaded near the surface and miss this effect. The high scattering levels in this wedge for frequencies greater than 500 Hz and depths less than 100 meters correspond to the resonance curves for scatterers with swim bladder radii of approximately .6 cm distributed throughout the upper 100 meters. These scatterers are not visible on the daytime diagram probably because they are then at depths greater than 100 meters - where their resonance is above the 1430 Hz limit of this analysis.

In addition to these small scatterers, somewhat larger ones in zone B with swim bladder radii about .75 to 1 cm have risen about 100 meters from their 200 to 300 meter daytime depth. At night they are filling in the diagram over the depth range 100 to 250 meters and frequency range 1200 to 1430 Hz. It is interesting that the expected wedge-shaped feature is missing for these scatterers. This may be because they have migrated only up to 100 meters and not all the way to the surface. The scatterers making up zones A and C have not migrated. Below zone D, and at frequencies less than 500 Hz, a region of low scattering strengths is still observed.

2.2 COMPARISON OF SITE 2 AND SITE 6

Site 6 is quite different from site 2. The daytime diagram, Fig. 4, is an average of seven shots over a 36 minute period. Zones A and C are missing, and zone B is about 100 meters shallower than at site 2. The nighttime diagram, Fig. 5, is an average of 5 shots over a 15 minute period and zones A and C are again missing. Above 100 meters, however, it does look similar to site 2 and is probably caused by .6 cm scatterers which can't be seen in the daytime when they migrate deeper than 100 meters. Since zone B scatterers are present in the daytime at depths of 100 to 200 meters and frequencies greater than 1100 Hz, it is hard to explain the lack of scattering in this region in the nighttime. One possible explanation is that at night the scatterers are not distributed throughout the water column, but have risen en masse to the surface, and cause the intense scattering from about 600-1100 Hz.

Because of the large differences between site 6 and site 2, temperature - salinity (T-S) diagrams were plotted in order to see if there were oceanographic differences between them. Fig. 7 shows that site 6 is composed of relatively cold, fresh Polar water as compared to site 2 (and the others). This difference in water masses (and associated biological scatterers) between the sites may explain the absence of zones A and C at site 6. Also blue whiting prefer water temperatures near $5°\pm3°$ and there was less of that at site 6 than site 2.

3.0 HORIZONTAL LINE ARRAY RESULTS

Figure 8 shows a typical ray trace for this data set at site 2. The 1st convergence zone (CZ) caustic is visible in the upper 1000 m of the water column. A very weak surface interaction is shown. A horizontal line array array pentagon data collection experiment was conducted 1 CZ away from site 2 during the day (see [6] for a similar experiment). Sources used were 0.818 kg SUS at 91 m depth. The five array headings are used to get rid of array ambiguity. The dB averaged and stacked results are shown for a 50 Hz band of energy around 1400 Hz, in the color plot of Fig. 6. The plot shows the five lines of bearing plotted to a distance of 37 km from the relative array position during each data acquisition. One can clearly see the strong return at the convergence zone predicted at 40-52 km. In addition, a close-in diffuse circular backscatter can be seen out to about 1/2

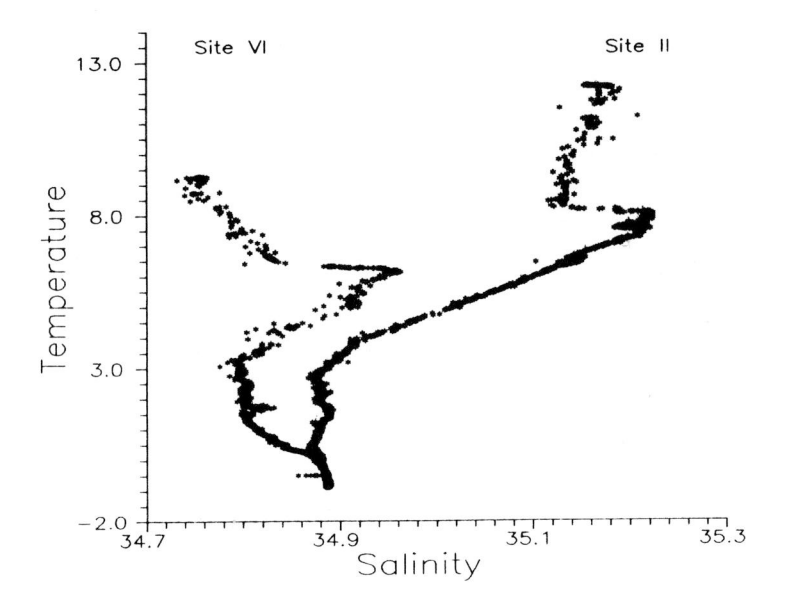

Figure 7. T-S diagram for sites 2 and 6 showing warm saline water and colder fresher water.

a CZ. The result shown in Fig. 6 was also seen for center frequencies of 900 and 1100 Hz, over wind speeds from 9-19 kts and was apparently independent of wind speed.

Finally Fig. 9 shows the result of comparing the beam time series of a broadside beam in the 1400 Hz frequency band with a prediction using GENERIC SONAR MODEL (GSM) [7]. The prediction was made using the measured CTD data at site 2 and the vertical volume scattering strength profile from Fig. 2 at 1400 Hz. GSM predicts that a small but noticeable convergence zone return will be caused by volume scattering and the data which overlay the prediction show that in fact the convergence zone return was slightly stronger than predicted. The inputs to GSM used to generate Fig. 9 included a Lambert's rule dependence for bottom scattering strength (coefficient =-32 dB), the Chapman Harris suface scatter formula for a 10 knot wind speed, a normal incidence bottom loss of 20 dB and since this was a monostatic prediction, a source depth of 95 m (mid-way between the actual source and receiver depth).

4.0 SUMMARY

A preliminary analysis of data from the GINRUNS cruise shows that: 1) the horizontal line array showed clear evidence of volume scattering in the daytime from 900 - 1400 Hz at 1 CZ (lower frequencies not measured); 2) the day - night migration effect was observed and gave increases at the shallower depths of 10 - 20 dB in scattering strength; 3) five zones or regions have been identified on the color scattering diagrams: four of

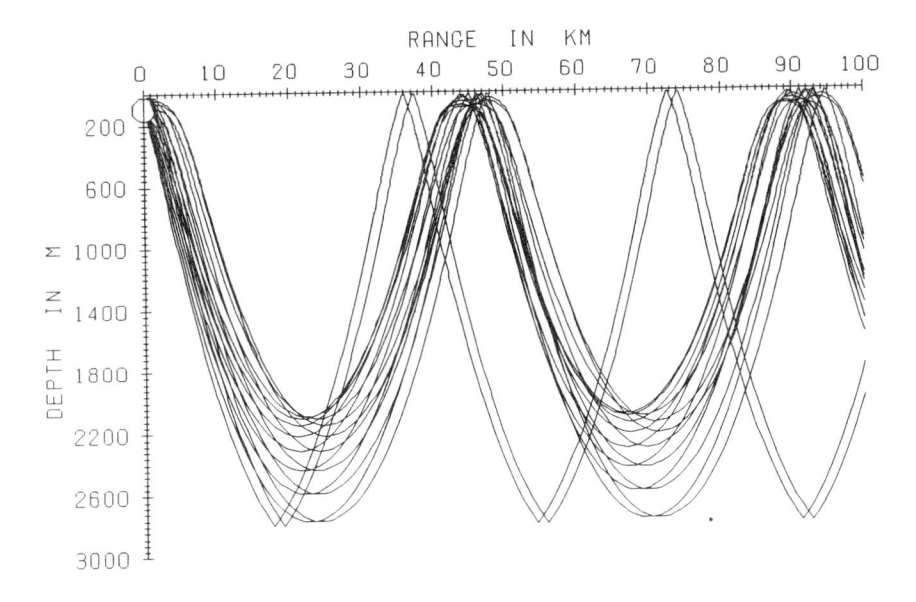

Figure 8. Estimated raypaths at site 2 from a source at 95 m showing CZ formation.

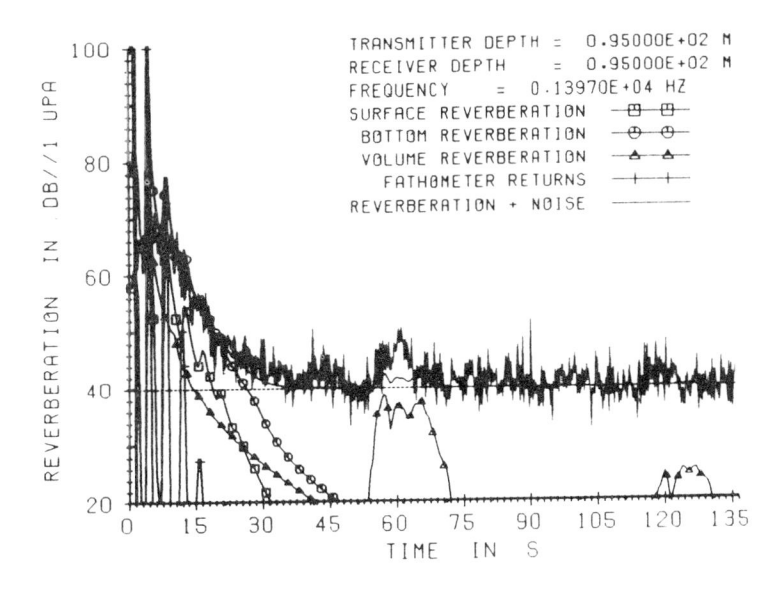

Figure 9. Comparison of broadside beam reverberation time series at site 2 (thick line) with predictions of GSM showing contributions of the surface, bottom and volume reverberation to the total time series.

high scattering strength and one of low scattering strength; and 4) except for a thin frequency independent layer near the surface the scattering is dominated by resonant scatterers (probably blue whiting and some redfish) and was important at frequencies from 500 to 2500 Hz.

REFERENCES

[1] Stockhausen, J.H. and Figoli, A. (1964) 'An upward-looking bistatic research system for measuring deep scattering layers in the ocean', SACLANTCEN TR-225. La Spezia, Italy, SACLANT ASW Research Centre, [AD 764-395].

[2] Akal, T., Dullea, R., Guido, G. and Stockhausen, J., (1993) 'Low-Frequency volume reverberation measurements', in Ellis, Preston and Urban (eds.), Ocean Reverberation, Kluwer Academic Publishers, Dordrecht.

[3] Andreeva, I.B. (1964) 'Scattering of sound in air bladders of fish in deep sound-scattering ocean layers.' Soviet Physics - Acoustics 10 (1), 17 - 20.

[4] Doutt, J.A. (1977) 'Broadband measurements of volume scattering strength in the Mediterranean', SACLANTCEN SR-17. La Spezia, Italy, SACLANT ASW Research Centre,.

[5] Thompson, C. and Love, R.H., (1993) 'Volume reverberation at mid frequencies in the Norwegian Sea', in Ellis, Preston and Urban (eds.), Ocean Reverberation, Kluwer Academic Publishers, Dordrecht.

[6] Preston, J.R., Akal, T. and Berkson, J.M. (1991) 'Analysis of backscattering data in the Tyrrhenian Sea', J. Acoust. Soc. Am. 87, 119-134 .

[7] Urick, R. (1983) Principles of Underwater Sound, 3rd Ed. McGraw Hill, New York, NY.

VOLUME REVERBERATION AT MID FREQUENCIES IN THE NORWEGIAN SEA

Charles H. Thompson and Richard H. Love
Naval Research Laboratory - Stennis Space Center
Stennis Space Center, MS 39529 U.S.A.

ABSTRACT

During July 1991, the Naval Research Laboratory (NRL) conducted volume reverberation measurements at seven locations in the Norwegian Sea between 63°N and 68°N as part of a multi-laboratory experiment directed by the SACLANT Undersea Research Centre. NRL obtained volume scattering strength versus depth profiles at frequencies between about 1 kHz and 10 kHz. During the day, scatterers were generally found between depths of 150 m and 500 m. Even though nights were short and the northern horizon was always visible except when obscured by cloud, scatterers were nearer the surface at night, extending from about 100 m to 400 m. Scattering strengths were high, with strongest scattering at about 2 kHz during the day and between 1 kHz and 2 kHz at night. Some geographic variation was observed, with scattering at the northernmost station being the weakest. Comparisons of the measurements to scattering strengths predicted from bioacoustic modeling of fisheries data have been made and indicate that the principal scatterers were most likely blue whiting.

1. INTRODUCTION

During a multi-laboratory exercise in July of 1991, the Naval Research Laboratory (NRL) measured volume scattering strengths at frequencies between about 1 kHz and 10 kHz at seven sites in the Norwegian Sea. The locations of these sites were: 1- 63° N / 2° W 2- 65° N / 0° W, 3- 66° N / 2° E, 4- 68° N / 5° W, 5- 65° N / 6° W, 6- 64° N / 8.3° W, and 7- 63° N / 6.5° W. Volume scattering strengths at freqencies below about 1.5 kHz were measured by the SACLANT Undersea Research Centre. Only NRL results are discussed in this paper.

2. MEASUREMENT METHOD

The volume reverberation measurements discussed in this paper are based on the method published by Machlup and Hersey[1] using small explosive charges and a directional receiver near the surface with its beam directed downward. The equations of Machlup and Hersey were modified to account for non-coincident source and receiver. The sound sources used were 180 g blocks of TNT, electrically detonated about 0.5 m below the surface. The receiver used was a thirty-two element line hydrophone, 1 m long, mounted along the axis of a conical reflector with a 2 m diameter base. By varying the number of active hydrophone elements, the receiver's 3-dB beamwidth can be maintained between 10° and 20° from 2.5 kHz up to 20 kHz. Below 2.5 kHz the main beam widens and performance degrades due to increasing sidelobes. When scattering strengths are high at the lower frequencies, the receiver can collect useful data down to about 800 Hz.

During this exercise, data were collected using sixteen of the receiver elements for frequencies from 5 kHz to 10 kHz and using the full aperture for frequencies of 5 kHz and below. Only data collected with the low frequency aperture are presented here. The output of the hydrophone was amplified by a differential amplifier with 20 dB gain and then bandpass filtered between 0.4 and 6 kHz with an additional 30 dB gain. Each volume reverberation record consisted of a short sample of noise immediately before each explosive charge, followed by the back-scattered signal. The records were digitally sampled at a rate of 20 kHz and stored on magnetic disks. A typical data collection sequence consisted of six to ten charges being fired and lasted about one and one-half hours. Each recorded signal was later digitally filtered into twenty-one 1/6-octave frequency bands and the voltage amplitude versus time envelope of each band's time series was calculated. Amplitude versus time envelopes at each frequency were averaged together and an average volume scattering strength versus depth profile for each frequency band was calculated for each data collection sequence. Scattering strengths were calculated using the equation:

$$S_V = 20 \log (V_{sig}) + 20 \log (t) - 10 \log E - 10 \log (1-\cos (\beta/2)) + \alpha ct + 10 \log \gamma - \text{gain} - \text{FFVS} - 171.1,$$

where S_V is the volume scattering strength in dB re $1\mu Pa^2$/Hz per m^3 at 1 m, V_{sig} is the noise-corrected voltage amplitude, t is time in seconds after the source detonation, E is the source energy per unit area in the 1/6-octave band in ergs/cm^2 at 100 m, β is the 3 dB beam width of the receiver, α is absorption in dB/m, c is the sound speed in m/sec, the γ term accounts for non-coincident source/receiver geometry, gain is the amplification of the recorded signal, FFVS is the free-field voltage sensitivity of the receiver, and 171.1 is a constant which includes the density of seawater and unit conversions for acoustic pressure and source energy.

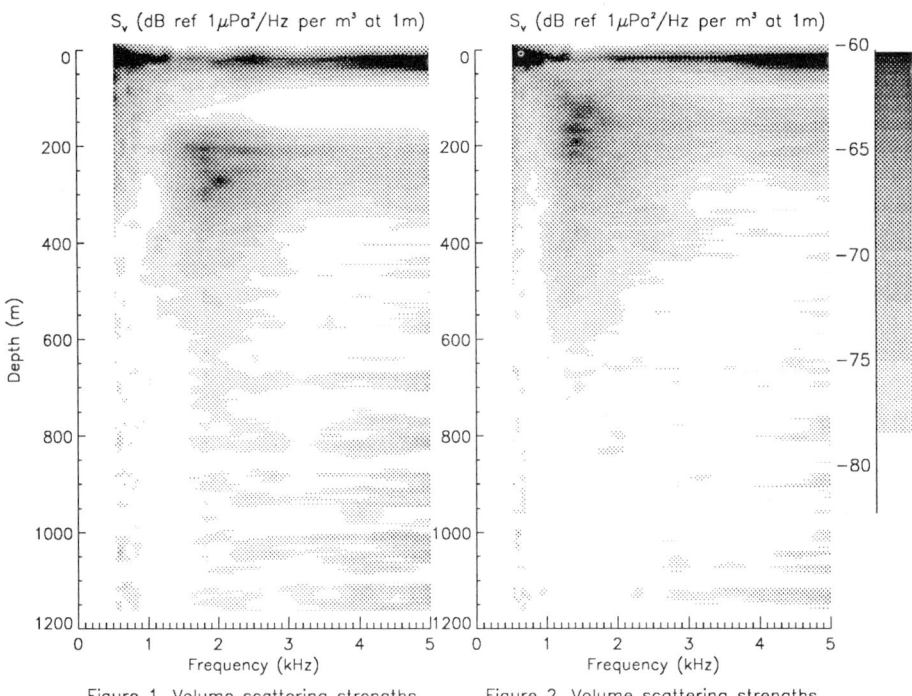

Figure 1. Volume scattering strengths at Site 2 during the day.

Figure 2. Volume scattering strengths at Site 2 during the night.

3. ANALYSIS

Scattering strength data were displayed versus frequency and depth with a color scale indicating S_V in dB. Shaded plots in this style are shown in Figures 1 through 4. The dynamic range of the data has been greatly reduced to be shown with greyscale shading. Each figure shows the initial blast from the charge and some surface scatter that obscures any volume scattering very near the surface. Parts of the data are also obscured by off-axis returns at the lower frequencies where the effect of the receivers side lobes is evident. The initial surface scatter appears to contaminate the data at increasingly greater depths as frequency decreases because the sensitivity of the receiver's main lobe decreases with respect to the side lobes. A similar effect can be seen below strong scattering from a layer. Off-axis returns from a layer arrive at the receiver slightly later than returns from directly below, creating a reverberation "tail" that extends to greater depths than those to which scatterers probably reside.

Figure 1 shows data from Site 2 during daytime and Figure 2 shows data from the same site at night. These two figures show a scattering layer between 120 and 390 m during the day, peaking around 2 kHz, that moved closer to the surface at night, between about 90 and 300 m, with a peak around 1.6 kHz. Figures 3 and 4 show daytime and nighttime data, respectively, from Site 4. Each of these two figures shows two distinct scattering layers. The shallower of the two layers lies between 150 and 335 m during the day, peaking at 1.8 kHz and between 125 and 300 m at night, peaking at 1.6 kHz, while the deeper layer extends to 500 m during the day and to 550 m at night. The upper range of the

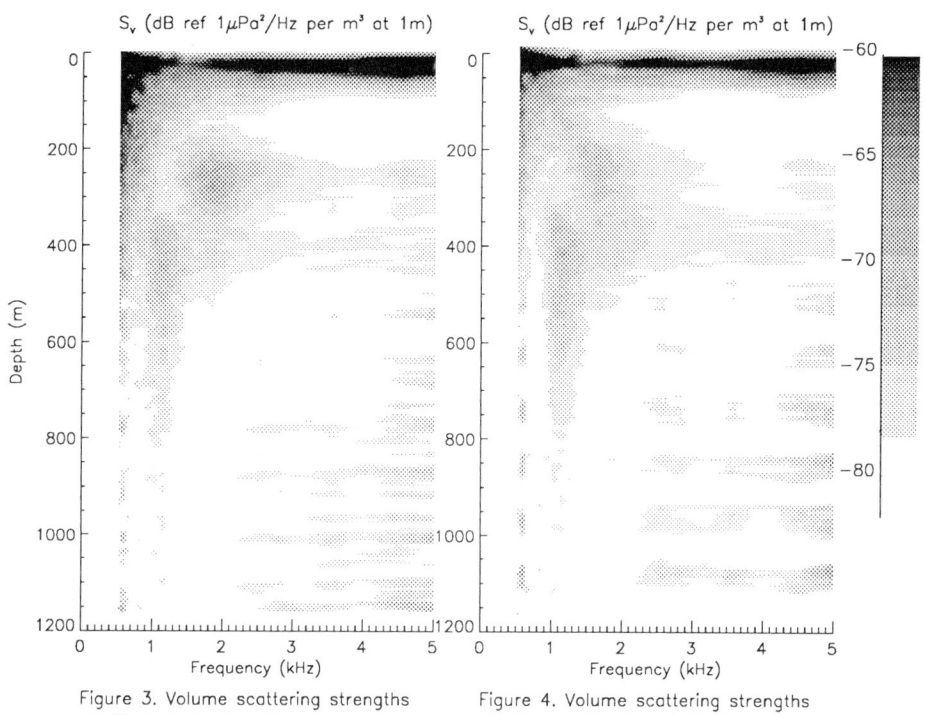

Figure 3. Volume scattering strengths at Site 4 during the day.

Figure 4. Volume scattering strengths at Site 4 during the night.

deep layer overlaps the bottom of the shallow layer indicating that at this depth the two groups of scatterers are mixed. The strongest scattering in the deep layer is at a lower frequency than that in the shallow layer, indicating that the resonant frequency is lower and that the fish are larger.

A closer examination of the data in Figure 1 reveals a layer similar to the deeper layer at Site 4, but weaker, centered at about 400 m depth and scattering most strongly at about 1.2 kHz. This layer was observed in data from most of the sites to some extent, but was usually weak, as in Figure 1, or at night, was within the reverberation tail from the strong shallow layer above it as in Figure 2.

From plots of S_V versus depth and frequency from each sequence, an analysis of the depth ranges of scattering layers was made. The scattering strength profiles were then integrated over the chosen layer depths to produce layer scattering strength (S_l) versus frequency curves. The sum of the layer strengths (the column scattering strength) at each site is shown in Figure 5. Multiple day and night sequences were conducted at Site 2; the curves shown are averages of the two day and two night sequences.

Both the daytime and nighttime curves from Site 4 show a pair of resonance peaks due to the two layers discussed above. These curves show Site 4, the northernmost site, to have the lowest scattering strengths over most of the frequency range measured. Scattering strengths below 1.2 kHz during daytime are among the highest, however, because of the low frequency peak in the deep layer.

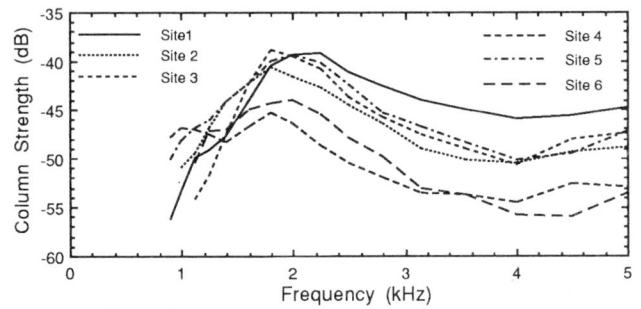

Figure 5(a) - Daytime column strengths at each site.

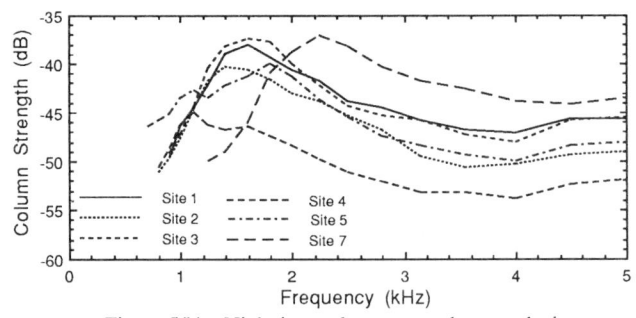

Figure 5(b) - Nighttime column strengths at each site.

Although nights were short and the northern horizon was always visible except when obscured by cloud, some upward migration of scatterers took place at night at all sites for which both day and night data were obtained. At most sites, the same number of layers was observed day and night. Site 5, however, was unique because a single layer that was observed in the day split to form two layers at night. A group of scatterers near the surface at night was responsible for the secondary peak in this curve at 1.2 kHz.

Although the curve from Site 6 is similar in shape to other daytime curves, the distribution of scatterers was quite different from other sites. At other sites during daytime the majority of the scatterers were deeper than 200 m. At site 6, however, the greatest density of scatterers was shallower than 200 m. Water temperature data taken during the exercise show that the temperature profile at Site 6 was different from the other sites, with colder water below 200 m. Scatterers at Site 6 may have moved shallower into water of a preferred temperature.

Site 7, the southernmost site, is unique among the nighttime curves with a resonance peak which is relatively high in frequency.

4. MODELING

Bioacoustic modeling was conducted to relate scattering strengths to potential scatterers. The strength of a layer of swimbladder-bearing fish is

$$S_l = 10\log\{\Sigma n(r,z)\sigma(f,r)\}$$

where S_l is the layer scattering strength in dB re $1\mu Pa^2/Hz$ per m^2 at 1 m, n is the number of swimbladders of radius r at depth z, and σ is the acoustic cross section of an individual fish at frequency f. Fish size is usually given in terms of length distributions, which can be related to distributions of swimbladder radii.

The Faeroese Fishery Laboratory conducted a limited acoustic survey of blue whiting near the Faeroe Islands during August 1991 [2]. They recorded 282,000 tons of fish, 75% of which were within 100 nm of the Islands and 80% of which were age 2 or less. These results differ significantly

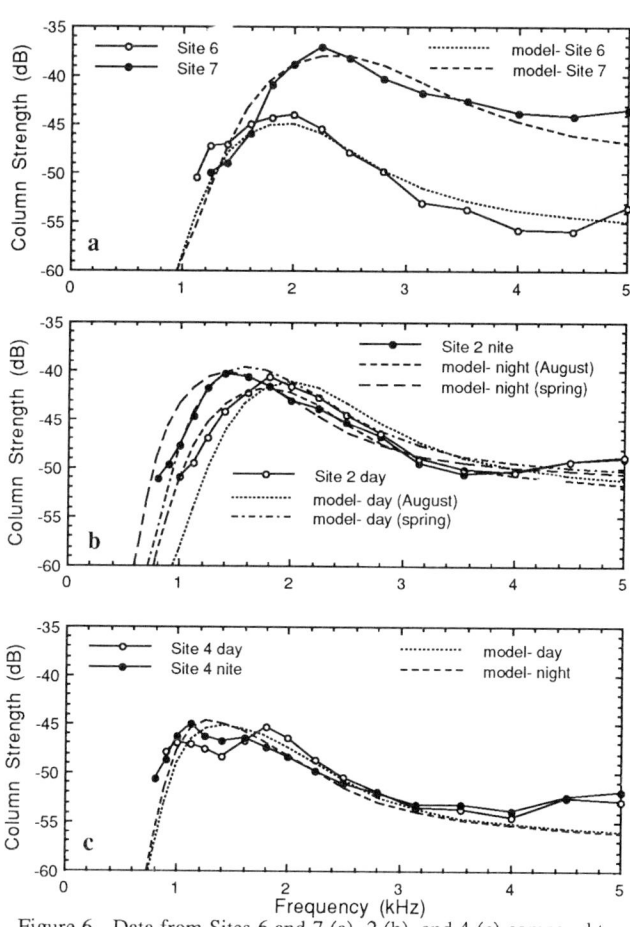

Figure 6 - Data from Sites 6 and 7 (a), 2 (b), and 4 (c) compared to model predictions from fisheries survey data.

from Russian/Norwegian surveys in March-April of 1991 that recorded 5,200,000 tons of blue whiting, 28% of which were age 2 or less [2]. This indicates that the Faeroese survey recorded little of the adult stock.

Figure 6 compares measured column strengths with model results [3]. The modeling used fish depths based on observed layer depths and varied the number of fish to fit measured S_l's. Figure 6a shows the results obtained for Sites 6 and 7: using the Faeroese August blue whiting length distribution gives a good match, indicating that juvenile blue whiting were predominant at these stations. Figure 6b shows the results from Site 2: a blue whiting length distribution that is intermediate to those obtained by the Faroese in August and the Russians and Norwegians in the spring would give the best fit, indicating a mixture of juveniles and adults at this station. At Site 4, the spring blue whiting distribution was used to model the shallow layer and, since redfish are found between 300 and 500 m at these latitudes in the Norwegian Sea, a redfish length distribution from a Norwegian survey in the eastern Norwegian Sea in October 1991 was used to model the deep layer [4]. The model results for each layer agree with the data reasonably well, indicating that redfish and adult blue whiting are responsible for the deep and shallow layers respectively. However, the model peaks are a little closer together and broader in frequency than the data so that, when the layers are combined, as in Figure 6c, only one peak is apparent in the total modeled layer strength, compared to two in the data.

5. CONCLUSIONS

These measurements show that volume scattering strengths can be quite high in the Norwegian Sea in the summer. Layer scattering strengths in excess of -45 dB were measured at frequencies from 1 kHz to 5 kHz, with peak layer strengths as high as -38 dB occurring between 1.5 and 2.5 kHz. Layer strengths were lowest at the northernmost site. Comparisons to fisheries data suggest that the primary scatterers were blue whiting, which were more numerous and smaller to the south than the north, with concentrations of redfish in some areas.

6. ACKNOWLEDGEMENTS

The authors would like to thank the officers and crew of the R/V LANGEVIN and the R/V ALLIANCE, John Preston, Chief Scientist on this experiment, Redwood Nero of NRL and personnel from the SACLANT Centre and GERDSM for assistance in the conduct of the measurements. This work was supported by Program Element 65857N of the Office of Naval Research, W. V. Harned and E. Estalote, Program Managers.

7. REFERENCES

1. Machlup, S. and J. B. Hersey "Analysis of Sound-Scattering Observations from Non-Uniform Distributions of Scatterers in the Ocean," Deep-Sea Res. 3, 1-22 (1955).

2. Monstad, T., Institute of Marine Research, Bergen, Norway, Personal Communication, 1992.

3. Love, R. H. "Resonant Acoustic Scattering by Swimbladder-Bearing Fish," J. Acoust. Soc. Am. 64(2), 571-580 (1978).

4. Nedreaas, K., Institute of Marine Research, Bergen, Norway, Personal Communication, 1992.

VOLUME REVERBERATION IN THE MARGINAL ICE ZONE OF FRAM STRAIT

Marcia A. Wilson and Richard H. Love

Naval Research Laboratory
Code 243, Bldg. 1005, Stennis Space Center
Mississippi 39529 United States of America

ABSTRACT Many measurements of volume reverberation have been conducted in the open oceans of the world. In May 1988, the Naval Research Laboratory at Stennis Space Center (NRL-SSC) had the opportunity to investigate volume reverberation in a unique region: the marginal ice zone (MIZ) of the Fram Strait between Spitsbergen and Greenland. Measurements were made from a drifting ice camp at three locations over a four day period. Volume scattering strength versus depth profiles were obtained at frequencies from 3.5 to 50 kHz using short CW pulses from a suite of downward and upward looking transducers. Results show that scattering layers occurred from about 100 to 200 m and 400 to 500 m at each location, with some variability in strength at the different locations. A comparison with volume scattering strengths reported from other cold water regions shows that values observed in the Fram Strait MIZ were lower than those reported at similar frequencies from open waters of the Norwegian Sea, Labrador Sea, and northern Baffin Bay, and comparable to those in the Denmark Strait, Davis Strait, Chukchi Sea MIZ, and northeast of Iceland.

1.INTRODUCTION

Volume reverberation measurements were made in May 1988 from an ice floe in the marginal ice zone (MIZ) of the Fram Strait by a team of scientists from the Naval Research Laboratory at Stennis Space Center (NRL-SSC). The experiment was located around 79°N and 0° to 3°W in the area where the West Spitzbergen Current meets the East Greenland Current, between Spitzbergen and Greenland. Volume scattering strength versus depth profiles were obtained at frequencies from 3.5 to 50 kHz using short CW pulses. Data were collected from May 15th to May 18th while the ice camp drifted westward about 80 km.

The Fram Strait MIZ is a unique region for several reasons. It is the only deep water channel between the Arctic Ocean and warmer waters to the south. The depth of the bottom in the Bering Strait is about 50 m, paths through the Barents sea must pass through areas around 300 m deep, and the Nares Strait between Baffin Bay and the Lincoln Sea is about 500 m deep. The Fram Strait, however has depths of 2500 to 3000 m. Strong currents bring polar water in from the north and North Atlantic water in from the south, as shown in Figure 1.[1] Eddies are common where these water masses meet and mix. There is a nearly permanent local gyre around the Molloy Deep near the experimental area. Temperature and sound speed profiles in the area have a large layer of low values near the surface, an abrupt increase around 100 m and smaller increases and decreases below this thermocline. Compared to other marginal ice zones, the ice edge in the Fram Strait does not change its position very much with the seasons. Ice is constantly drifting down from the north and melting as it reaches the polar front.[2]

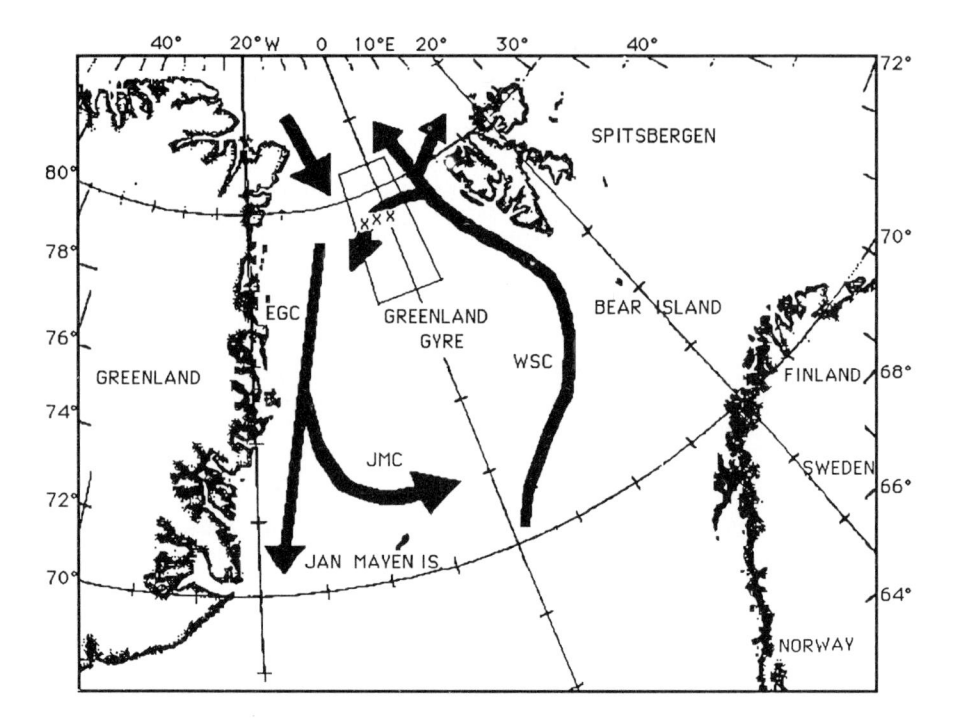

Figure 1. Locations of experimental stations shown in relationship to the prevailing currents in the area. EGC is East Greenland Current, JMC is Jan Mayen Current, WSC is West Spitsbergen Current.

2. THEORY AND METHODS

2.1 Biological Data:

At the lower end of our frequency range volume scattering in the ocean is caused primarily by gas-filled swimbladders of pelagic fish. At very high frequencies, however, the shells of planktonic animals become important sources of reverberation. Little information is available on fish in the Fram Strait, but sightings of seals, polar bear and walrus during the experiment indicated that under the ice there was a food source that was most likely fish. Two similar fish species, *Boreogadus saida* and *Arctogadus glacialis*, commonly called polar cod, are the main food sources for seals and polar bears. These fish have been found as far north as 84°N. One year old polar cod off the Labrador coast are about 9 cm long, but those from the arctic do not grow as fast.[3] Northeast Arctic cod, caught to the west and south of Spitzbergen north of Bear Island, consume young polar cod, redfish, capelin, and deep sea shrimp, 5 to 9 cm in length, and herring and haddock, 10 to 14 cm long.[4] Capelin, herring, and haddock are not found as far north as the MIZ 88 experiment site. Most adult redfish migrate into the Barents Sea in the spring and summer, but many larval and young redfish may be carried as far as the Fram Strait in the West Spitsbergen Current. Polar cod inhabit all parts of the Arctic Ocean and are known to spawn between December and February off Spitzbergen and in the southeast Barents Sea. Although they prefer cold water, polar cod may be at any depth seeking out the available food.[5] They have been caught in the Fram Strait, Svalbard area, and western Barents Sea.[6] Therefore, since they are known to be found in the experimental area and they have swimbladders, polar cod are potential sources of volume scattering.

Plankton surveys in the Fram Strait indicate that the most common zooplankton are the copepods *Calanus hyperboreus*, *Calanus finmarchicus*, and *Calanus glacialis*.[2] Also, polar cod caught near the under ice surface in the Fram Strait had mainly calanoid copepods, and amphipods in their stomachs.[7] Large numbers of copepods can be strong scatterers, as can aggregations of diatoms, ciliates and flagellates which are the main food sources for *Calanus*,[8] but only at much higher frequencies than were used in the MIZ 88 experiment. Small pteropods which are common in the Chukchi Sea MIZ possibly could be resonant scatterers at 30 to 50 kHz in the Fram Strait MIZ as well.[9]

2.2 Acoustic Measurements and Model:

The sources for this experiment were a set of 5 transducers generating 10 or 40 ms pulses at frequencies between 3.5 and 50 kHz. They were placed in the water just below the ice for downward looking measurements and at 65 m for looking up. For downward looking data, the first layer starts at 35 m, after the surface returns diminish. Upward looking data were obtained only at Station 2 from 3.5 to 12 kHz. Scattered signals were received by the same transducer, log amplified and analog recorded for further processing at NRL-SSC. About 10 pulses were averaged for each frequency and pulse length. Volume scattering strength, as a function of depth, was obtained using the following equation:

$$Sv = 20 \log V - SL - FFVS - 10 \log m - 10 \log(1 - \cos(b/2)) + 20 \log t + act - Gain + 20.8$$

where Sv is volume scattering strength, V is voltage level of the received signal, SL is source level, FFVS is free field voltage sensitivity of the transducer, m is pulse length, b is the -3dB beam width, t is time, a is the attenuation coefficient, and c is sound speed.[10] Peaks in these profiles indicate the depths at which scatterers are most numerous. Integrating the scattering strength over a selected depth range gives a layer strength or column strength. Column strengths were calculated from the depth at which volume scattering data began to the bottom of the deepest scattering layer in the upper 800 m.

A swimbladder scattering model was employed to estimate sizes of fish responsible for observed scattering.[11] The length of a fish that would produce its maximum scattering strength at a given frequency and depth was estimated using the equation:

$$L = q[(z+10)/10]^{1/2} / \pi f$$

where L is the fish length, q is the ratio of the fish length to the effective radius of the swimbladder, z is the depth in meters, and f is the frequency in kHz. Selecting the right q value requires some knowledge of the type of fish that are found in the experiment area. A thin fish would have a larger q value than a fat one. For polar cod the q value is about 25.

3. RESULTS AND DISCUSSION

Figure 2 shows the Station 2 sound speed profile next to a scattering strength vs. depth profile. The upper scattering layer is above the thermocline in cold polar water brought

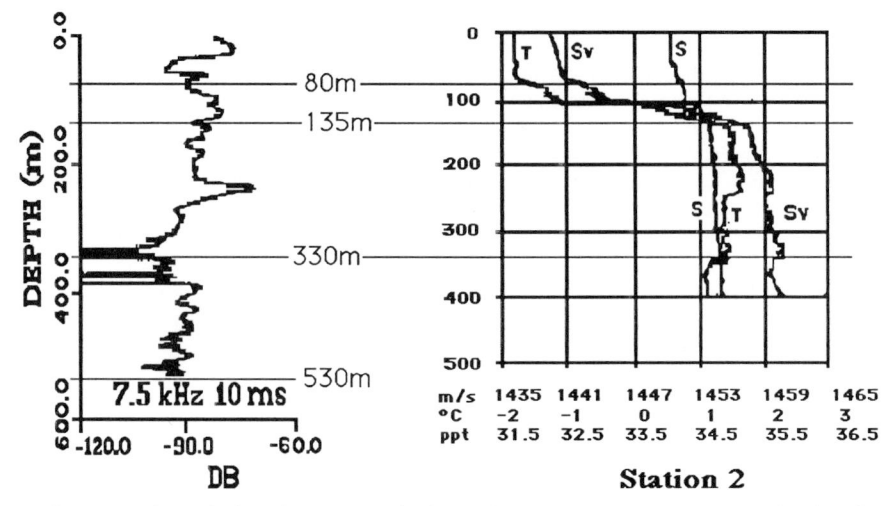

Figure 2. Layer boundaries chosen to coincide with temperature structure at Station 2

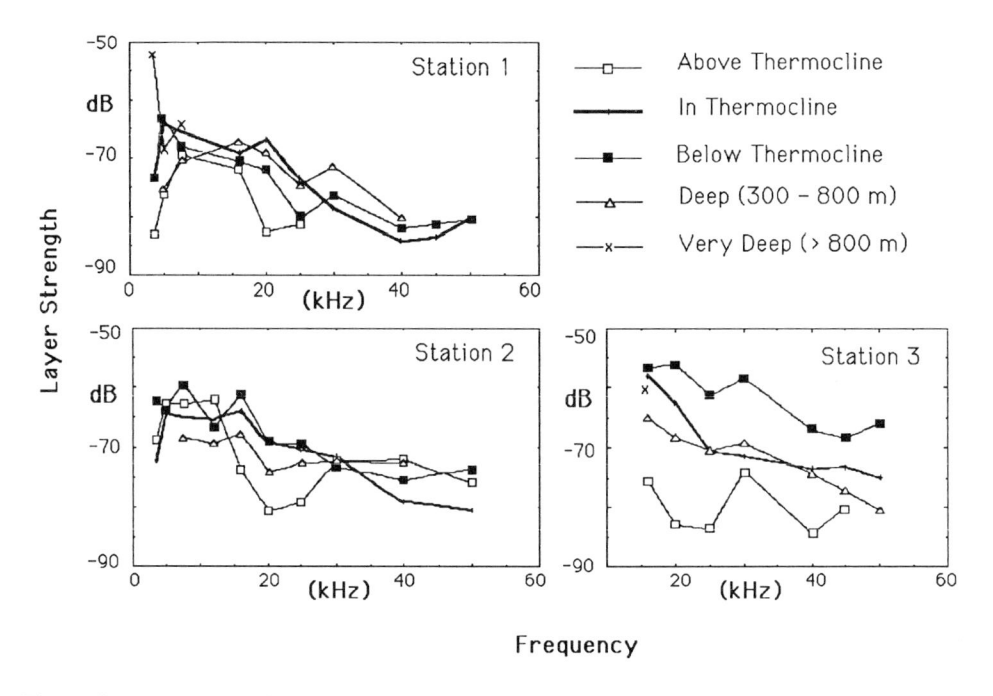

Figure 3. Layer strength vs. frequency for layers related to temperature profile.

from the north in the East Greenland Current. The second layer is in the thermocline where polar water is mixing with warmer Atlantic Intermediate Water and the temperature increases from -1.6° to +1.5°. The next layer is below the thermocline to 330 m. At 7.5 kHz there is also a deep layer from 330 to 530 m. Layer boundaries differ only slightly with frequency and station. Layer strength versus frequency data are shown in Figure 3. Values above the thermocline are notably higher when upward looking data are included. Lower frequencies generally have higher layer strengths. For Station 1, at the three lowest frequencies, there are five layers: four similar to those at Station 2 and one strong layer deeper than 800 m. Station 3 had a very deep layer also, but only at 16 kHz, which was the lowest frequency measured at that location. Above the thermocline, layer strengths for Stations 1 and 2 peaked between 5 and 16 kHz. If this scattering is from polar cod, they are between 1 and 4 cm long. That would indicate mainly 3 to 6 month and perhaps one year old fish. Stations 2 and 3 also have a layer strength peak at 30 kHz. This may be scattering from pteropods or larval fish about 7 mm long with an effective scattering radius about 0.3 mm. Polar cod 5 to 6 cm long could provide scattering layers at the depths and low frequencies of the in thermocline and below thermocline layers of Stations 1 and 2. The 300 to 800 m layer peaked at 16 kHz for all three stations. This could be caused by polar cod 3 to 4 cm long, or perhaps young redfish or Atlantic cod.

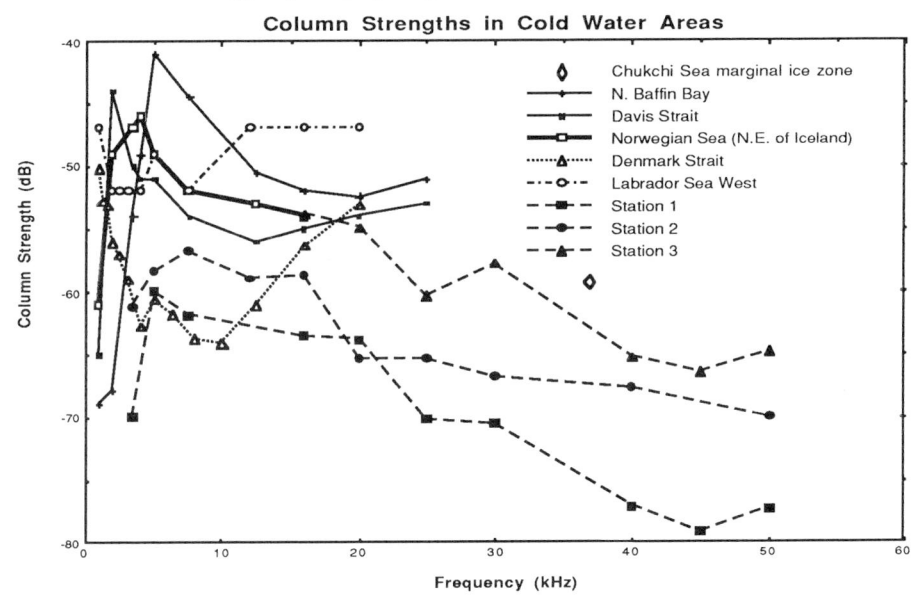

Figure 4. Column Strengths for the upper 800 m in the Fram Strait marginal ice zone and for other cold water areas.

Figure 4 shows column strengths for the Fram Strait and other cold water areas for comparison. Column strengths in the Fram Strait increased as the ice floe drifted west in the MIZ. Levels at Station 1 and 2 are similar to those in other cold water straits and lower than those in open water. Station 3 data matched some Norwegian Sea data from northeast of Iceland at 16 kHz, the only frequency at which they overlapped. Data from other parts of the Norwegian Sea were generally higher. The Chukchi Sea MIZ data at 38 kHz was about the same as Station 3 data at 30 kHz, but higher than that for 40 kHz.

4. CONCLUSIONS AND SUMMARY

Column strengths in the Fram Strait MIZ varied between -50 and -80 dB in the 3.5 to 50 kHz range with a general decrease from low to high frequencies. Levels increased as the ice camp moved toward the Greenland continental shelf. The variations of layer strength with frequency indicate more than one scattering species. Young polar cod from 2 to 6 cm long are the most likely scatterers in the lower part of the frequency range. The peak layer strengths at 30 kHz and above may be from larval fish or perhaps pteropods.

5. ACKNOWLEDGEMENTS

This research was supported by the Office of Naval Technology. Program management was provided by Dr. Edward R. Franchi of Naval Research Laboratory, Stennis Space Center under Program Element 0602435N, Project Number RJ35I21. NRL SSC Contribution PR92:005:243. Approved for public release; distribution is unlimited.

6. REFERENCES

1. Meredith, R.W., P.J. Bucca, and K. McCoy (1989) "Environmental Measurements and Analysis: Arctic Acoustics Experiments in the Marginal Ice Zone" NORDA Report 210.
2. Smith, S. L., W.O. Smith, L.A. Codispoti, and D.L. Wilson (1985) "Biological observations in the marginal ice zone of the East Greenland Sea", Journal of Marine Research, 43, pp. 693-698.
3. Scott, W.B. and M.G. Scott (1988) Atlantic Fishes of Canada, Can. Bull. Fish. Aquat. Sci.,219, pp. 260-270.
4. Mehl, S. (1989) "The Northeast Arctic cod stock's consumption of commercially exploited prey species in 1984-1986" in G. Hempel(ed.), Oceanography and Biology of Arctic Seas, ICES,Rapp. P.-v. Réun. Cons. int. Explor. Mer., 188, pp. 185-205.
5. Farquhar, G.B. (1990) "Biological Scattering in the Barents Sea", internal report Naval Oceanographic and Atmospheric Research Laboratory (NOARL) Liaison Office.
6. Monstad, T. and H. Gjøsæter (1987) "Observations on Polar Cod (*Boreogadus saida*) in the Barents Sea 1973 to 1986", ICES, C.M.1987/G:13.
7. Lønne, O.J., and B. Gulliksen (1989) "Occurrence and ecological importance of sympagic fauna in the Fram Strait, Svalbard area, and western Barents Sea" in G. Hempel(ed.), Oceanography and Biology of Arctic Seas, ICES, Rapp. P.-v. Réun. Cons. int. Explor. Mer., 188, p. 170.
8. Barthel, K-G. (1989) "Feeding of Calanus in the Greenland Sea" in G. Hempel(ed.), Oceanography and Biology of Arctic Seas, ICES, Rapp. P.-v. Réun. Cons. int. Explor. Mer., 188, p. 173.
9. Hansen, W.J. and M.J. Dunbar, (1970) "Biological Causes of Scattering in the Arctic Ocean" in G.B. Farquhar (ed.) Proceedings of an International Symposium on Biological Sound Scattering in the Ocean, Maury Center for Ocean Science, DON,Washington, D.C., pp. 508-526.
10. Urick, R.J. (1983) Principles of Underwater Sound, McGraw-Hill Book Company, New York.
11. Love, R.H. (1978) "Resonant Acoustic Scattering by Swimbladder-bearing Fish," J. Acoust. Soc. Am. 64, pp. 571-580.

Section 7

Signal Processing Issues

ANALYTICAL DESCRIPTION AND EXPERIMENTAL RESULTS FOR REVERBERATION-RESISTANT ACOUSTIC TRACKING SIGNALS

Thomas J. Curry and Thomas A. Casey
Naval Undersea Warfare Center Division, Newport
Newport, Rhode Island 02841-5047 USA

ABSTRACT The emerging need to design and operate transportable undersea tracking systems in other than deep water scenarios is discussed in the context of the performance evaluation of undersea systems. In particular, a class of differentially encoded binary phase shift keyed (DPSK) signals are examined in regard to its ability to provide resistance-to-boundary interaction while still meeting the needs of the underwater tracking system. Receiver operating characteristics are presented for a variety of DPSK signals with emphasis on their resistance to interference that is characteristic of boundary reverberation. The parameters of the DPSK signals examined include the code length, bandwidth, and duration of the signal. Analysis and simulation predict, and experimental results verify, that longer code words provide a significant improvement in interference immunity. Acknowledgement is extended to M. Amaral, S. Jarvis and D. Moretti for their assistance in producing results.

1. INTRODUCTION

The changing geopolitical environment has greatly influenced the thrusts of undersea systems research, development, test, and evaluation (RDT&E). The current attention on contingency and limited objective encounters carried out in littoral waters has placed greater emphasis on conducting exercises in realistic oceanographic environments that more closely match those where the undersea system is likely to be deployed.

Testing and evaluating the performance of undersea systems in the above environments requires portable (or transportable) test and evaluation (T&E) systems that can perform satisfactorily in the presence of reverberation characteristic of the shallow water environments. Underwater tracking systems for test and evaluation typically utilize active acoustic sources on the targets to be tracked. The transmissions from the targets are received on spatially separated receivers that are placed so that time-of-arrival (TOA) measurements at the receivers can be used to determine the spatial coordinates of the target.

Because of multiple access requirements and the need to minimize bandwidth, the binary phase shift keyed (BPSK) tracking signal has been the signal of choice for several tracking problems, and has been used successfully in several deep water tracking ranges including those in the Atlantic and the Pacific. This class of tracking signal has enjoyed considerable success in deep water; however, its performance potential in the presence of reverberation which is characteristic of shallow water environments has not been examined prior to the work reported herein.

In the following paper, the structure of the BPSK signal processor is first described followed by an analytical prediction of performance of several variants of the BPSK processor. Next, the performance is analytically assessed in the presence of modeled shallow water reverberation. The results of experiments conducted in Narragansett Bay and Block Island Sound off the coast of New England are used to evaluate the performance of the BPSK system in these highly reverberant environments. These results are then compared with those predicted analytically.

2. DESCRIPTION OF THE BPSK SYSTEM

The BPSK signal can be thought of as a sinusoidal carrier multiplied by a binary baseband waveform of ±1's, which achieves the logic levels of 0 and 1 being encoded in the phase of the sinusoidal carrier. The conventional BPSK system uses a fully coherent detector, and for the purposes of multiple access and TOA measurement, the recovered baseband code in this system is processed by a matched filter.

The coding used is chosen for desirable auto-correlation and cross-correlation properties so that timing ambiguities for the signal of interest are minimized (i.e., low sidelobes of the autocorrelation function), and the probability of incorrect target identification is minimized (i.e., low cross-correlation peaks).

In underwater acoustic tracking applications, the signal is typically only present about 1 or 2% of the time; consequently, fully coherent detection has not been historically used. Rather, a differentially coherent system (DPSK) of the type shown in Figure 1 below has been used to recover the baseband code for subsequent processing.

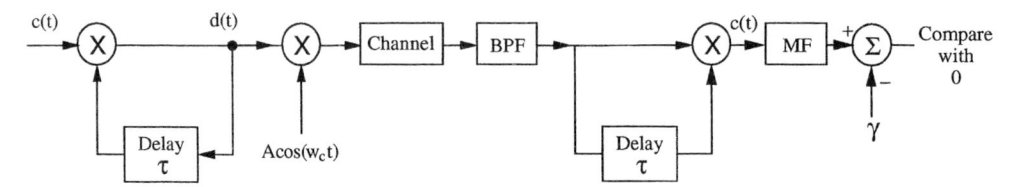

Figure 1. DPSK System.

In addition to eliminating the local oscillator in the coherent system, this scheme tends to be more tolerant of Doppler and other phase non-stationarities that occur over periods of time longer than the code length. However, its performance is inferior to that of the coherent system because the DPSK receiver essentially recovers the baseband code using a noisy reference signal.

The application of these detectors to the transportable T&E scenario also places a premium on power consumption, volume, and weight. For these reasons, zero-crossing detectors or hardlimiters are currently used in deep underwater tracking systems to reduce computational load. Another benefit of the hardlimiter is that the receiver becomes constant false alarm rate (CFAR), allowing the use of fixed thresholds.

An additional PSK receiver configuration (referred to as the hardlimited bandpass correlator (HLBPC)) is basically the coherent PSK system with the addition of a hardlimiter at the output of the bandpass filter. The addition of these limiters results in additional degradation with respect to that incurred by using the differentially coherent scheme instead of the fully coherent approach. Each of these variants is analyzed below in order to set a baseline against which the performance of these systems in reverberation can be compared.

3. NOISE-LIMITED ANALYSIS OF THE BPSK SYSTEM

The following analysis assumes that the BPSK signal is corrupted with white Gaussian noise, and that all filters are normalized to have unity dc gain. It is further assumed that the time-bandwidth product is sufficiently large to permit a Gaussian assumption at the output of the matched filter. In all noise-limited cases, the analytical approach is to determine the mean and variance at the output of the matched filter for both the noise-alone case and when the signal is aligned in the filter. In some cases where nonlinear devices (delay-multiply loop and hardlimiters) are added, simulation is used to determine the means and variances.

The mean and variance at the output of the matched filter at the time when the signal is perfectly aligned can be found directly for the coherent system. For the DPSK system without hardlimiters, the delay-multiply loop adds considerable complexity to the analysis; and as it will be shown, this loop also accounts for a significant part of the performance degradation caused by using the DPSK system.

To illustrate the impact of the delay-multiply loop, it is convenient to relate the signal-to-noise ratio (SNR) at the output of the bandpass filter (BPF) to that of the baseband signal into the matched filter (MF). Defining SNR1 as the ratio of signal power at the output of the BPF to the noise power at the same point and SNR2 as the square of the mean at the input to the matched filter divided by the variance at that same point, then it is found that for the coherent system that SNR2 = 2 • SNR1 and for the DPSK system that SNR2 = 2 • SNR1^2/(2 • SNR1 + 1). At high SNR1, the coherent system performs 3 dB better than the DPSK system; at low SNR1, there is increasing loss as SNR1 decreases. Figure 2 compares these two SNRs over the range indicated.

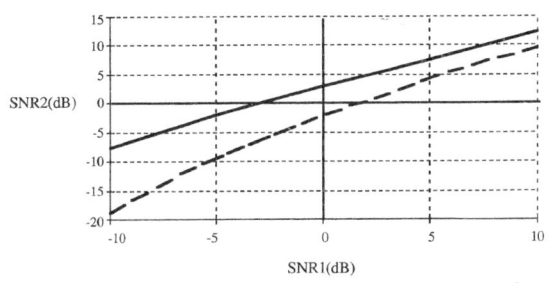

Figure 2. Comparison of Coherent PSK and DPSK Performance.

Addition of the hard limiters can further degrade performance because of the non-linear signal and noise interactions. While a detailed dissertation on the BPSK systems discussed above is not the primary focus of this paper, the performance of the variants of the BPSK system in noise-limited environments provides a baseline against which the performance can be compared for the reverberation/interference cases. Using the methodology outlined earlier, the noise-limited performance of the BPSK systems was determined as a function of input signal-to-noise ratio. The coherent case provides the best performance followed by the hardlimited bandpass correlator and the DPSK system without hardlimiters.

From a performance point of view, the choice is clear; however, recall that other considerations such as power consumption, size, and weight are also of importance for the application, and can make the suboptimum configurations of interest. With these considerations in mind, the focus of the remainder of this report will be on the hardlimited bandpass correlator (HLBPC) and the hardlimited differentially coherent phase shift keyed (HLDPSK) systems.

4. EFFECTS OF INTERFERENCE

Of interest in this analysis is the effect of code-word length on the ability of the DPSK system to reject reverberation. Increasing the code length can be accomplished by decreasing the bit duration, or by simply adding more bits of the same length with decreased amplitude to maintain constant signal energy for all codes.

It is known that the performance of the optimal detector in white noise depends only on the signal energy and noise spectral density; consequently, it is expected that fixed energy signals of any code length will have equivalent performance in noise for the coherent system. As a baseline for comparison, Figure 3 shows the performance of the candidate systems for different fixed-energy code lengths in noise. The values of SNR1 are all measured in the 3-kHz bandwidth of the 19-bit code.

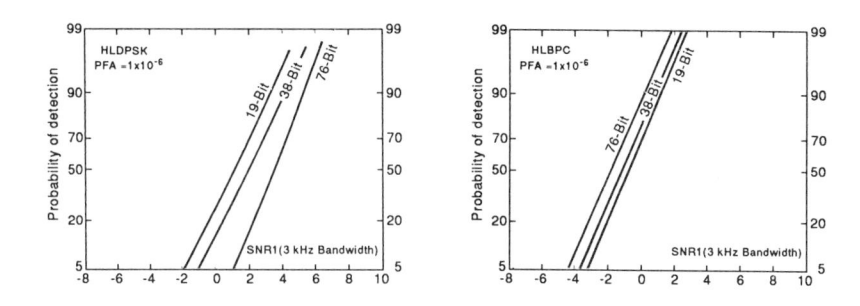

Figure 3. Fixed Energy Codes in Noise.

For the HLBPC, all codes perform equally well. Note that the longer codes incur a penalty in performance for the HLDPSK system, which is caused by the nonlinear delay-multiply loop adversely reacting to the decreased SNR1 at the bit level for the longer codes.

The signal structures analyzed above are examined in the presence of modeled reverberation. The modeled reverberation uses the approach developed by Estes and Fain[1] which is outlined below. The model utilizes time-varying impulse responses and windowing techniques to determine the response of an underwater channel to any input signal. Statistical descriptors of the complex impulse responses and the continuous wave (CW) response of the underwater acoustic channel of interest are used to specify the channel impulse responses which modify the input signal. The process is then replicated to estimate statistics of the received signal.

Figure 4 depicts the basis of the model, where the source and the receiver are in general moving, indicating that there are several acoustic paths that couple the receiver and the transmitter.

$r(t) = \sum_i \sum_n \alpha_i(t - n\tau_{ci}) S(t - \tau_i(t)) * h_{in}(t)$
(path) (coherence times)

$\alpha_i(t)$ = Raised cosine aperture for the ith path

$h_{in}(t) = \bar{h}_i(t) + \delta h_{in}(t)$

$\tau_i(t)$ = Delay model for ith path

τ_{ci} = Coherence time for ith path

$\bar{h}_i(t)$ = Mean impulse response of ith path

$\delta h_{in}(t)$ = Non-stationary random component of ith path @ $n\tau_{ci}$

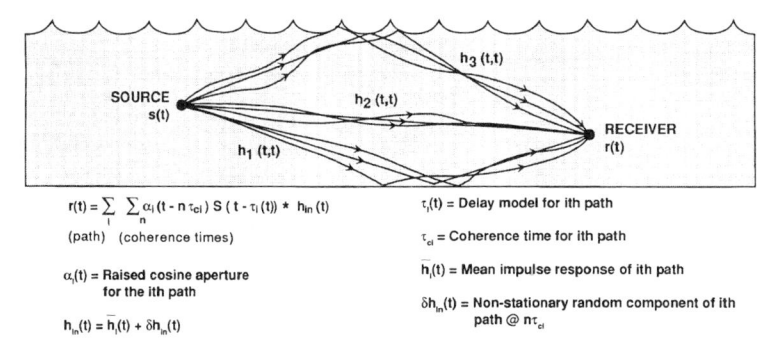

Figure 4. Model for Estes-Fain Channel.

These paths are defined on the basis of their average Doppler shift and are further characterized by a source-to-receiver time delay, $T_i(t)$, which can either be calculated based upon a range model for the ith path, or determined experimentally. Each of these paths is described by a nonstationary random impulse response with a stationary mean. The time over which the nonstationary component of the ith path varies is called the coherence time, T_{ci}, while the duration of this same path is called the reverberation time, T_i. An example of the input and output of this reverberation model is shown in Figure 5.

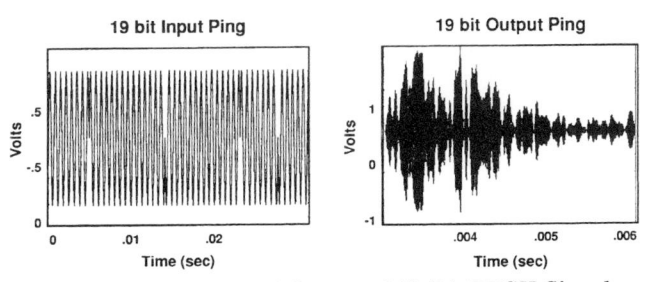

Figure 5. Input and Output of 19-Bit DPSK Signal.

Using the above approach, short and long duration CW burst signals were used to probe the propagation channel to determine T_{ci} and T_i for the reverberation model. These data were gathered in a shallow water environment characterized by a depth of about 20 meters in Narragansett Bay with transmitter and receiver separation of about 500 meters. Simulated signals were applied to the model using the experimentally derived parameters, and the modeled output signals were applied to the hardlimited DPSK and BPC systems. The performance of the systems was assessed, and the results are summarized in Figure 6 for both the HLDPSK and HLBPC systems using fixed-energy signals.

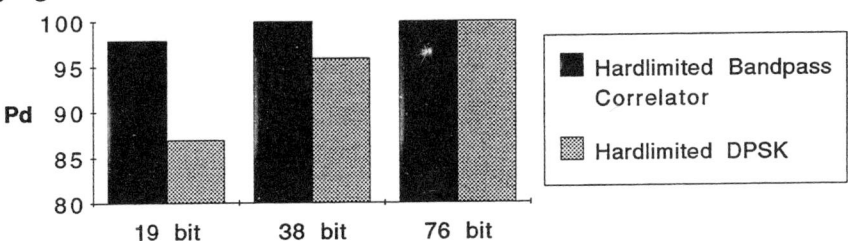

Figure 6. Performance Simulation with Fixed Energy Signals Generated from Estes-Fain Model.

5. EXPERIMENTAL RESULTS

With the above analytical descriptions and predictions of performance as a basis, two experiments were conducted off the coast of the northeast USA. The first of these (described above) was used to provide channel measurements for the Estes-Fain reverberation model; however, real BPSK data was collected and processed with the previously described BPSK systems. The second experiment was similar to the first, but was conducted in a different location in deeper water (65 meters) and longer horizontal range (1850 meters) south of Block Island off the southern coast of Rhode Island.

The experimental data was appropriately digitized and applied to both systems. The thresholds were set to achieve the desired false alarm rate in noise, and detection probability was estimated by counting the number of threshold crossings divided by the total number of signal transmissions.

Figure 7 shows the experimental results for the Bay Tests using several signal formats, along with the results for the modeled reverberation.

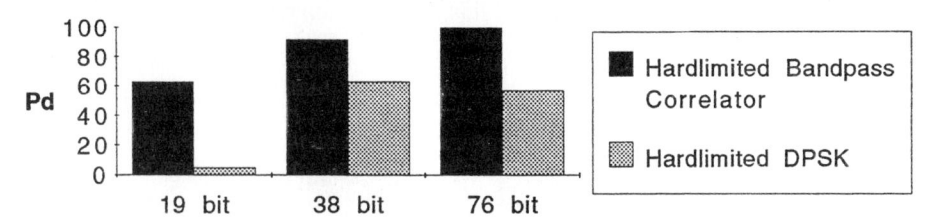

Figure 7. Performance of Fixed Energy Signals in Narragansett Bay.

Similarly, the results for the Block Island Test are shown in Figure 8.

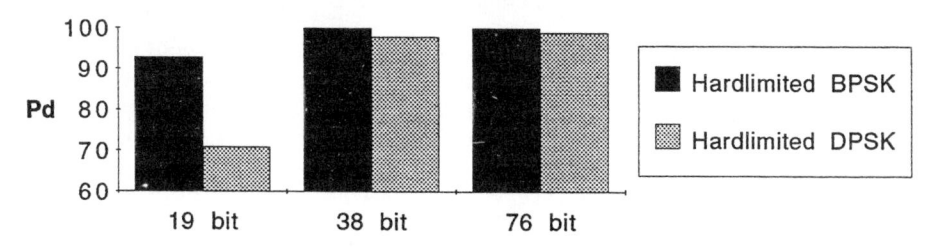

Figure 8. Performance of Fixed Energy Signals in Block Island Sound.

Experimental results from both tests, while different in detail, show remarkably similar trends and behavior. The Bay test environment was much more severe than for the Block Island tests in regard to reverberation because of the extreme shallow depth in the Bay test, as well as the presence of underwater pilings that may have further corrupted the received signal in the Bay tests. Note that longer fixed-energy codes generally produce better results, and that the HLBPC does quite well in both environments for the longer codes.

6. SUMMARY AND CONCLUSIONS

The requirements for testing and evaluating undersea systems were outlined in the context of the changing world environment. Emphasis on portability and operation in shallow water environments focused attention on the performance of underwater tracking systems in reverberation-limited situations characteristic of shallow water. Because of multiple access and time of arrival measurement requirements, BPSK tracking systems were of particular interest. Toward that end, analyses, simulation, and empirical investigations were undertaken on several variants of the BPSK tracking system.

Both analytical and experimental results with the HLDPSK and the HLBPC show that longer time-bandwidth product signals result in improved performance in reverberation.

7. REFERENCE

[1] L. Estes and G. Fain, "Submarine Underwater Communications Project," Final Report prepared under contract N66604-91-M-CN81, NUWC, Feb. 28, 1992.

BROADBAND ADAPTIVE BEAMFORMER FOR LINEAR-FREQUENCY-MODULATION ACTIVE SIGNALS

James C. Lockwood
Surveillance Department
Naval Command, Control and Ocean Surveillance Center
Research, Development, Test and Evaluation Division
San Diego, California 92152-5000

ABSTRACT A novel signal processing approach is described that allows broadband sonar echoes to be optimally beamformed without first frequency-decomposing the signal waveform. Use is made of the property of the LFM waveform that a time shift results in a constant frequency difference relative to the original waveform. Prior to beamforming, echoes are demodulated by multiplying by an LFM reference. This process reduces the echoes from point targets into narrowband signals, which are beamformed using narrowband techniques. Results of a computer simulation are presented that illustrate both conventional and adaptive beamforming results.

1. INTRODUCTION

An efficient and robust method of adaptively beamforming broadband active sonar signals, which will be called ALMA (Active LFM Mixer Adaptive) beamforming, is described. The technique makes use of the property of the Linear-Frequency-Modulation (LFM) waveform that a time shift results in a constant frequency difference with respect to the original waveform. Prior to beamforming, echoes are demodulated by multiplication by an LFM reference, producing narrowband target signals in which range has been mapped into frequency shifts. The LFM waveform suppresses reverberation by providing temporal (range) resolution equal to the inverse bandwidth. In addition, adaptive beamforming applied to each time resolution cell suppresses reverberant features from directions other than that to which the array is steered. The use of the LFM waveform, and demodulation prior to beamforming, makes possible the application of narrow-band ABF techniques. The LFM preprocessing permits adaptation based on much more local parameter estimation than other frequency-domain techniques.

The ALMA beamforming process has been simulated using an IBM-compatible personal computer to explore the performance potential of the technique. The simulation comprised calculating demodulated array element responses to a hypothetical field of point scatterers, representing targets and reverberant features, insonified by an LFM signal. The resulting cross-spectral matrices were averaged, with the addition of an assumed white-noise component, and the ALMA beamforming weights where calculated by matrix inversion. Results are presented that demonstrate the potential performance of the technique, as well as limitations associated with correlation among multiple scatterers at the same range. These results depict ALMA beamforming as a powerful new technique for suppressing reverberant features in a wide range of active sonar applications.

2. BACKGROUND

In applying an ABF algorithm to active signals, one seeks to adapt to a process that is nonstationary with a signal-resolution-like time scale. Frequency domain approaches similar to ones developed for passive transients [1] have been proposed. However, their ability to adapt to local conditions is limited by the processed block size and by the nature of coded-pulse waveforms, such as LFM, that are used in active sonar.

Linear-Frequency-Modulation signals have long been used as search waveforms in sonar systems. An advantage of LFM in this context is the fact that it has a degree of doppler tolerance. That is, the doppler-distorted waveform will still correlate with the undistorted replica over a reasonable range of doppler frequency shifts. In some applications, LFM has been replaced by hyperbolic frequency modulation (HFM) waveforms, which have the property of doppler invariance. One application of LFM is the continuous transmission frequency modulation (CTFM) sonar, which has been used in fish-finding and obstacle avoidance applications. In the CTFM sonar, a very long LFM sweep is transmitted continuously. The angular sweep of a receiving hydrophone gives the target echo a pulse length equal to the time it takes the beamwidth of the receiving hydrophone to sweep past the target. The target echo is mixed with the transmitted sweep and the (constant) difference frequency is detected. The frequency of the demodulated target echo determines the target range. This feature of the LFM waveform, that the frequency difference between a point-target echo and a replica of the transmitted signal is proportional to the time (range) offset, is used in ALMA beamforming to compress *pulsed*-LFM target echoes to narrowband signals prior to beamforming.

3. TECHNICAL DISCUSSION

Consider an LFM signal of duration T reflected from a stationary point target at range R and angle θ. For simplicity a line-array receiver is assumed, with the origin at the first element, and the angle $\theta=0$ broadside to the array as depicted in Fig. 1. The transmitted signal is represented, during the on-time of the pulse by

$$f(t) = \sin \omega_0(1+at)t \ .$$

Let t=0 represent the arrival time of a signal from range R_0 at the first element of the array. Then the arrival time of a signal from range R is $\delta t = 2(R-R_0)/c$, where c is the sound speed. The arrival time of a signal from (R,θ) at the n-th element of the array (periodic for illustration) is

$$\Delta_n = \delta t - (n-1)(d/c)\sin\theta,$$

where d is the element spacing. The LFM signal arriving at the nth element may be written

$$f_n(t) = \sin\omega_0[1+a(t-\Delta_n)](t-\Delta_n) .$$

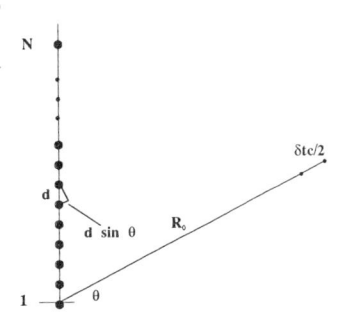

Figure 1. Array geometry.

The element signals are multiplied by f(t). Then the signal from a point target becomes
$$g_n(t) = f(t)f_n(t) = \sin\omega_0[2a\Delta_n t + \Delta_n(1-a\Delta_n)] + \text{Sum-Frequency Terms}.$$

The signals are converted to an analytic representation

$$z_n(t) = g_n(t) + i\,\hat{g}_n(t) ,$$

in which case the point-target signal becomes

$$z_n(t) = (1/i)\exp i\omega_0[2a\Delta_n t + \Delta_n(1-a\Delta_n)] .$$

Because the relative delays among the elements result in the target having different frequency shifts, it is necessary to frequency align the target signals for each steering direction prior to beamforming. This is accomplished by shifting each element's spectrum $Z_n(\omega)$ by $-2\omega_0 a(n-1)(d/c)\sin\theta$. The signal from range R is thus made to appear at each element at the same frequency $2\omega_0 a\delta$. The element signals at each frequency can now be beamformed to steering direction θ using a steering vector with components

$$S_n = \exp i\omega_0\Delta_n(1-a\Delta_n) ,$$

where Δ_n may be rewritten

$$\Delta_n = \omega/2\omega_0 a - (n-1)(d/c)\sin\theta .$$

For each steering direction and for each range cell, the analysis is now formally the same as narrow-band beamforming. The cross-spectral matrix is defined with elements

$$R_{mn}(\omega) = Z_m(\omega)Z_n(\omega)^* .$$

It is observed that the output power for a conventional beamformer steered to θ is

$$P_{CBF} = \mathbf{S}^*\mathbf{R}\mathbf{S} .$$

For the minimum-energy beamformer, weights **W** are sought such that the output power

$$P_{ME} = \mathbf{W}^*\mathbf{R}\mathbf{W}$$

is minimized subject to the constraint that signal gain in the steering direction θ be unity

$$\mathbf{W}^*\mathbf{S} = 1 \ .$$

The matrix **R** defined above yields the instantaneous power output. In order to obtain a non-singular solution, weights are calculated that are optimum in an ensemble-average sense using an average over K frequency (range) cells

$$\overline{\mathbf{R}} = \frac{1}{K} \sum_{k=1}^{K} \mathbf{R}(\omega_i + k\Delta\omega) \ ,$$

where i is the index of the initial frequency in the average, and Δω is the frequency bin width. The well-known solution [2] for the weights

$$\mathbf{W} = \overline{\mathbf{R}}^{-1}\mathbf{S} \ / \ (\mathbf{S}^*\overline{\mathbf{R}}^{-1}\mathbf{S})$$

applies. However, unlike the narrow-band case, a different, suitably-frequency-shifted cross-spectral matrix must be used for each steering direction. The application of ABF for arbitrary broadband waveforms as discussed in [1] has an analogous requirement.

For the ALMA beamformer, the foregoing analysis is applied once for each frequency, that is, each range-resolution cell. This process may be interpreted as effectively a time-domain ABF implementation, wherein the frequency averaging of the sample cross-spectral matrices amounts to a cross correlation using a block of K range cells.

The ALMA beamformer process will adapt to field conditions much more local to the range cell processed than will a frequency-domain implementation along the lines discussed in [1]. Frequency-domain approaches to active ABF have the fundamental limitation that adaptation is based on parameters estimated over the entire temporal block processed, which for broadband waveforms is typically long compared to the temporal (range) resolution. For echoes from coded-pulse waveforms, such as LFM, the domain over which parameters are estimated is governed by the pulse length, which is typically even longer. To achieve local adaptation with coded-pulse waveform requires pulse compression prior to beamforming, which is a feature of the ALMA approach.

4. RESULTS

The process described above has been simulated using an IBM-compatible PC. Results for two different fields of scatterers are now presented. The array modeled is a 21-element line with elements spaced at half the wavelength at the top of the signal band. The range-resolution-cell width was equal to twice the element spacing.

The first field is depicted in Fig. 2. Fifteen 10-dB scatterers are spaced at 20° intervals and two range cells apart to test the ability of the ALMA beamformer to achieve full signal gain against well separated targets. The size of the box is 4000 array lengths square. Figure 3 shows typical results obtained with this field. The figure shows pairs of conventional and adaptive response patterns for a range cell containing a 10-dB scatterer and an adjacent cell containing no scatterer. In Fig. 3b, it is seen that the full signal gain is achieved. In the absence of a scatterer, the maximum response (Fig. 3d) is about 13 dB lower.

The second field is depicted in Fig. 4, where "X" indicates a scatterer with relative strength greater than 0 dB, "o" indicates between 0 and −6 dB, and scatterers below −6 dB are marked by a single dot. Figure 4 has two scatterers in each of 15 adjacent range cells. At the far right are depicted 13-dB scatterers in adjacent range cells with 1° angular separations. The remaining scatterers are in the same 15 range cells. The left-most has a level of 15-dB. The others are on 5° angular separations and drop by 3-dB

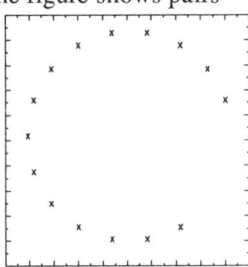

Figure 2. 10-dB Scatterers at discrete ranges

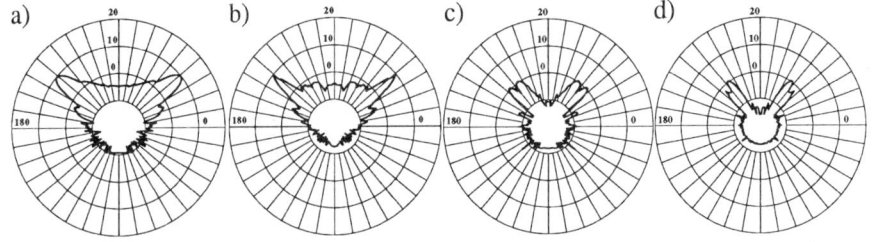

Figure 3. Conventional a) and adaptive response b) for range cell with 10-dB scatterer and conventional c) and adaptive response d) for adjacent cell with no scatterer.

steps going clockwise. The size of the box is the same as in the previous example. The purpose of this distribution was to test the ability of the ALMA beamformer to distinguish weak targets in the presence of strong interferors at the same range. Figure 5 shows examples of the adaptive response for 4 different target levels in the presence of 13-dB interferors. The heavy circle in each polar plot indicates the 0-dB level as a point of reference. Clearly the target in each of the first three examples and possibly in the fourth would be detectable above the general sidelobe levels. However, it is noted that the full signal gains have not been achieved. This is due to the fact that the target and interferor signals are correlated. Figure 6 shows results for a 4-dB target that illustrate

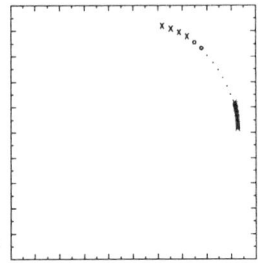

Figure 4. 13-dB Scatterers interfering at same ranges with various target levels

the effects of averaging on signal gain. The three plots are the conventional response and two adaptive responses exemplifying cases when 9 range cells and 3 range cells, respectively, were averaged in computing the weights. The 9-cell case (Fig. 6b) corresponds to the same conditions as Fig. 5. The first thing to notice is that Fig. 6b shows a

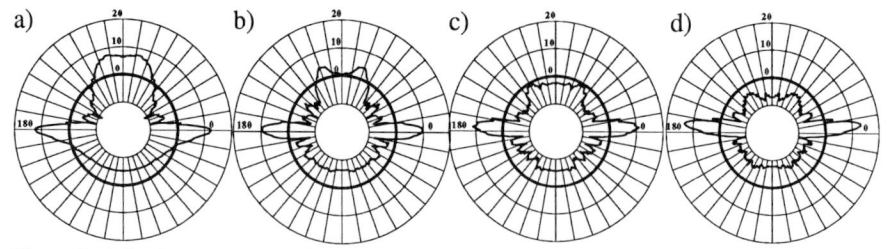

Figure 5. Adaptive response against 13-dB interferor and a) 13-dB target b) 7-dB target c) 1-dB target and d) −5-dB target.

clear gain in detectability relative to the conventional response shown in Fig. 6a. Second, it is clear that 9-cell averaging has improved signal gain relative to the 3-cell case shown in Fig. 6c. A possible explanation for this improvement with averaging is that the weights must suppress signals from all the averaged range cells, which eventually overtaxes the available degrees of freedom.

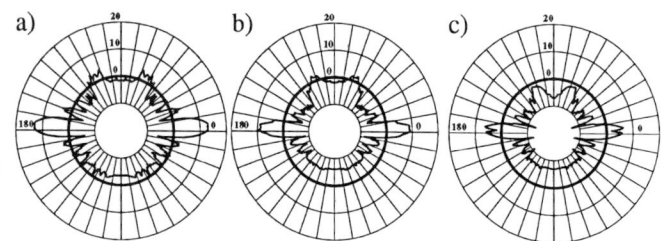

Figure 6. Responses against 13-dB interferor and 4-dB target: a) conventional b) adaptive with 9 cells averaged c) with 3 cells averaged.

CONCLUSIONS

A novel signal processing approach has been described and results of a simulation presented. The results show that the ALMA beamformer is a powerful new technique for suppressing reverberant features interfering with targets. This, as well as other active ABF techniques, is limited by the correlation among multiple scatterers within the same range cell. Sufficient averaging over range in computing adaptive weights has been shown to reduce, but not eliminate problems due to signal correlation. Because the field at each range cell may be independent of all the others, range averaging does not necessarily reduce cross-correlation products as would the analogous time averaging in a passive implementation. How to implement ALMA beamforming in a way that minimizes this difficulty remains a significant research issue. Still, the results to date clearly show a substantial expected performance improvement over conventional beamforming.

1. Reuter, M., "Issues and Methods of Broadband Array Processing," Naval Ocean Systems Center Technical Document 2086 (April 1991).

2. Cox, H., Zeskind, R.M. and Owen, M.M., "Robust Adaptive Beamforming," IEEE Trans on Acoustics, Speech and Signal Processing **35** 1365-1376 (1987).

BROADBAND PULSE DISTORTION:

Waveform Design Issues for Active Systems

FRANK. J. RYAN and EDWARD F. RYNNE

Naval Command, Control and Ocean Surveillance Center
RDT&E Division (NRaD)
San Diego, CA 92152-5000

25 May 1992

Abstract. In this paper, we study the effects of the ocean waveguide on waveforms used in active acoustic localization and detection systems. A broadband normal mode code was developed for use in a Fourier synthesis technique to predict the pressure time series and correlation function of pulsed waveforms. A full suite of signal processing and analysis codes were developed that allow waveforms to be injected into a realistic ocean waveguide environment from a vertical aperture source and propagated to long ranges. The model was used to study the effects of pulse distortion in a convergence zone environment and the impact on replica correlator performance. The predicted correlation structure was studied in terms of correlator loss, maximum active correlator width, and the number of correlator peaks. Comparisons were made between various waveforms, ocean environments, and other factors. We find that observed signals may differ dramatically from input signals after being propagated through the ocean guide.

1. Introduction

Active sonar systems often transmit broadband waveforms, i.e., waveforms having a bandwidth that is a large fraction of the center band frequency. One rationale for using a broadband transmit signal is to produce a Doppler-invariant waveform that provides accurate range localization. The signal processing of a broadband pulse is often done using a matched filter which is simply the correlation between a time-delayed replica of the transmitted signal $S(t)$ and the received echo waveform $P(t)$. Replica correlation is optimum processing for a maximum output signal-to-noise ratio criterion assuming the received echo signal is simply the time-delayed transmit pulse plus additive Gaussian noise. Such processing implicitly assumes a linear, time-invariant underwater acoustic channel with linear phase characteristics.

For a narrow band waveform, such as used in radars, the conventional practice is to represent the Doppler distorted signal as a frequency shifted version of the original waveform. This assumes that dispersive propagation effects on the signal are minimal. For a broadband waveform which propagates in a dispersive media such as the ocean, this approach does not work. Instead, the exact description of the Doppler effect as a dilation of the signal time base must be used (Williams and Battestin, 1976). This pulse distortion can be most prevalent in convergence zone (CZ) regions, and often exhibits marked spatial variation.

The purpose of this study is to examine in detail the temporal characteristics of broadband pulses in CZ environments with the goal of improving system performance. To help understand and quantify the effects of broadband pulse distortion, a broadband propagation model was developed to allow parametric studies of environmental effects on pulse distortion. This model has sufficient fidelity to accurately simulate the received pressure time series of realistic pulse waveforms after they have propagated to multi-CZ ranges. Knowing the received pressure time series it is a simple matter to compute the replica correlator and study its spatial properties.

2. Source Model

Consider, for example, a vertical line array (VLA) source consisting of N omnidirectional transducers, each emitting a pressure field p_0, and spaced a distance d apart. The normalized output pressure of the array is then the product, $p(N) = p_0 D(\theta)$, of the elemental pressures p_0 and the directivity index of the array $D(\theta) = \sin(N\phi)/(N\sin\phi)$, where the steering vector is $\phi = (\sin\theta - \sin\theta_0)k_0 d/2$.

In practice, the free-field radiation patterns of individual transducers are modified when placed in an array due to mutual interference or coupling between the transducer elements. The resulting source radiation pattern is then changed from that predicted using free-field radiation patterns. The array mutual interaction is frequency dependent and affects the broadcasted waveform, tending to raise sidelobe energy levels.

In this study we focus on arrays comprised of flextensional type transducers, consisting of a pre-stressed elastic shell driven by a ceramic stack assembly. When multiple transducers are driven in an array, they interact. These interactions arise from the mutual radiation impedance between elements of the array that changes the radiation loading on the surface of the transducer element leading to a change in the displacement degrees-of-freedom of the elastic shell. This in turn changes the electrical impedance seen by the power supply driving the array. To simulate the array interactions, we employ a three part model consisting of 1) a finite element structural model for the shell and ceramic driver bar, 2) an acoustic radiation model, and 3) a lumped-circuit model of the ceramic stack assembly driver.

If $\mathbf{U}$ represents the shell displacement vector, then the linear elasticity problem is defined by

$$[-\omega^2 \mathbf{M} + \mathbf{K}]\mathbf{U} = \mathbf{F}^r + \mathbf{F}^d, \qquad (1)$$

where $\mathbf{M}$ is the mass matrix, $\mathbf{K}$ is the stiffness matrix, $\mathbf{F}^r$ is the vector of acoustic radiation load forces, and $\mathbf{F}^d$ is the load vector of driving forces produced by the ceramic stack assembly on the flextensional shell. For harmonic motion, the surface normal velocity vector $\mathbf{V}$ is related to the displacement vector by $\mathbf{V} = i\omega \mathbf{C}^T \mathbf{U}$, where $\mathbf{C}$ is the coupling matrix.

The flexing shell radiates acoustic energy and the surrounding water imposes a radiation reaction on the shell. An acoustic radiation model based on the Combined Helmholtz Integral Equation Formulation, **CHIEF**, is used to model the radiation loading of the water on the transducer (Schenck and Benthien, 1989 and Benthien, 1990). The acoustic pressure field $\mathbf{P}$ couples to the elastic shell through the load vector $\mathbf{F}^r$, which in matrix form is $\mathbf{F}^r = -\mathbf{CDP}$, where $\mathbf{D}$ is a diagonal matrix. Following Pritchard (1960), the pressure is related to the surface normal velocity via the impedance relation $\mathbf{P} = \mathbf{ZV}$, with $\mathbf{Z}$ the mutual radiation impedance matrix

$$Z_{mn} = i\Re(Z_0)\frac{e^{ik_0 d_{mn}}}{k_0 d_{mn}}, \quad k_0 = \frac{\omega}{c_0}$$

between the mth and nth elements having center separations d_{mn}. The self radiation impedance of a single transducer in the free-field is denoted by Z_0, and is computed using CHIEF. Combining the elastic structure and acoustic models yields the following set of equations which relate the driving forces $\mathbf{F}^d$ and the normal surface velocities $\mathbf{V}$:

$$v_m = i\omega \mathbf{C}^T \left(-\omega^2 \mathbf{M} + \mathbf{K}\right)^{-1} \left[\sum_{n=1}^{N} \mathbf{C} d Z_{mn} v_n + f_m^d\right], \quad m = 1, 2, \ldots, N. \quad (2)$$

The equations for the jth ceramic stack assembly, which relate the input voltage e_j and current i_j to the stack end velocity v_j and corresponding drive force f_j^d, are (Ding et al. ,1971)

$$\begin{pmatrix} i_j \\ f_j^d \end{pmatrix} = \begin{pmatrix} h_{11} & h_{12} \\ h_{21} & h_{22} \end{pmatrix} \begin{pmatrix} e_j \\ v_j \end{pmatrix} \quad (3)$$

Collocation of the ceramic-bar, flextensional shell, and radiated pressure equations along with appropriate boundary conditions at their respective interfaces leads to a single matrix equation, the T-matrix, that combines the shell displacement coefficients with equivalent mechanical impedances and relates the normal surface shell velocity vector to the input drive voltage vector:

$$v_j(\omega) = \sum_{k=1}^{N} t_{jk}(\omega) e_k(\omega), \quad j = 1, 2, \ldots, N, \quad (4)$$

when the transducers are part of an array. The $N \times N$ complex T-matrix $\mathbf{T}(\omega) = \{t_{jk}\}$ is a function of the type of transducer and the array geometry. Thus a single set of T-matrices can be used for multiple ocean environments and pulse types.

Having solved the element interaction problem, via the T-matrix, all that remains is to compute the far-field radiated pressure of the array. Near each transducer, we assume that the ocean is locally homogeneous with density ρ_0 and sound speed c_0. In the far field ($r \gg \lambda$), a multipole expansion of the acoustic radiation from each transducer is made, keeping only the lowest order term. This corresponds to an *effective* fluctuating-volume monopole source having a mass flux $Q(t) \doteq \rho_0 A v(t)$ given by the product of the area, A, and radial surface velocity v. For harmonic time variation $v(t) = v_0 \exp(-i\omega t)$, the resulting radiated pressure from the monopole is just (Ross, 1976)

$$p_{mono}(r,t) = -i\frac{\omega Q_0}{4\pi r} e^{i\omega(r/c_0 - t)}, \quad Q_0 \doteq \rho_0 A v_0. \quad (5)$$

We see from Eq. (5) that the monopole radiated pressure is directly related to the effective transducer normal velocity $v(t)$. Accordingly, each transducer element in the source array is modeled by an equivalent "effective" monopole singlet element having a prescribed radial surface velocity $v(t)$. This is termed *velocity control* since the transducer radial surface velocity is specified explicitly. Alternatively, the drive voltage $e(t)$ to each transducer in the array may be specified and the radial velocity determined implicitly; this is termed *voltage control*.

3. Broadband Propagation Model

The above source model defines the broadband source field of the array. Our goal is to propagate this broadband sound pulse to long ranges in a dispersive ocean waveguide and then examine its temporal and spatial characteristics. In this study we treat the ocean medium as being horizontally stratified and use an azimuthally symmetric cylindrical coordinate system: $\mathbf{r} = (r, z)$.

The source array is modeled as a collection of monopole radiators with appropriate time varying amplitudes, $S(t) = \sum_n S_n(t)\delta(\mathbf{r}_n)$, and the resulting pressure field from the nth source, P_n, is governed by the acoustic wave equation:

$$\nabla^2 P_n(\mathbf{r},t) - \frac{1}{c(z)^2}\frac{\partial^2 P_n(\mathbf{r},t)}{\partial t^2} = S_n(t)\delta(\mathbf{r} - \mathbf{r}_n), \qquad (6)$$

where $c(z)$ is the depth dependent sound speed. Instead of solving Eq. (6) directly in the time domain for the pressure field P_n, it proves efficacious to first transform to frequency space using temporal Fourier transforms defined by:

$$\tilde{P}_n(\mathbf{r},\omega) = \int_{-\infty}^{\infty} P_n(\mathbf{r},t)e^{+i\omega t}\,d\omega, \quad \tilde{S}_n(\omega) = \int_{-\infty}^{\infty} S_n(t)e^{+i\omega t}\,d\omega. \qquad (7)$$

Since the source waveforms S_n (and by extension the pressures P_n) are real causal signals, this leads to constraints on the spectral components: $\tilde{P}_n(\mathbf{r},-\omega) = \tilde{P}_n^*(\mathbf{r},\omega)$, $\tilde{S}_n(-\omega) = \tilde{S}_n^*(\omega)$.

Taking the temporal Fourier transform of Eq. (6), we first compute the point source Green's function $G_\omega(\mathbf{r}|\mathbf{r}_n)$ that is the solution to

$$[\nabla^2 + k^2(z)]G_\omega(\mathbf{r}|\mathbf{r}_n) = -4\pi\delta(\mathbf{r} - \mathbf{r}_n), \quad k(z) = \omega/c(z). \qquad (8)$$

Knowing $G_\omega(\mathbf{r}|\mathbf{r}_n)$, the total pressure P is obtained by convolving with the spatial source distribution and then taking an inverse Fourier transform:

$$\begin{aligned} P(\mathbf{r},t) &= \frac{1}{\pi}\Re \int_0^\infty \sum_n \tilde{P}_n(\omega)e^{-i\omega t}\,d\omega, \\ &= \frac{1}{\pi}\Re \int_0^\infty \sum_n \tilde{S}_n(\omega)b_n(\omega)G_\omega(\mathbf{r}|\mathbf{r}_n)e^{-i\omega t}\,d\omega. \end{aligned} \qquad (9)$$

The $\{b_n(\omega)\}$ are optional complex weighting factors that take into account transducer mutual interactions and the source array radiation pattern.

To solve Eq. (8), we make use of separation of variables methods to expand the Green's function in a discrete modal series:

$$G(r,z,z_0) \approx \sqrt{2\pi i}\sum_m \frac{\phi_m(z)\Phi_m}{\sqrt{r\lambda_m}}e^{i\lambda_m r}, \qquad (10)$$

where $\{\phi_m\}$ are the ortho-normal depth eigenfunctions, $\{\lambda_m\}$ the corresponding modal eigenvalues, and Φ_m is the modal source function. The $\{\lambda_m\}$ and $\{\phi_m\}$ in general are complex and take into account losses due to ocean surface roughness and absorption of sound by a geoacoustic bottom. The mode sum in Eq. (10) includes high-order 'leaky'-modes that correspond to large grazing angle ray-equivalents at both the surface and bottom. For an N-element VLA source with the kth transducer located at z_k, the *effective* modal array source function Φ_m is defined by $\Phi_m(\omega) \equiv \sum_{k=1}^{N} b_k(\omega) \phi_m(z_k)/B$, with $B^2 = \sum_{k=1}^{N} |b_k|^2$. The $\{b_k(\omega)\}$ are complex source weights that incorporate the T-matrix effects and complex phasors needed to form a vertical beam in a particular direction.

For a particular signal $S(t)$ with Fourier transform $\tilde{S}(\omega)$, the time dependent pressure field from the source array is then just

$$P(\mathbf{r},t) = \sqrt{\frac{2i}{\pi r}} \Re \int_0^\infty \tilde{S}(\omega) \sum_m \frac{\phi_m(z)\Phi_m}{\sqrt{\lambda_m}} e^{i(\lambda_m r - \omega t)} d\omega. \tag{11}$$

At long range, the pulse propagation time T_{obs} to the observation point r_{obs} may exceed the available time window: $T_{obs} \gg T_w = N\Delta t$. The analysis time window is therefore shifted by applying a linear phase shift, $\exp(-i\omega T_{off})$, to each spectral FFT bin. The time offset is $T_{off} \approx r_{obs}/v_g$, where v_g is the nominal pulse group velocity computed using ray methods.

The Fourier transforms were implemented using fast Fourier transform (FFT) algorithms. For the results presented here, the frequency band of interest spanned 225–275 Hz, for which a sampling frequency $f_s = 600$ Hz is adequate. An $N = 2^{14}$ point FFT was used, yielding a frequency resolution of $\Delta f = 600/N \approx 0.037$Hz. The corresponding time resolution is then $\Delta t = 1/f_s \approx 1.7$msec, and the total analysis time window is $T_w = N\Delta t \approx 27$s. This frequency sampling requires full-wave pressure fields to be generated for 2730 frequencies at each range/depth grid point.

The Green's function G_ω was computed with a modified version of the ATLAS normal mode model (Ryan, 1989). At the higher frequencies, aproximately 1000 modes were required. The calculations were made on a network of SUN system SPARC computers and took 60 cpu hours to generate the modal pressure database.

After the modal pressure database had been constructed, it was used by the pulse synthesis program PT4 (Rees, 1990). The PT4 program executes interactively on a Sun4 workstation and can display the input and output time series and spectra, construct several varieties of input waveforms or use user-defined ones, construct replica correlators and autocorrelators, add Gaussian noise, and in general provides the user with the tools to investigate waveguide propagation effects on broadband signals.

4. Examples

The broadband propagation characteristics (i.e., signal transfer function) of different ocean water mass structures, source array configurations, and pulse waveforms were examined in this study. A representative example for a deep water CZ region of the NE Pacific is covered next.

The scenario employed a VLA-source steered to an elevation angle of 8-degrees. The source consisted of flextensional transducers, and mutual interactions were included using the T-matrix described earlier. The source transmitted a 2-second hyperbolic frequency modulated (HFM) waveform having the functional form[Chou 1988]

$$p(t) = \begin{cases} \cos\left[\dfrac{2\pi}{b}\ln(T_0 + bt) + \phi_0\right], & \text{for } 0 \leq t \leq T \\ 0, & \text{for } t < 0, t > T, \end{cases}$$

where b and T_0 are expressed in terms of the pulse start frequency f_1, stop frequency f_2 and duration T as: $T_0 = 1/f_1$, and $Tb = (f_1 - f_2)/(f_1 f_2)$. The canonical HFM pulse used in this study had $f_1 = 225$ Hz, $f_2 = 275$ Hz, and $T = 2$s. Because the pulse frequency structure remains essentially the same no significant differences are expected for longer pulses.

A rectangular range/depth observation window was selected that covered the second convergence zone (CZ), and a dense grid of points (ranges 90–119.5 km in increments of 250m, and depths 60–280m in 10m steps) were selected at which the pressure field and corresponding correlator output would be computed. Two representative points are chosen to present detailed waveform characteristics: point-1 is at range 94.5km and depth 190m, and point-2 is at range 109.0km and depth 190m.

The replica correlator is computed as a product in frequency space of the propagated pulse and the transmitted waveform. This Fourier technique is doubly convenient when using the Fourier synthesis technique, as the frequency space representations of the input and propagated signals are immediately available. The correlator loss CL is defined in a manner analogous to transmission loss TL. Let $C_{\max}$ be the maximum value of the replica correlator output in the available time window, then $CL \equiv -10\log C_{\max}$.

Figure-1 shows the pressure time series (sampled at 1.7 msec intervals), the corresponding pulse power spectrum (sampled at 0.037 Hz intervals), and the correlator output for the two points described earlier. Note the strong pulse distortion exhibited at point-2 and the splitting of the correlator output. The spatial variation of the CL in the observation window is shown in Fig.-2. Note the strong spatial variation in the replica correlator output.

5. Conclusions

A suite of codes has been developed which can propagate broadband waveforms over long ranges with high fidelity. Propagated waveforms may be processed with a replica correlator, and the correlator output studied in terms of correlator loss, total active correlator width, and number of correlator peaks, in addition to presenting the explicit correlation function. In addition, this method allows one to construct a filter matched to the impulse response of the ocean channel in situations where there are time and frequency spreads in the propagated signal. In a sense, this is an extension of conventional narrowband matched field processing to broadband waveforms.

A novel feature of this work is the inclusion of transducer mutual interaction effects into the calculations.

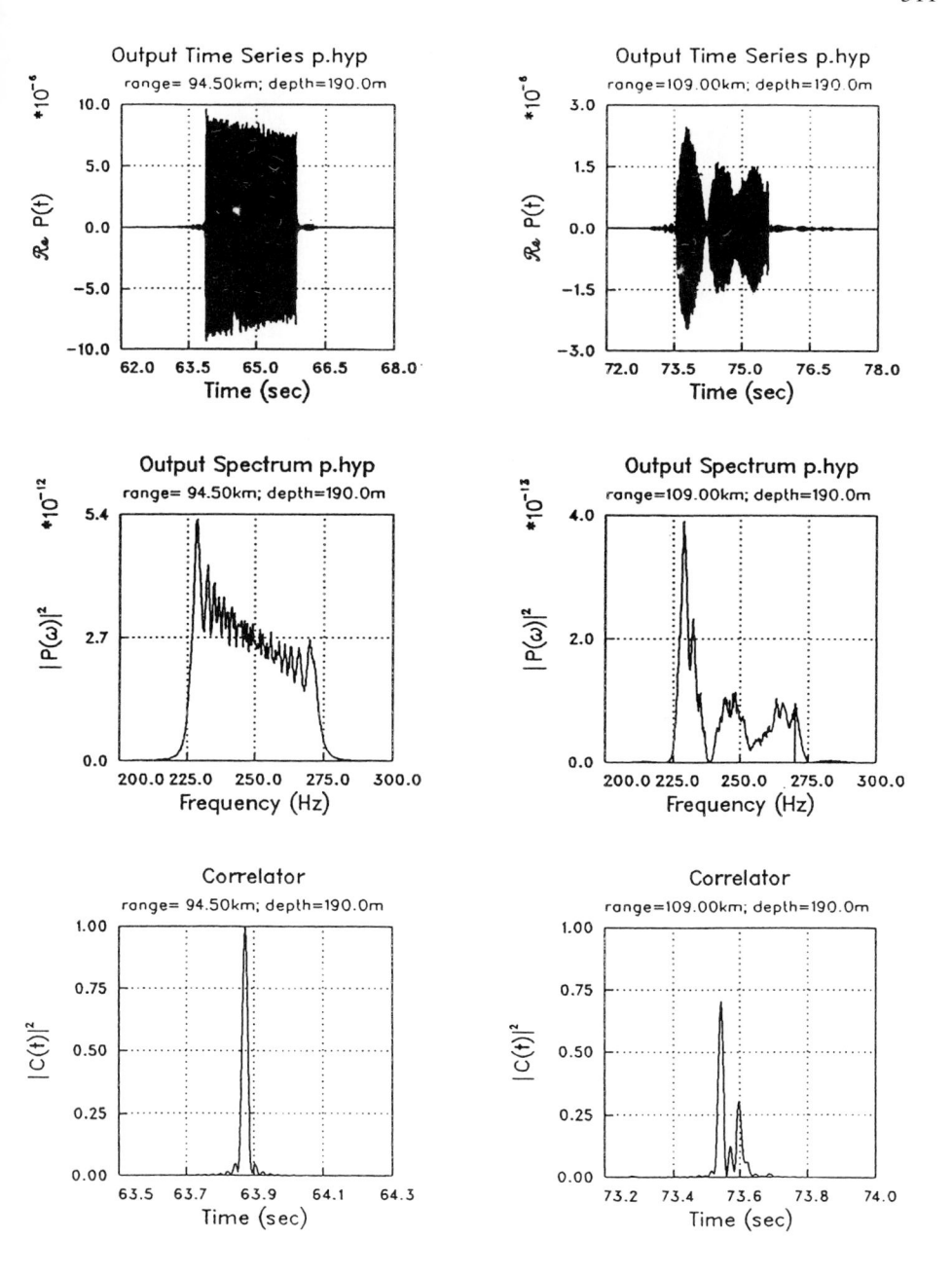

Fig. 1. Predicted observed signal characteristics at the two test points for the Pacific profile, a source beamforming angle of +8 degrees, and the HFM source pulse. The predicted time series, power spectrum, and replica correlator are displayed for Point 1 in the left column and Point 2 in the right column.

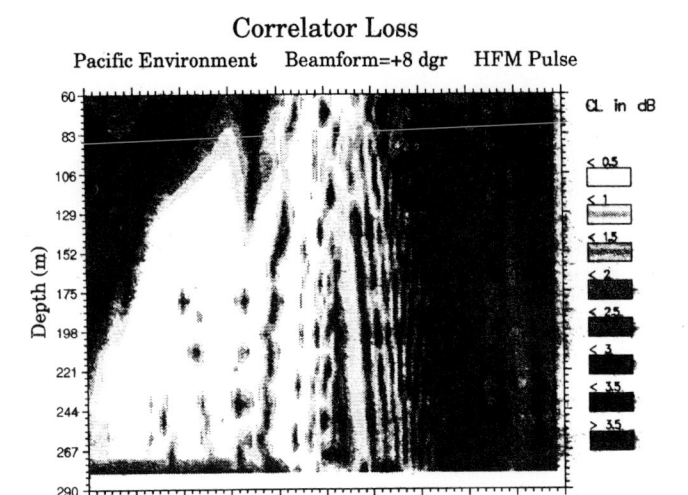

Fig. 2. Predicted CL throughout the target area for the Pacific profile, a source beamforming angle of +8 degrees, and the HFM source pulse. Regions of low CL have the least signal distortion.

6. Acknowledgements

We thank D. Barach and D. Gillette for supplying the T-matrices used in the analysis, and D. Rees for developing the pulse synthesis and graphics codes.

References

Benthien, G.W.: 1990, 'Numerical Modeling of Array Interactions', in *Power Transducers for Sonics and Ultrasonics*, B.F. Harmonic, O.B. Wilson, and J.N. Descarpigny eds., Springer-Verlag, 109–124.
Brigham, G.A.: 1974, 'Analysis of the class-IV flextensional transducer by use of wave mechanics', *J. Acous. Soc. Am.* **56**, 31–39.
Chertock, G.: 1964, 'Sound radiation from vibrating surfaces', *J. Acous. Soc. Am.* **36**, 1305–1313.
Chou, S.I.: 1988, 'Hyperbolic frequency modulated (HFM) waveforms and wavetrains' **TR-1259**, Naval Ocean Systems Center, San Diego, CA.
Ding, H., McCleary, L. and Ward, J.: 1971, 'Computerized Sonar Transducer Analysis and Design Based on Multiport Network Interconnection Techniques', **TP-228**, Naval Undersea Research Center, San Diego, CA.
Pritchard, R.L.: 1960, 'Mutual acoustic impedance between radiators in an infinite rigid plane', *J. Acous. Soc. Am.* **32**, 733–737.
Rees, C.D.: 1990, frequency synthesis computer program **PT4**, Naval Ocean Systems Center, San Diego, CA.
Ross, D.: 1976, *Mechanics of Underwater Noise*, Pergamon Press: New York, Chap. 4.
Ryan, F.: 1989, normal mode computer program **ATLAS**, Naval Ocean Systems Center, San Diego, CA.
Schenck, H.A. and Benthien, G.W.: 1989, 'Numerical Solution of Acoustic-Structure Interaction Problems', **TR-1263**, Naval Ocean Systems Center, San Diego, CA.
Veenkant, R.L.: 1977, 'Investigation of the propagation stability of a doubly spread underwater acoustic channel', *IEEE Trans. Acoust., Speech, Signal Proc.* **ASSP-25**, 109–116.
Williams, R.E. and Battestin, H.F.: 1976, 'Time coherence of acoustic signals transmitted over resolved paths in the deep ocean', *J. Acous. Soc. Am.* **59**, 312–328.

ACTIVE MATCHED FIELD PROCESSING FOR CLUTTER REJECTION

L. B. Dozier
H. A. Freese
Science Applications International Corporation
1710 Goodridge Drive
McLean, Virginia 22102
U. S. A.

ABSTRACT. Traditionally, matched field processing schemes have been used passively. However, in this paper we outline an approach to extending such schemes for active use. We use the aperture of a vertical array to take advantage of the inherent modal character of the echo and bottom reverberation. When the matched field beamformer assumes a knowledge of the correct target position, significant rejection of the unwanted clutter (reverberation) is obtained, as demonstrated by a numerical computation of array gain of the echo against reverberation.

1. Introduction

Passive matched field processing (PMFP) is a beamformer which does not assume a locally homogeneous plane-wave signal field, but rather a modelable complex pressure field with non-plane-wave multipath interference. The signals received at an array are correlated with modeled signals from point sources at assumed locations within a search space. The success of PMFP depends on the ability to accurately predict the acoustic field at the array for all possible source locations of interest.

In this paper we consider *active* matched field processing. Now the desired signal is echo from a target at an unknown position. The masking field (clutter) includes not only the noise of the passive case, but also the reverberant field from the surface and bottom; in this first study we consider only bottom reverberation. Also in contrast to PMFP, the signal is not usually single frequency (CW), although the bandwidth may be quite narrow. The general scheme, then, is to assume various target positions in a search space and select as the true target position the one which shows the highest maximum output of the beamformer over time. Since the full solution to this problem will involve massive computation, in this first study we make several simplifying assumptions. After outlining the full method for a monostatic geometry, we consider only a small time segment received at the array, and assume that the echo and reverberation reaching the array at that point came from the same range relative to the array. We also assume that the beamformer is perfectly matched to the target (as if its position were known), and compute the gain against clutter in that case.

2. Active Matched Field Formulation

Let $r(f,z_l)$ and $e(f,z_l)$ be the reverberation and echo observed at frequency f at the lth hydrophone of a vertical array. We apply a space-frequency filter (beamformer) described by w(f,z$_l$) to the data. The time history of the beamformer output is then given by

$$b(\tau) = \sum_{l} \int df\, e^{i2\pi f\tau}\, w^*(f,z_l)\, [\, r(f,z_l) + e(f,z_l)\,]$$

The mean square value of the beamformer output is then given by

$$\langle |b(\tau)|^2 \rangle = \sum_{l,k} \iint df\, d\hat{f}\, e^{i2\pi(f-\hat{f})\tau} w^*(f,z_l)[e(f,z_l)e_k^\dagger(\hat{f},z_k) + \langle r(f,z_l)r^*(\hat{f},z_k)\rangle] w(\hat{f},z_k)$$

where we have assumed that the reverberation is a zero mean random process and the target echo is deterministic and predictable. Assuming that the source and the vertical array are collocated, the reverberation can be written as

$$r(f,z) = \iint \rho\, d\rho\, d\theta\, T_r(f,z,\zeta_s,\zeta_r,\rho,\theta) F(f)$$

where F(f) is the source spectrum and the subscripts r and s refer to the depth of the reverberant region and the source respectively. The variables ρ and θ refer to a cylindrical coordinate system centered at the source and thus describe the distance and the bearing of the reverberator from the source or the vertical array. We now write the transfer function T$_r$ as the following double sum over the normal modes which represents the forward and backward waves and the effects of mode coupling induced by the reverberator. Here, the complex function Γ represents the mode coupling induced by the scattering event. For bottom or surface scattering Γ is assumed to be a random function which depends on the location of the scatterer and the frequency. We shall assume that the scattering is spatially uncorrelated and that its frequency correlation is large compared to the bandwidth of the transmitted signal. For a target, the scattering will be treated deterministically so that Γ^t (target scattering function) could be computed by a boundary integral over the surface of the target.

$$T_r = \sum_{m,n} \phi_m(\zeta_s)\phi_m(\zeta_r)\, e^{-i(k_m - i\alpha_m)\rho}\, \Gamma_{m,n}\, \phi_n(\zeta_r)\phi_n(z)\, e^{-i(k_n - i\alpha_n)\rho} / \rho \sqrt{k_m k_n}$$

We now approximate the exponential expressions involving the horizontal wavenumber, which is a function of frequency, in the following way using the modal group velocities:

$$-ik_n\rho = -ik_n^0\rho - i2\pi(f-f_0)\rho/c_n$$

The expression for the transfer function can therefore be written in the following way:

$$T_r = \sum_{m,n} A_{m,n}(\zeta_s,z) E_{m,n}(f,\rho) \frac{C_{m,n}(\zeta_r,\rho)}{\rho} \Gamma_{m,n} ,$$

by grouping terms defined by

$$A_{m,n}(\zeta_s,z) = \frac{\phi_m(\zeta_s)\phi_n(z)}{\sqrt{k_m k_n}} \quad ,$$

which depends only on the source depth and the depth z of the hydrophone;

$$E_{m,n}(f,\rho) = e^{-i2\pi(f-f_0)\rho(\frac{1}{c_m}+\frac{1}{c_n})} \quad ,$$

which depends only on the propagation distance to the scatterer and the separation (f-f$_0$) of the frequency f from the center frequency f$_0$; and

$$C_{m,n}(\zeta_r,\rho) = \phi_m(\zeta_r)\phi_n(\zeta_r) e^{-(\alpha_m+\alpha_n)\rho} e^{-i(k_m^0+k_n^0)\rho} \quad ,$$

which depends only on the depth of the scatterer and the propagation distance to the scatterer. Note that we have assumed that the scattering function (coupling matrix $\Gamma_{m,n}$) is independent of frequency, although it need not be.

We next evaluate the expectation of the reverberation term. We assume that the scattering function has zero mean, that the reverberations arising from different regions are uncorrelated, and that the scattering function is spatially homogeneous. Also, we assume that the scattering function for one pair of modes is statistically independent of the scattering function for any other pair. These assumptions allow us to simplify the expectation of the scattering term to the following:

$$\langle \Gamma_{m,n}(\rho,\theta) \Gamma_{\hat{m},\hat{n}}^+(\hat{\rho},\hat{\theta}) \rangle = S_{m,n} \delta_{m,n} \delta_{\hat{m},\hat{n}} \quad ,$$

where $S_{m,n}$ is the scattering kernel which describes how the energy in one mode couples to another. This in turn allows us to write the equation for the space-frequency correlation of the reverberation as follows:

$$\langle r(f,z) r^+(\hat{f},\hat{z}) \rangle = 2\pi \sum_{m,n} \int d\rho A_{m,n}(\zeta_s,z) A_{m,n}^+(\zeta_s,\hat{z}) E_{m,n}(f,\rho) E_{m,n}^+(\hat{f},\rho) |C_{m,n}(\zeta_r,\rho)|^2 S_{m,n} F(f) F^+(\hat{f})$$

Similarly, we can write the target response as (Γ^t is the target scattering function)

$$e(f,z) = F(f) \sum_{m,n} A_{m,n}(\zeta_s,z) E_{m,n}(f,\rho_t) \frac{C_{m,n}(\zeta_t,\rho_t)}{\rho_t} \Gamma_{m,n}^t \quad .$$

Note that the scattering function for the target is deterministic. We are now in a position to compare the reverberation rejection performance, or gain, of the vertical array to that of the average single sensor performance in the following sense. The gain, G, against the clutter is defined as the signal-to-clutter ratio at the output of the beamformer to the average signal-to-clutter ratio of a single hydrophone. Mathematically, G is defined as

$$G = \frac{|b_t(\tau)|^2 / \langle |b_r(\tau)|^2 \rangle}{\sum_l |p_t(\tau,z_l)|^2 / \sum_l \langle |p_r(\tau,z_l)|^2 \rangle}$$

where p is the Fourier transform of e or r.

In this study, we assume that the beamformer outputs are perfectly matched to the expected target return, i.e., w is proportional to e. Thus, we are not carrying out a search in target space, but rather computing the peak gain. Also, we make some additional simplifications that make the evaluation of the above expressions computationally easier but yet retain much of the underlying physics. First, we assume that the signal bandwidth is much smaller than the reciprocal of the received time spread of all energy scattered at a specified range from the source or array. The signal is also assumed long enough so that there is significant overlap between the target and clutter return coming from the same range cell as the target. Finally, we assume that the signal is still short enough so that we can associate a range with a central arrival time τ_0 of this energy via $\tau_0 = 2\rho_t/c_g$, where c_g is an average group velocity. We thus expect that we can develop a coarse time history of the reverberation in this manner, but we mainly want to study the effect that the vertical aperture has on the gain against the reverberation. Given the above assumptions, analysis yields the single sensor echo and reverberation terms as

$$\sum_l |p_t(\tau_0,z_l)|^2 = \sum_l |\sum_{m,n} A_{m,n}(\zeta_s,z_l) \frac{C_{m,n}(\zeta_t,\rho_t)}{\rho_t} \Gamma^t_{m,n}|^2$$

for the echo and

$$\sum_l \langle |p_r(\tau_0,z_l)|^2 \rangle = 2\pi\Delta\rho \sum_l \sum_{m,n} |A_{m,n}(\zeta_s,z_l)|^2 |C_{m,n}(\zeta_r,\rho_r)|^2 S_{m,n}$$

for the reverberation. In this last expression $\Delta\rho$ represents the range interval which corresponds to the pulse length.

For the beamformer outputs the corresponding terms are

$$|b_t(\tau_0)|^2 = D [\sum_l |\sum_{m,n} A_{m,n}(\zeta_s,z_l) \frac{C_{m,n}(\zeta_t,\rho_t)}{\rho_t} \Gamma^t_{m,n}|^2]^2$$

for the echo (D is a normalizing constant), and for the reverberation,

$$\langle |b_r(\tau_0)|^2 \rangle = 2\pi D \Delta \rho \sum_{l,k} [\sum_{m,n} A^+_{m,n}(\zeta_s,z_l) \frac{C^+_{m,n}(\zeta_t,\rho_t)}{\rho_t} \Gamma^{*t}_{m,n}]$$

$$\cdot [\sum_{m,n} A_{m,n}(\zeta_s,z_l) A^+_{m,n}(\zeta_s,z_k) |C_{m,n}(\zeta_r,\rho_r)|^2 S_{m,n}] [\sum_{m,n} A_{m,n}(\zeta_s,z_k) \frac{C_{m,n}(\zeta_t,\rho_t)}{\rho_t} \Gamma^t_{m,n}]$$

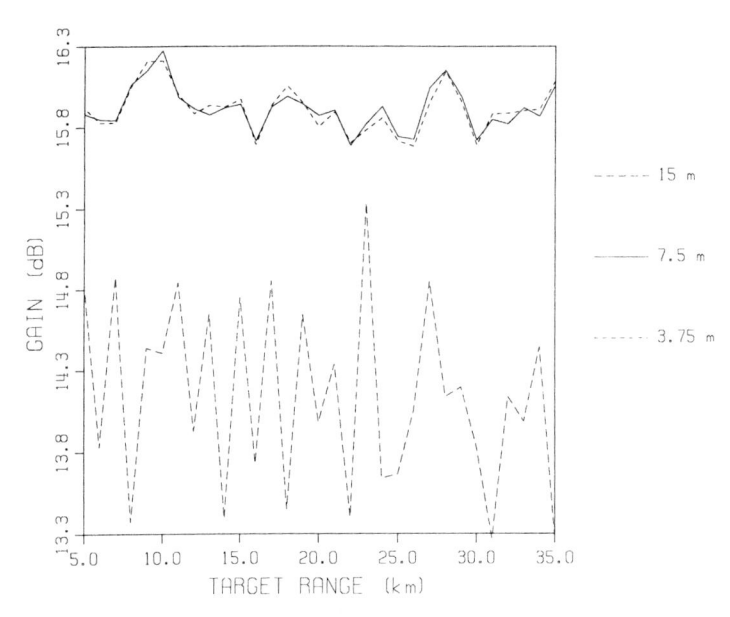

Fig. 1: Gain of echo against clutter for three array element spacings. Each array spans the entire water column.

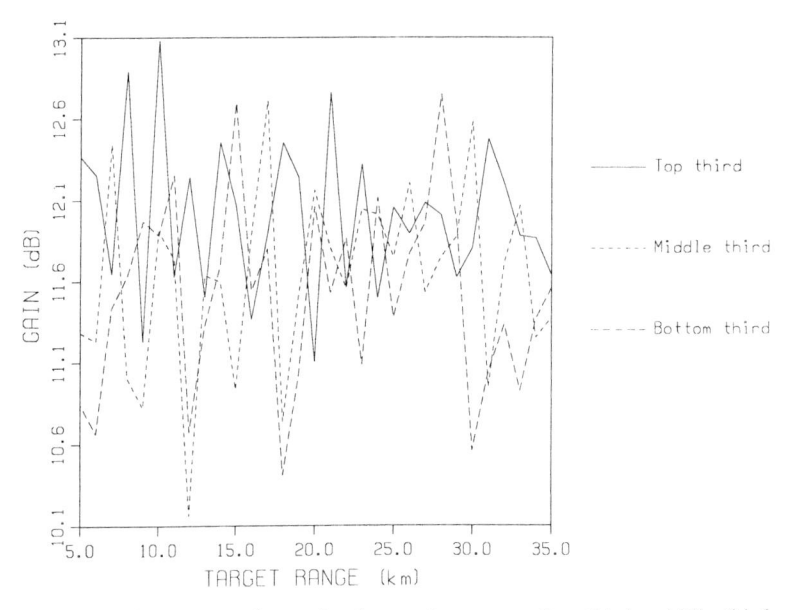

Fig. 2: Gain of echo against clutter for three subapertures (top third, middle third, and bottom third of water column) for one wavelength element spacing.

Note that if there were no depth dependence of the modes, so that the A_{mn} were constants, G would be unity (no gain against clutter). On the other hand, in the limiting case where the A_{mn} are "uncorrelated" at different hydrophone depths, i.e., $\phi_m(z_k)\phi_n(z_l)=0$ unless k=l, the ideal gain G=N is obtained. To the extent that more energy is contained in the (more oscillatory) higher modes, this condition should tend to be better satisfied.

3. Numerical Simulation.

We chose a sound speed profile from a continental slope region, at a depth of 500 m, where the sediments would most likely be thin and rough basement scattering would predominate. The profile is sharply downward refracting away from the surface and then is roughly isovelocity from about 300 m to 500 m. We chose a frequency of 200 Hz. A source was placed at a depth of 100 m, and vertical arrays of various apertures were placed at the same range as the source. A target was placed at a depth of 100 m, at varying ranges from the source. Pure backscatter from the target (no mode coupling) was assumed; note that G is independent of target strength. We used a layered normal mode code (using Airy functions) to compute the depth functions. We assumed a very rough bottom, with scattering governed by Lambert's Law using the ray-equivalent angles of the modes. Note that G is independent of the Lambert's Law constant. Figure 1 shows the results of computing the gain G as a function of the range (sampled at 1 km intervals) of the target from the source for three different arrays with equally spaced elements, all spanning virtually the entire water column, from 10 m to 490 m. The arrays have spacings of one-half acoustic wavelength (3.75 m, 129 elements), one wavelength (7.5 m, 65 elements), and two wavelengths (15 m, 33 elements). It is seen that for spacings down to a wavelength, G is generally within about 2 dB of the ideal array gain of 10 log N, where N is the number of elements (10 log 65 = 18.1 dB, 10 log 33 = 15.2 dB), but that sub-wavelength spacings offer no additional gain for the increased number of elements. Figure 2 then shows G for three sub-aperture arrays of 22 elements, all with one-wavelength spacing, and spanning the top, middle, and bottom third of the water column, respectively. The top aperture, being farthest away in depth from the bottom, on average shows somewhat higher gain (i.e., ability to reject bottom clutter) than the other apertures, being almost always within 2 dB of the ideal gain of 13.4 dB. However, even for the other two apertures, G is almost always within 3 dB of the ideal gain.

4. Summary.

We have outlined an active matched field processing scheme and shown that it has the potential to yield gain against clutter not far from the ideal array gain. In the future, we plan to carry out the large-scale computations required to integrate over large areas of the bottom and surface, to include broadband effects, to add more realistic target and boundary scattering kernels, and to include bistatic geometries.

5. Acknowledgements. This work was supported by DARPA and OPNAV under contract MDA972-91-C-0068.

REVERBERATION SUPPRESSION AND MODELING

D.W. Tufts, *University of Rhode Island, Kingston, RI, U.S.A.*

D.H. Kil, *Lockheed Sanders, Inc., Nashua, NH, U.S.A.*

R.R. Slater, *NRL, SSC, MS, U.S.A*

Abstract: We apply the Principal Component Inverse (PCI) method of rapidly adaptive interference suppression and signal detection to reverberation suppression and modeling. We present preliminary results for the challenging example in which the transmitted pulse is a two-second burst of hyperbolic frequency-modulated (HFM) sinusoid sweeping from 210 Hz to 280 Hz. We demonstrate improved detection performance after reverberation suppression, as compared to conventional matched filtering. We provide a physical interpretation of the principal components by presenting corresponding conventional and high-resolution temporal frequency/spatial frequency analysis of sets of snapshots from a subarray.

1 Introduction

In this paper we apply a method of interference suppression–the Principal Component Inverse (PCI)–for estimating, removing, and modeling of reverberation. It is a method of interference suppression and signal detection that adaptively nulls the interference by exploiting any low-rank structure of matrices that are formed from the samples of the interference. Such low-rank structure, whether present naturally or artificially induced by transformation, is used by making a reduced-rank approximation to a natural or transformed data matrix to obtain an estimate of the interference due to reverberation. The interference estimate, here the estimate of reverberation waveform from beams or from individual hydrophones, is subtracted from the observed data and the residual is matched filtered. The reverberation estimate itself is used in model testing and validation.

The major advantages of the PCI method are (1) the adaptation is rapid, allowing one to follow the fluctuations in reverberation frequency and direction, and (2) information for both reverberation suppression and signal detection are obtained from the same data set. No reverberation-only training data is needed.

Reverberation data from the Acoustic Reconnaissance Cruise of the Acoustic Reverberation Special Research Program (ARSRP) [1] were analyzed using the PCI method. For the case in which a two-second burst of a hyperbolic frequency-modulated (HFM) sinusoid is transmitted, short segments (about 1/6 sec. long) of reverberation can be used to form matrices that have the needed low-rank structure. The time intervals over which a sequence of reverberation is estimated is much shorter than the duration of the transmitted pulse.

The SRP data acquisition and processing system [1] consists of a vertical line array (VLA) with ten coherent sources for transmission and a horizontal line array (HLA) of 128 hydrophones for reception. The HLA was towed at a depth of about 170 m during data collection. The array section proper begins at 868 m from the tow point, and the distance of the center of the array from the transmitting source is 1062 m. The ship was moving at a speed of 3.3 knots and a heading of 300 degrees true, with its projector array steered downward by 6 degrees. For the data analyzed here, the transmitted signal was an HFM upsweep located at 0948Z on Track 5 of the ARSRP and with pulse identification number SPSS053. Samples 2263-2272, 2266-2275, 2269-2278, and 2272-2281 were used in tests shown later in this paper. It is given by the following expression:

$$s(t) = w(t)\sin(a\ln(1 - kt)), \qquad 0 \leq t \leq T \qquad (1)$$

where

$$k = (f_2 - f_1)/(f_2 T) \qquad f_1 = \text{initial frequency}$$
$$a = (-2\pi f_1)/k \qquad f_2 = \text{final frequency}$$
$$T = \text{waveform duration} \qquad w(t) = \text{window function.}$$

The HFM waveform parameters were: $f_1 = 210$ Hz, $f_2 = 280$ Hz, $T = 2$ sec. yielding $k = 1/8$ and $a = -3360\pi$. The window function $w(t)$ is nearly rectangular with some reduction at the beginning and at the end of the 2 sec. interval. Each hydrophone output was band-pass filtered by a filter having center frequency of 250 Hz and a bandwidth of 128 Hz and then shifted to baseband. The sampling frequency of the complex-valued baseband signal was 128 Hz. 0 Hz and 128 Hz of the complex envelope correspond to 186 Hz and 314 Hz of the original real-valued signal, respectively.

2 The Principal Component Inverse (PCI) Method of Interference Suppression

In [2, 3, 4, 5, 6, 7, 8, 9] Kirsteins and Tufts have considered the problem of rapidly adaptive signal detection when the noise plus reverberation matrix is approximately low rank and only a limited amount of data is available. The resulting method, called the Principal Component Inverse (PCI) method, is now motivated and described.

Let us assume that Y, the interference data matrix alone, without perturbation by signal or background noise components,

$$Y = [\mathbf{y}_1|\mathbf{y}_2|\cdots|\mathbf{y}_K] \tag{2}$$

is strongly low rank. The i-th column of Y could, for example, correspond to interference samples from the i-th hydrophone of an array. In the PCI method, the interference is first regarded as a data component to be enhanced. This interference estimate is subtracted from the observed data and the residual is then matched filtered. More specifically, the steps are as follows:

1) **Estimating the Interference Component**

   An estimate of the m-th column of the interference-only matrix Y, $\hat{\mathbf{y}}_m$, is

$$\hat{\mathbf{y}}_m = \tilde{U}_s \tilde{U}_s^H \tilde{\mathbf{y}}_m, \tag{3}$$

   where columns of the matrix $\tilde{U}_s$ are the principal left singular vectors of the total data matrix $\tilde{Y}$, consisting of interference, signal and noise components

$$\tilde{Y} = [\tilde{\mathbf{y}}_1|\tilde{\mathbf{y}}_2|\cdots|\tilde{\mathbf{y}}_K] \tag{4}$$

   and the projection matrix $\tilde{U}_s \tilde{U}_s^H$ in Eq. (3) can be thought of as a matrix filter that enhances the interference part of the data.

2) **Removing the Interference Component**

   The residual vector $\mathbf{r}$, after removal of the estimated interference from the m-th column is

$$\mathbf{r} = \tilde{\mathbf{y}}_m - \hat{\mathbf{y}}_m = \tilde{\mathbf{y}}_m - \tilde{U}_s \tilde{U}_s^H \tilde{\mathbf{y}}_m \tag{5}$$

3) **Matched Filtering the Residual r**

   The PCI test statistic is

$$z = \left|\mathbf{s}^H \mathbf{r}\right|. \tag{6}$$

   where $\mathbf{s}$ is the signal shape vector, say, the samples of the HFM signal.

4) Compare z Against Threshold

To test for the presence of the signal component we compare the value of the test statistic with a threshold value that has been set for a specified probability of false alarm.

The structure of the PCI estimate of the interference can be justified using some results of Claus et al. [10] and Liu and Nolte [11] on detection of a signal of known form but random phase, in correlated gaussian noise. Claus et al. showed in their analysis that the optimum processor removes an estimate of the interference from the data. Those eigencomponents corresponding to large eigenvalues of the noise covariance matrix R are scaled more in that removal process. This can be seen by writing the likelihood ratio test statistic for the case of Gaussian reverberation and noise as follows [10]:

$$\left|\mathbf{s}^H R^{-1}\mathbf{x}\right| = \sigma^2 \left|\mathbf{s}^H(I - P)\mathbf{x}\right|, \tag{7}$$

where $\mathbf{x}$ is the observed data, and $\mathbf{s}$ is the hypothesized signal vector. P is the matrix filter that operates on the data vector $\mathbf{x}$ to produce the estimate of the interference vector:

$$P = \sum_{k=1}^{r} \frac{\lambda_k}{\lambda_k + \sigma^2} \mathbf{v}_k \mathbf{v}_k^H. \tag{8}$$

The $\{\lambda_k\}$ and $\{\mathbf{v}_k\}$ are, respectively, the r principal eigenvalues and eigenvectors of R, and σ^2 is the variance of the background white noise component. When the interference is strongly low rank, i.e., $\lambda_k \gg \sigma^2$, then P is approximately the following projection matrix:

$$P \approx \sum_{k=1}^{r} \mathbf{v}_k \mathbf{v}_k^H. \tag{9}$$

Hence the matrix $I - P$ in Eq. (2) corresponds to the projection onto the orthogonal complement of the interference space. The premultiplication of the data vector $\mathbf{x}$ by the matrix $(I - P)$ can be considered as a generalized null-steerer.

Because we work with short segments of non-stationary data it is more efficient to find the vectors $\mathbf{v}_k$ and the reverberation estimate $P\mathbf{x}$ directly from the data using the Singular Value Decomposition (SVD) of the data matrix rather than by forming an estimated reverberation covariance matrix. The data matrix can be a matrix of snapshots collected from the hydrophone array, or overlapping segments of data from a single hydrophone or beam can be used to form the rows of a matrix.

For a single hydrophone or beam the steps are as follows: Blocks of N samples from a single hydrophone's output are taken and a forward-backward data matrix is formed [6, 12]. Next, its low-rank approximation is computed. The rank is decided by first computing the SVD of the data matrix. Then the sum of squared singular values (starting from the smallest singular value) is computed till this sum exceeds a particular threshold. For example, if after adding the square of the k-th singular value the sum exceeds a fixed threshold, the rank is chosen as k. An estimate of the interference is obtained using the diagonal averaging procedure [6]. Since this method is known to yield better signal estimates in the middle of the data record [13], only these estimates are retained, which necessitates the use of overlapping blocks to obtain contiguous blocks of interference estimates. In the final step of the above procedure, the estimated interference is subtracted, the residual is matched filtered, and its output compared against a threshold to test for presence of a weak signal.

The threshold mentioned in the previous paragraph for rank selection is set high enough that it is unlikely that target returns at their levels predicted by propagation and target-strength models at the current range, ambient noise, or weak reverberation components would cause the threshold to be exceeded. That is, for cases in which beamforming and matched filtering are nearly optimum without the use of adaptive reverberation suppression, the threshold is unlikely to be exceeded and the reverberation is not estimated.

3 Experimental Results

Various computer experiments were done using the ARSRP data. These were done to test the hypotheses that matrices formed from short-duration segments of reverberation are approximately low rank and to test the ability to detect weak signals buried in strong reverberation by the matrix-based reverberation suppression method of the previous section.

An example of the above interference estimation procedure using data from a broadside beam formed from 126 uniformly spaced hydrophones is shown in Fig. 1(a). The solid line shows a portion of the real reverberation plus a simulated weak signal, while the dashed line shows the reverberation estimate, which was obtained using the procedure described in the previous two paragraphs; the approximating rank was unity. In Fig. 1(b) the estimated reverberation in the absence of the weak signal is shown. It is seen that the effect of the weak signal on the interference estimate is quite minimal, and the interference estimate remains quite good.

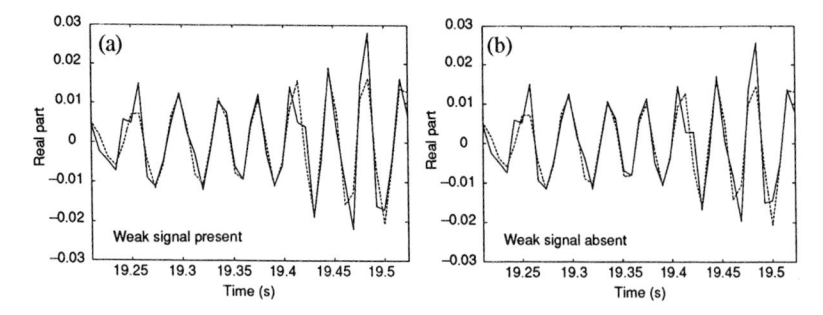

Fig. 1 (a) Portion of broadside reverberation plus an artificially added weak signal (solid line) and its estimate (dashed line) using rank 1 approximation in the presence of weak signal. (b) Portion of broadside reverberation (solid line) and its esimate (dashed line) using rank 1 approximation in the absence of weak signal.

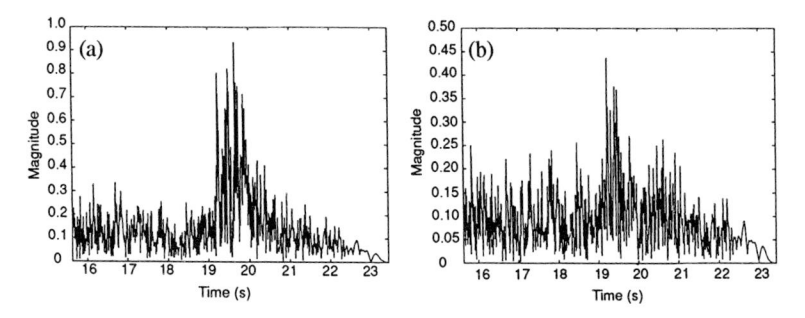

Fig. 2 (a) Magnitude of matched filter output of beamformed data prior to reduced-rank reverberation cancellation. Nearby false peaks obscure the true location of the weak signal. (b) Magnitude of matched filter output after reduced-rank reverberation cancellation. The peak location now corresponds to the true delay of the weak signal.

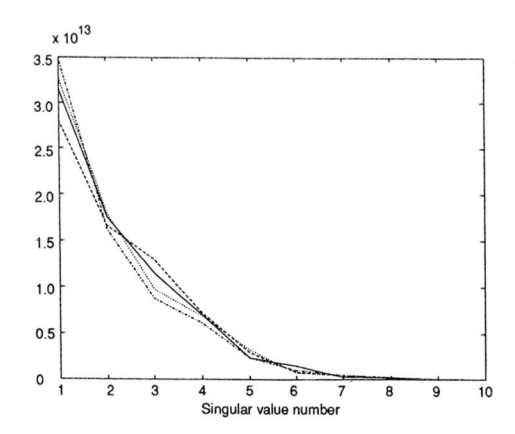

Fig. 3 Squared singular values of four different 10 x 10 matrices of the SRP data. The four time-overlapped data matrices which were used to compute these squared singular-values are also used to compute the results of Fig. 4.

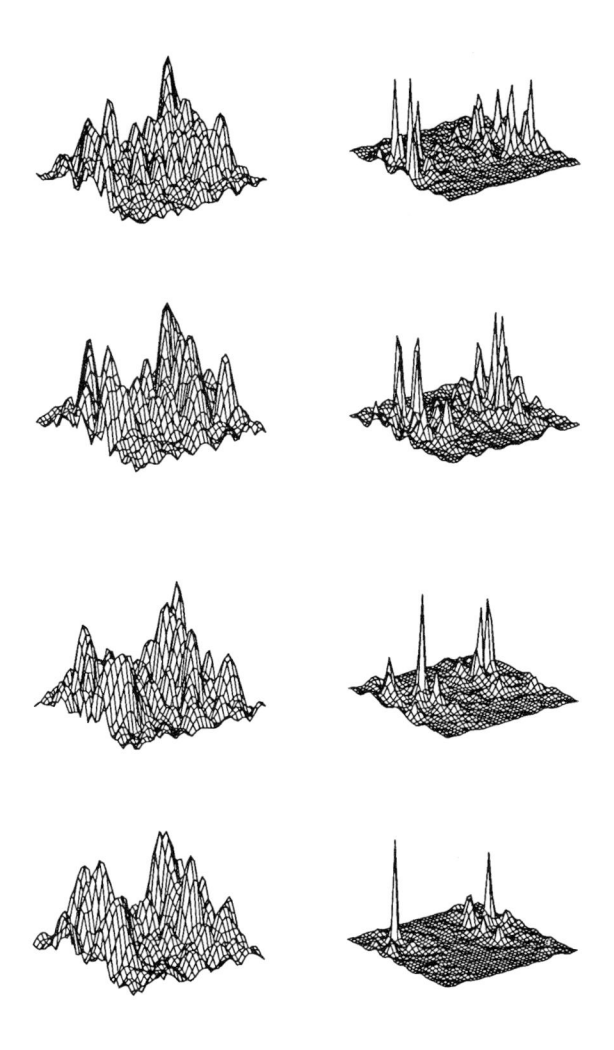

Fig. 4 Magnitude plots of 2D-DFTs (1st column) and 2D, reduced-rank Linear Prediction using an 8 x 8 window (2nd column). The time overlap between successive plots down either column is seven-sample period (7/128 sec.). The axis going left represents the temporal frequency and the axis going right is the spatial frequency.

A portion of the set of snapshots from the array containing strong reverberation was considered. To it array snapshots corresponding to a weak signal coming from broadside was added. The signal strengths were chosen such that direct matched-filtering of a broadside, time-delay-formed beam was not sufficient to detect the weak signal echos. To improve signal detectability, the matrix-based reverberation estimation procedure described in Section 2 was used for interference suppression prior to matched filtering. Fig. 2(a) shows the magnitude of matched filter output of received data prior to reverberation cancellation. The presence of many false peaks resulted in poor detection performance. Fig. 2(b) shows the magnitude of the matched filter's output after reverberation cancellation. Its peak corresponds to the true location of the weak signal. Our ability to detect the weak signal has improved. The reason appears to be that there are only few strong reverberation components in the directional beam and it is then more likely that we can suppress reverberation without significantly cancelling the desired signal.

In a second series of experiments, data matrices were formed from sets of ten snapshots from a subarray of ten consecutive hydrophones of the array. Each snapshot is a set of ten simultaneous samples, one sample from each of the ten hydrophones. A ten-by-ten matrix is formed by using a group of ten consecutive snapshots as rows of the matrix. The snapshots are spaced by 1/128 of a second.

Four successive matrices are formed by removing the three oldest of the ten snapshots(rows) at the top of the previous matrix and inserting three new snapshots at the bottom of the matrix to obtain a new ten-by-ten data matrix. Four such matrices were formed to compute the results of Figs. 3 and 4.

Two-dimensional, spatial frequency/temporal frequency analysis results are shown in Fig. 4. The left side of Fig. 4 from top to bottom shows the magnitude of the two-dimensional Fourier transform of each of four successive matrices of subarray snapshots. The right side shows the corresponding results obtained using two-dimensional low-rank, linear-prediction estimation of spatial and temporal frequencies [14]. Fig. 4 shows that six components explain much of the energy. This is consistent with the results of Fig. 3 which shows four superimposed plots of squared-singular values for the four consecutive matrices. Much work remains to connect these principal components of reverberation, which are obtained using matrix approximation, with the important components obtained from physical modeling [18].

4 Principal Components and Reverberation Models

Much has been learned about low frequency reverberation, especially about power density spectra and about the distribution of short-time reverberation energy in time and angle. Examples are the results from Project Artemis as summarized in [15] and the work on modeling reverberation [16, 17, 18].

The models that have evolved from such studies provide information about the expected variation in short-time energy at the output of a matched filter due to reverberation. Currently they do not provide information about the number of principal components, or more importantly, the percentage of the total number of components which are likely to be required for good representation and hence good suppression of the reverberation. Further work is needed to relate the important predicted physical components of reverberation, such as the stronger multipath arrivals from reflecting sites in a given direction, to the important approximating components formed from the data by matrix analysis using the SVD.

ACKNOWLEDGEMENTS

We are grateful to Chintana Griffin, Tim Pierce, and to C.S. (Ramli) Ramalingam for their help in processing of the ARSRP data and in the preparation of the manuscript.

References

[1] "Acoustic Reverberation Special Research Program," Technical report, Office of Naval Research, 1991.

[2] I. P. Kirsteins and D. W. Tufts, "On the probability density of signal-to-noise ratio in an improved adaptive detector," in *IEEE Proc. ICASSP-85*, (Tampa, FL), pp. 572–575, Apr. 1985.

[3] I. P. Kirsteins and D. W. Tufts, "Rapidly adaptive nulling of interference," in *High Resolution Methods in Underwater Acoustics*, (M. Bouvet and G. Bienvenu, eds.), New York: Springer-Verlag, 1991.

[4] I. P. Kirsteins and D. W. Tufts, "Signal enhancement by low-rank approximation to a data matrix with application to adaptive detection," *IEEE Trans. Aerospace and Electronic Systems*, to appear 1993.

[5] C. D. Melissinos, D. W. Tufts, and I. P. Kirsteins, "Very fast adaptive array processing," in *25th IEEE Conf. on Decision and Control*, (Athens, Greece), Dec. 1986.

[6] D. W. Tufts, R. Kumaresan, and I. Kirsteins, "Data adaptive signal estimation by singular value decomposition of a data matrix," *Proc. IEEE*, vol. 70, pp. 684–685, June 1982.

[7] D. W. Tufts, I. P. Kirsteins, and R. Kumaresan, "Data-adaptive detection of a weak signal, single frequency, plane-wave signal in noise and strong, unidirectional interference," in *Statistical Signal Processing*, (E. J. Wegman and J. G. Smith, eds.), New York: Marcel Dekker Inc., 1984.

[8] D. W. Tufts, I. P. Kirsteins, and P. F. Swaszek, "Detection of signals in the presence of strong, signal-like interference and impulse noise," in *Topics in Non-Gaussian Signal Processing*, (E. J. Wegman, S. C. Schwartz, and J. B. Thomas, eds.), New York: Springer–Verlag, 1988.

[9] R. J. Vaccaro, D. W. Tufts, and G. F. Boudreaux-Bartels, "Advances in principal component signal processing," in *SVD and Signal Processing*, (F. Deprettere, ed.), pp. 115–146, The Netherlands: Elsevier Science Publishers, 1988.

[10] A. Claus, T. Kadota, and D. Romain, "Efficient approximation of a family of noises for application in adaptive spatial processing for signal detection," *IEEE Transactions on Information Theory*, vol. IT-26, pp. 588–595, 1980.

[11] C. Liu and L. W. Nolte, "Performance evaluation of array processors for detecting Gaussian acoustic signals," *IEEE Trans. Acoust., Speech, Signal Processing*, vol. ASSP-28, pp. 328–333, June 1980.

[12] D. W. Tufts and R. Kumaresan, "Estimation of frequencies of multiple sinusoids: Making linear prediction perform like maximum likelihood," *Proc. IEEE*, vol. 70, pp. 975–989, Sept. 1982.

[13] A. A. Shah and D. W. Tufts, "Estimation of the signal component of a data vector," in *IEEE Proc. ICASSP-92*, (San Francisco, CA), pp. 393–396, Mar. 1992.

[14] A. Efron and D. W. Tufts, "Estimation of frequencies of multiple two-dimensional sinusoids: Improved methods of linear prediction," in *Proc. ICASSP-85*, (Tampa, FL), pp. 1777–1779, Mar. 1985.

[15] B. B. Adams, D. T. Deihl, L. B. Palmer, and J. T. Warfield, "Long-range monostatic reverberation in the deep ocean," *U.S. Navy Journal of Underwater Acoustics*, vol. 24, pp. 215–242, Apr. 1974.

[16] C. Eckart, "The scattering of sound from the sea," *J. Acoust. Soc. Am.*, vol. 25, pp. 566–570, 1953.

[17] D. D. Ellis and D. V. Crowe, "Bistatic reverberation using a three-dimensional scattering function," *J. Acoust. Soc. Am.*, vol. 89, pp. 2207–2214, 1991.

[18] J. W. Caruthers and E. J. Yoerger, "ARSRP reconnaissance results and BISSM modeling of direct path backscatter," *Presented at the Ocean Reverb. Symp.*, (LaSpezia, Italy), May 1992.

SUB-BOTTOM SCATTERING AT LOW FREQUENCIES

Dr. James H. Wilson
Neptune Sciences, Inc.
PO Box 1235
San Clemente, CA 92672

Dr. Marshall Bradley
Mr. Melvin Wagstaff
Planning Systems, Inc.
115 Christian Lane
Slidell, LA 70458

ABSTRACT Operation of active sonar systems at low frequencies has been of high interest to many research efforts this past year. Bottom and sub-bottom reverberation can degrade the performance of active sonar systems at these frequencies. In this paper, the results of data analysis of a 12 element vertical receive array with bistatic, low frequency, impulsive source are given. The bottom and sub-bottom multipath arrivals are detected and analyzed. A frequency domain, reverberation subtraction technique is developed that uses Inverse Beamforming (IBF) and an innovative covariance matrix to FFT approximation technique. The seismic data shows numerous multipath arrivals that include a number of strong sub-bottom/surface reflected paths interfering with sub-bottom reflected paths. The sub-bottom frequency attenuation of multipaths over the frequency range from 12 Hz to 63Hz is clearly shown in the data. The bottom/sub-bottom reverberation technique is applied to the data and results are given.

The CAPARAY (ray theory) propagation model is employed to simulate the bottom and sub-bottom multipath arrivals. Use of range-independent propagation models is suitable, since the ocean bottom and sub-bottom in this area is nearly a flat layered of "pancake" geological structure. Preliminary model/data comparisons of the seismic data are discussed.

1. INTRODUCTION

ASW R&D efforts are currently being directed toward both active acoustic systems and shallow water areas. The new diesel submarines being acquired by third world nations are very quiet and do not spend a great deal of time in deep water areas within missile range of the United States. Active acoustic systems operating at low frequencies in shallow water will be significantly impacted by bottom and sub-bottom reverberation. Also, in seismic signal processing, bottom reverberation is the "signal" in ASW terms, and being able to subtract unwanted multipath arrivals or multiples that are not single reflections from the bottom or sub-bottom layers improves seismic images of bottom sublayer structure.

This paper describes an integrated process to model bottom reverberation using the CAPARAY model, identify the multipath arrivals (or multiples in seismic signal processing terms), and then subtract the unwanted multipath bottom reverberation. The CAPARAY model assumes a horizontally layered sub-bottom or "stacked pancake" bottom structure. While there are some ocean areas that don't fit the stacked pancake or layered bottom characteristic, vast areas of the ocean bottom/sub-bottoms are approximately this way. In addition, CAPARAY is used only to identify multipaths to subtract and an innovative frequency domain beamforming method[2], called the Fourier Integral method (FIM) is used to do the actual subtraction. The inverse FFT of a covariance matrix (outer product of FFTs) is not a trivial operation in this case because a subtraction between covariance matrices produces a covariance matrix that is not an outer product of individual hydrophone FFTs. An innovative, robust algorithm to perform this inverse FFT approximation is shown to involve solution of non-linear equations.

The ultimate objective of this research is to identify individual bottom reverberation multipaths with CAPARAY, automatically subtract the unwanted energy in the frequency domain, then inverse FFT the data to hydrophone time series with the unwanted energy subtracted.

2. Bottom and Sub-Bottom Reverberation Subtraction Techniques

The signal processing schematic to subtract bottom reverberation is shown in Figure 1. There are two reasons for performing the subtraction in the frequency domain for this application. First, the solution to the Inverse Beamforming (IBF) Integral Equation solved by Dr. Wilson of Neptune Sciences, Inc. can only be implemented in the frequency domain. The IBF solution is the most natural plane wave density estimate of the acoustic field, as shown in a recent paper[2] by Nuttall and Wilson and can not be formulated in the time domain. Second, an incoherent noise normalization technique is applied independently in each frequency bin to remove unwanted, incoherent noise before using IBF to subtract the coherent noise.

As Figure 1 shows, an outer product of the FFTs of the individual hydrophone time series forms the covariance matrix of all hydrophone pairs in the array. The covariance matrix in each frequency bin is input to IBF and the plane wave density of the bottom reverberation acoustic field is estimated as a function of frequency, azimuth, and time. The CAPARAY model described in the next section is used to identify the specific bottom reverberation multipaths to subtract. Peak energy at each

Deriving Time Series Data from Cross Spectral Matrix (CSM)

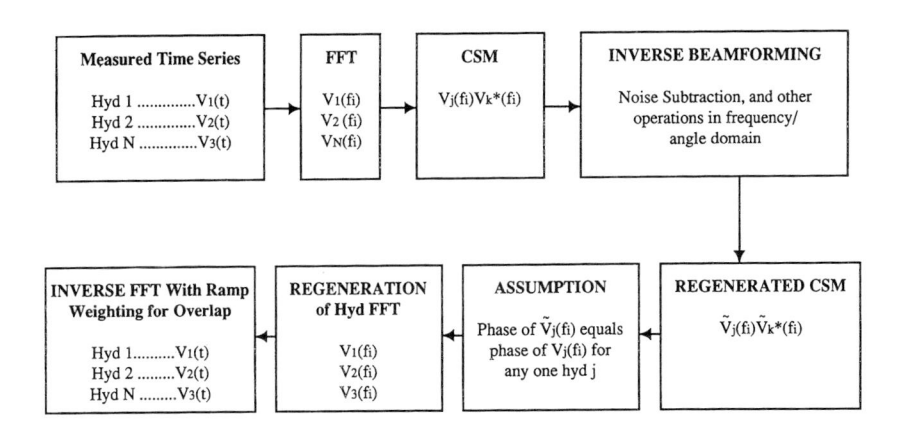

Figure 1. IBF Reverberation Subtraction Technique

FFT time epoch is determined from the measured covariance matrix and IBF is used to subtract the unwanted energy in each frequency bin. The IBF plane wave estimation is made for each arrival to be subtracted and the covariance matrix is regenerated from IBF for each plane wave. This IBF regenerated covariance matrix is subtracted from the original or measured covariance matrix and the "subtracted" covariance matrix that results must be used to form an estimate of "subtracted" FFTs so that an inverse FFT can be performed to regenerate time series with the unwanted bottom reverberation subtracted. Two problems arise at this step. First, the absolute phase in each FFT time epoch is lost when the covariance matrix is formed (relative phase information among hydrophone pairs is maintained). This is solved by a phase fit algorithm that minimizes the least square error between the phases of the hydrophone FFTs immediately preceding and following the FFT time epoch in which the subtraction was performed. The second, and more serious, problem is that the "subtracted" covariance matrix that resulted from the subtraction of the IBF regenerated covariance matrix from the measured covariance matrix is <u>not an exact</u> outer product of FFTs - although <u>both</u> the measured covariance matrix and IBF regenerated covariance matrix are <u>exact</u> FFT outer products of FFTs. A set of non-linear equations has to be solved to find the "best" or least square error FFTs whose outer product approximates the subtracted covariance matrix. A separate set of equations had to be solved for FFT phase and amplitude. Space does not allow detail discussion of these equations but they are specified by setting the unknown phase of the "subtracted" FFT's equal to the <u>average</u> relative phase and <u>average</u> amplitude of the covariance matrix for a line array of M equally spaced elements as follows:

$$|\overline{Si}| = \frac{1}{M} \sum_{j=1}^{M-1} \frac{1}{|si|} [|sij|] \qquad (1)$$

$$\overline{\Phi}_i = \frac{1}{M-1} \sum_{j=1, j \neq i}^{M} [\Phi_i - \Phi_j] + \Phi_j \qquad (2)$$

Where $|\overline{Si}|$ is the magnitude of the subtracted FFT elements $\overline{Si}$, $|Sij|$, is the magnitude of the subtracted covariance matrix element for hydrophones i and j, $\overline{\Phi}_i$, is the phase of the subtracted FFT elements $\overline{Si}$, and $[\Phi_i - \Phi_j]$ is the relative phase of the covariance matrix Sij.

To illustrate the subtraction technique, data from a 48 element bottom planar array are used with an impulsive source. Time series data are shown in Figures 2 and 3 and the large direct arrival near one second in Figure 2, a higher frequency arrival near 3 to 4 seconds in Figure 3, and a coherent multipath arrival near the end of the time series at very low frequency in Figure 3 were required to be subtracted. Using a .25 second FFT with no overlap, these interference terms are shown in the IBF output in Figure 4 in the direction of the source. The IBF or plane wave energy is shown in 19 time epochs of .25 seconds each and in 32 frequency bins of 4Hz each from 0Hz to 128Hz in Figure 4. The IBF frequency domain subtraction technique was applied to the data in Figure 2, 3, and 4 and the corresponding figures after subtraction are shown in Figures 5, 6 and 7. Figure 5 shows that the unwanted energy has been subtracted from the IBF output while Figures 6 and 7 show the time series after these three unwanted sources of energy have been subtracted. The results are exceptionally good but it is emphasized that the subtraction technique has many innovative facets that can not be described in detail here. The two points emphasized are: IBF (and not conventional beamforming) must be used; and factorizing an arbitrary covariance matrix into an optimum (least square error) outer product of FFTs required solving the set of non-linear equations given in Equations (1) and (2). This proved to be a very difficult problem.

In the next section, data from a 12 element bottom, vertical array is used to model multipath arrivals using CAPARAY.

3. Modeling Bottom and Sub-Bottom Arrivals Using CAPARAY

Figure 8 shows the broadband output for 12 equally spaced hydrophones comprising the vertical linear receive array. The horizontal source-receiver separation is 324.6 meters and the source depth is 6.1 meters. The lowest trace in Figure 8 is the voltage output of the shallowest phone, located at a depth of 197.9

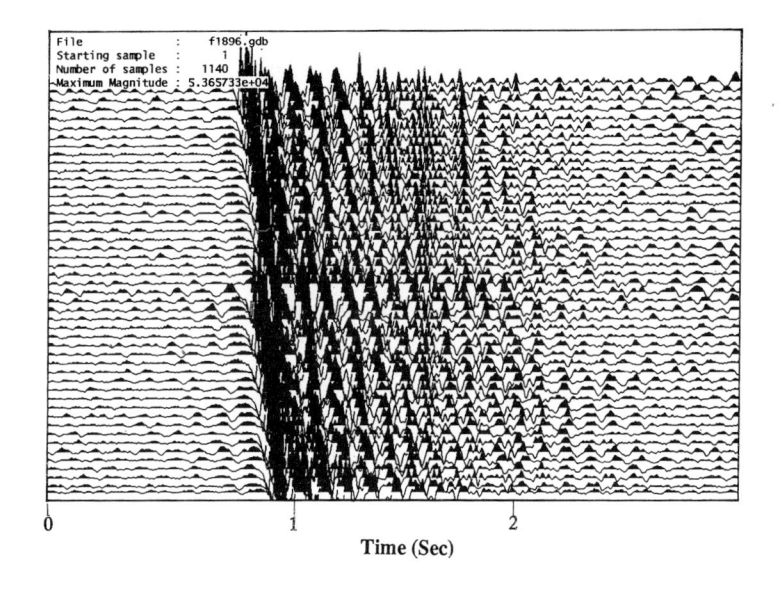

Figure 2. Planar Array Time Series Before Subtraction

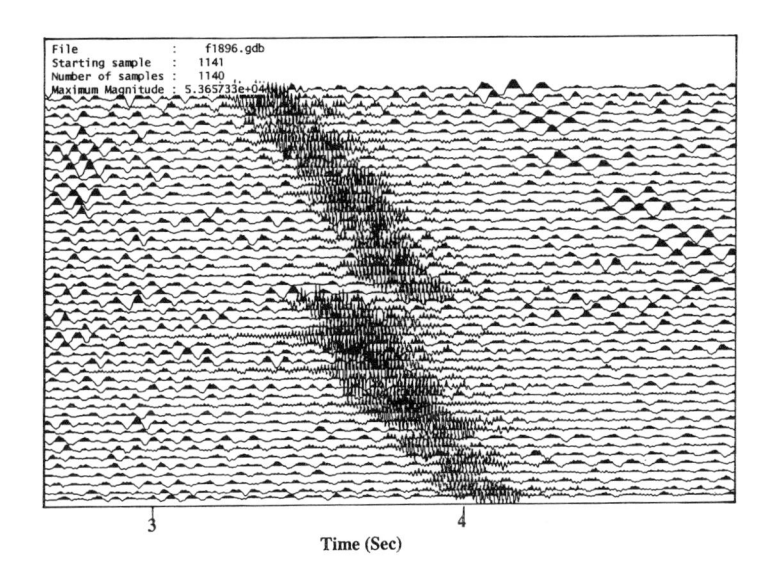

Figure 3. Planar Array Time Series Before Subtraction (cont.)

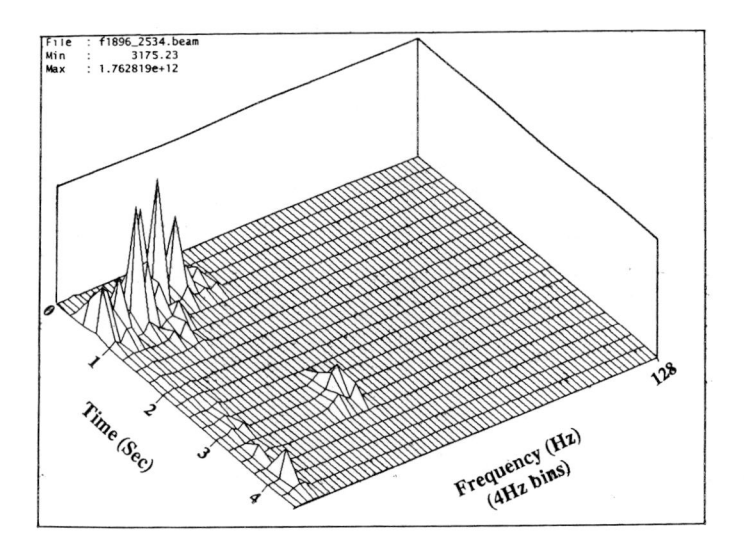

Figure 4. IBF Frequency Time Surface in Source Direction Before Subtraction

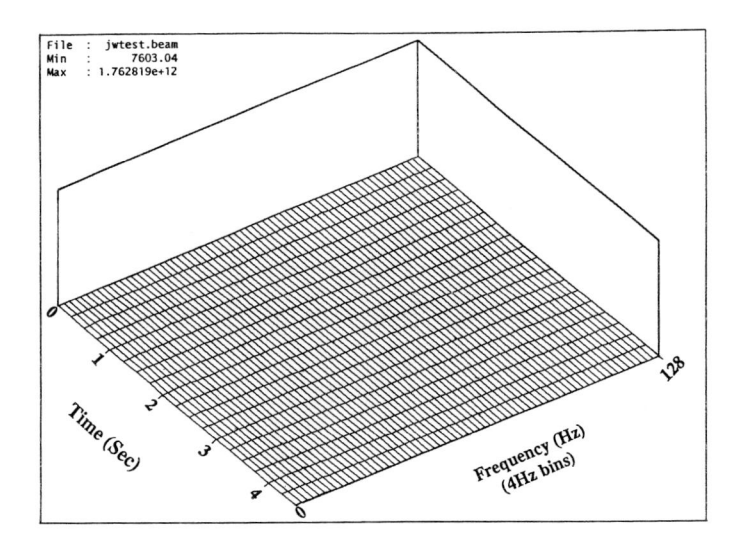

Figure 5. Frequency Time Surface in Source Direction After Subtraction

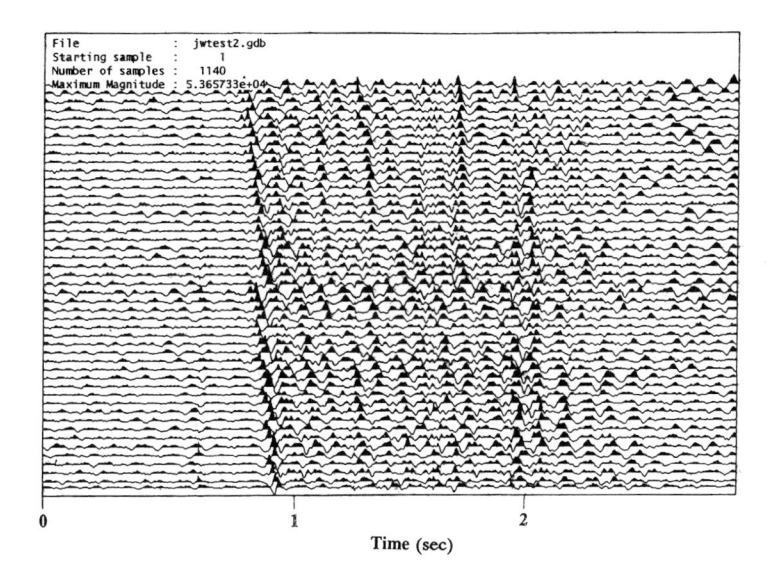

Figure 6. Planar Array Time Series After Subtraction

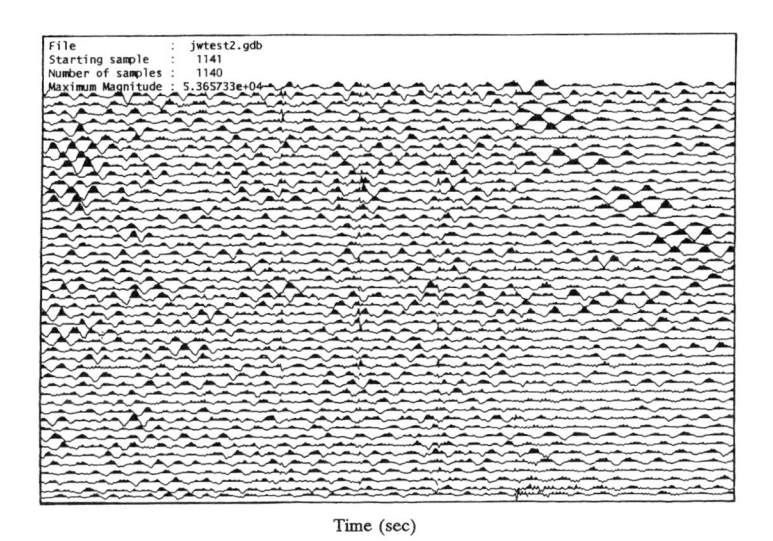

Time (sec)

Figure 7. Planar Array Time Series After Subtraction (cont.)

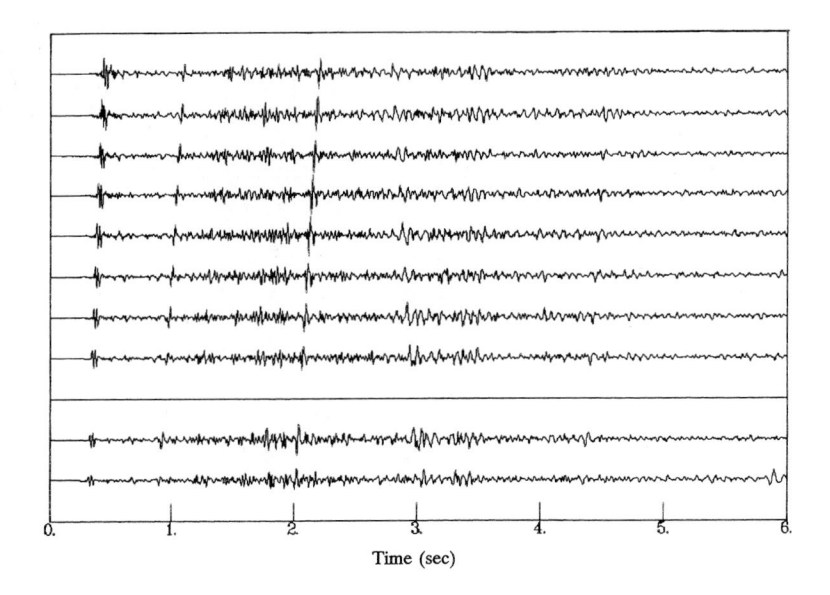

Figure 8. Measured Seismic Data

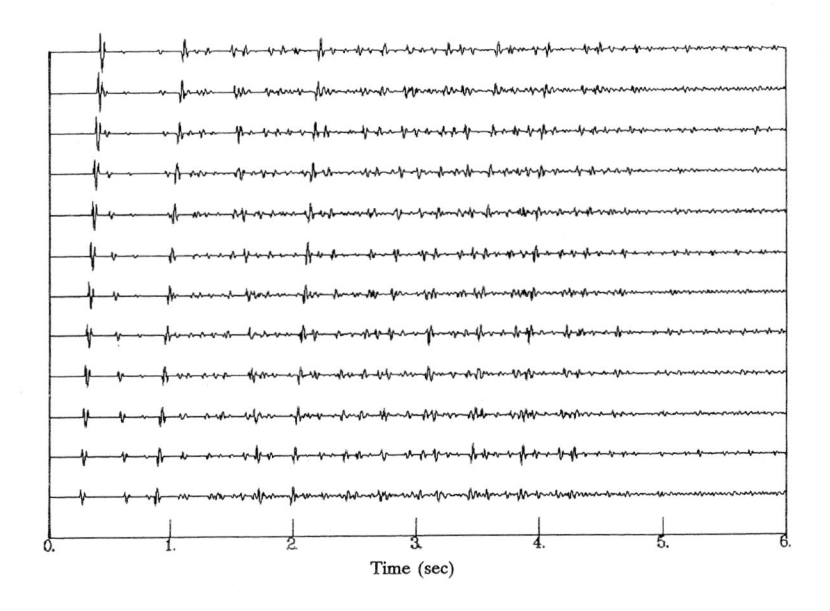

Figure 9. Simulated CAPARAY Data

meters, and the upper trace is the voltage output of the deepest phone, located at a depth of 532.9 meters. The voltage outputs shown in Figure 8 contain a gain proportional to time squared in order to compensate for spherical spreading and to facilitate interpretation of later arrivals. The third and twelfth phones were not operational during this shot sequence.

For the first phone, a combination of direct and surface reflected path arrivals is clearly evident at about 0.3 seconds. The weak bottom reflection arrival at about 0.7 seconds indicates a relatively low impedance contrast across the ocean bottom. The arrival at about 0.9 seconds is the first bottom-surface reflection. The applied gain makes this arrival appear stronger than it actually is. The next significant event, at about 2 seconds, indicates a strong impedance contrast between layers 4 and 5. Similar though weaker events can be found at 3.3 and 4.3 seconds. Since the arrival time for these events increase as a function of phone depth it is inferred that these arrivals are bottom-surface reflections. With a few exceptions, sub-bottom reflections do not readily appear in the data.

In order to further analyze the data presented in Figure 8, synthetic hydrophone outputs have been generated using the CAPARAY model[1]. CAPARAY is a range-independent eigenray model that simulates the propagation of broadband signals which interact with a layered ocean bottom. Profiles of compression velocity, shear velocity, compression attenuation, shear attenuation and density can be specified for each bottom layer. An efficient technique is used to find all eigenrays that propagate from a source to a receiver, including those with multiple reflections, refractions and conversions within a layer.

For each requested source-receiver location combination, CAPARAY computes the travel time, arrival amplitude and phase shift of all eigenrays up to a specified energy loss level. At short ranges and high propagation angles where caustic effects are not important, the receiver pressure level due to a source which emits a pulse $S(t)$ is:

$$S(t) = \sum A_n S(t-t_n) \cos(\Phi_n)$$

where A_n is the amplitude, t_n is the travel time and Φ_n is the phase of the nth eigenray which propagates from the source to receiver. The source function was not available and we have assumed that the source emits a pulse of the form:

$$S(t) = \begin{cases} \exp(-t/t_o) \sin(2\pi t/t_o), & 0<t<t_o, \\ 0, & \text{otherwise,} \end{cases}$$

where t_o the pulse length, is 0.05 msec. Estimates of the compression velocity and density with in the ocean bottom are shown in Table 1. Compression velocity (V_p)

estimates and bottom layer depths (Z) were supplied with the seismic data tape. Shear velocity (V_s) estimates were obtained by dividing the compresion velocities by 1.5. Layer density (ρ) was assumed to begin at 1.5 gm/cc and to gradually increase. Densities were adjusted to obtain appropriate reflection amplitudes. Compression and shear attenuation were not used as these variables are difficult to obtain and impedance contrasts can serve adequately to produce the desired relative amplitude and phase terms.

Table 1

Z(m)	V_p(m/s)	V_s(m/s)	ρ (gm/cc)
538.6	1507.3	0.0	1.0
692.0	1583.4	1055.6	1.5
950.8	1644.3	1096.2	1.62
1218.8	1766.1	1177.4	1.74
1505.0	1827.1	1218.0	1.86
2010.5	1948.8	1299.2	2.22
2540.4	2070.6	1380.4	2.34
2991.0	2192.4	1461.6	2.46
3466.1	2314.3	1542.9	2.65
3965.5	2436.1	1624.01	2.77
half-space	2557.9	1705.3	2.89

A CAPARAY prediction corresponding to the measured data shown in Figure 8 is presented in Figure 9. The modeled data shown in Figure 9 also contains a gain proportional to time squared. Many features delineated in the measured data are reproduced in the simulated data. Some phase information is not present, though this is expected due to inadequate modeling of the source. One feature that we have not been able to model is a strong return at 3.0 seconds on the first trace of the measured data. This return behaves as a bottom-surface reflection across all twelve phones but does not appear in the modeled data.

4. CONCLUSIONS

Two conclusions result from the analyses in this paper. First, IBF can be used to subtract unwanted bottom reverberation arrivals using a frequency domain technique with exceptionally good results. Second, the CAPARAY model can be used to model bottom and sub-bottom multipath arrivals as a way to identify specific multipath arrivals to be subtracted. CAPARAY currently assumes layered or "pancake" bottom layering structures.

5. References

1. Westwood, E. and Vidmar, P. (April 1987) 'Eigenray finding and time series simulation in a layered-bottom ocean', Journal of Acoustical Society of America, 81(4), 912-924.

2. Nuttal, A. and Wilson, J. (October 1991) 'Estimation of the acoustic field directionality by use of planar and volumetric arrays via the Fourier series method and Fourier integral method', Journal of the Acoustical Society of America, 90(4), 2004-2019.

BROADBAND DETECTION AND CLASSIFICATION OF UNDERWATER SOURCES

Azizul H. Quazi and Albert H. Nuttall
Naval Undersea Warfare Center
New London, Connecticut 06320 USA

ABSTRACT: The radiated signals from sources are of great importance for passive sonar, which is designed to exploit peculiarities of this form of signal and distinguish it from ambient noise, in which it is normally observed. The radiated signals usually consist of both narrowband and broadband components. In this paper, we describe techniques that exploit the broadband signals for detection and possible classification of the underwater sources.

INTRODUCTION: Ships, submarines, and torpedoes create and generate underwater sound. The principal sources of radiated acoustic noise-like signals are the propulsion system, propellers, auxiliary machines, and hydrodynamic effects [1,2]. The acoustic spectra of these noise-like signals typically contain a broad continuous spectral component as well as narrowband sinusoidal components, referred to as discrete lines or tonals. Traditionally, these narrowband tonals have been instrumental for passive detection and classification of underwater sources. However, if the tonals are reduced or disappear, it will be difficult to detect and classify based solely on these tonals. An alternative way to address the problem of detection and classification is through broadband signals. In this paper, we describe techniques that will lead to detection and possible classification of underwater sources; however, the classification work that has been initiated will not be discussed in detail here.

DETECTION AND CLASSIFICATION SYSTEM DESCRIPTION: A passive noise-like signal may be characterized as a finite duration of a stationary or nonstationary process. This type of random signal is very often of unknown duration, bandwidth, time location, and frequency shift, and occurs in the presence of colored noise. The detection of this type of signal is often accomplished in practice by filtering the received waveform to the frequency band of interest, squaring the filter output, and time-weighting this quantity prior to accumulation and threshold comparison. The performance of this processor depends on the spectra of the received signal and noise processes and on the transfer characteristics of the input filter. The uncertainty about the signal location and duration will mandate that the receiver observation time be lengthened, in order to ensure capture of any impingent signal energy. Similarly, lack of detailed knowledge of signal spectral characteristics or frequency shift will often dictate that a fairly broad (mismatched) filter passband be utilized in order that the signal be passed (when present).

Since the signal time location and duration are often unknown, the receiving filter must be kept open at all times, thereby allowing the filter output noise to build to its full strength steady-state value.

By contrast, a gated input signal leads to a filter output component which gradually builds up in time and decays slowly after the signal is turned off. This nonstationary effect, coupled with the particular weighting employed during the observation interval, influences the performance of the detector in a complicated fashion. In addition, it will not suffice to resort to Gaussian approximations for the processor output, because the number of samples employed is not large enough to utilize the central limit theorem, especially on the tails of the distribution (small false alarm probabilities) where we are interested. This need is addressed in this paper, for the case of Gaussian input signals in Gaussian noise; we provide exact performance of the mismatched energy detector for a variety of signal and noise characteristics and selection of parameters.

The detector of interest is shown in figure 1 below. We will discuss this detection scheme and provide performance results in terms of probability of detection versus probability of false alarm, as a function of the input signal-to-noise ratio, number of samples, and other associated parameters.

The detector of interest is the digital processor depicted in figure 1; time sampling increment Δ is arbitrary but should be approximately matched to the coherence time of the input signal and filter. The digital input sequence $x(k\Delta)$, filter impulse response $\Delta\, h(m\Delta)$, and summer weights $w(k\Delta)$ are all real. Input $x(k\Delta)$ consists of either noise-alone or a gated-signal plus noise, according to the model

$$x(k\Delta) = \left\{ \begin{array}{c} n(k\Delta) \\ \text{or} \\ s(k\Delta)\, g(k\Delta) + n(k\Delta) \end{array} \right\} \quad \text{for all } k.$$

Input noise $n(k\Delta)$, which is present for all time, is a stationary zero-mean discrete Gaussian sequence with covariance

$$R_n(m\Delta) = \overline{n(k\Delta)\, n(k\Delta - m\Delta)} \quad \text{for all } m, k,$$

where an overbar denotes an ensemble average. The underlying input signal $s(k\Delta)$ is also a stationary zero-mean discrete Gaussian sequence with covariance

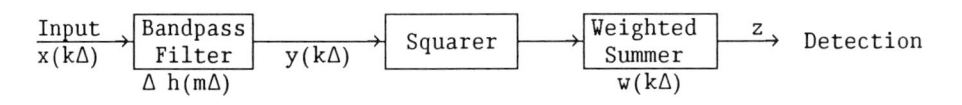

Input $x(k\Delta)$ → Bandpass Filter $\Delta\, h(m\Delta)$ → $y(k\Delta)$ → Squarer → Weighted Summer $w(k\Delta)$ → z → Detection

Figure 1. Block Diagram for Detection

$$R_s(m\Delta) = \overline{s(k\Delta) \, s(k\Delta-m\Delta)} \quad \text{for all } m, k.$$

However, input signal samples s(kΔ) are gated by function g(kΔ) which is nonzero only in the limited region given by

$$g(k\Delta) = 1 \quad \text{for} \quad k_a \leq k \leq k_b, \quad \text{and zero otherwise.}$$

This results in a gated burst of stationary signal sequence s(kΔ) being input to a digital filter Δ h(mΔ); the input signal starting and ending times $k_a\Delta$ and $k_b\Delta$, respectively, are generally unknown, except in an approximate fashion. This generality allows for consideration of input signals of unknown location and duration at the detector input. The filter output y(kΔ) can be conveniently broken into signal and noise components and is available by means of discrete convolution

$$y(k\Delta) = \Sigma_m \, \Delta \, h(m\Delta) \, x(k\Delta - m\Delta) = y_s(k\Delta) + y_n(k\Delta)$$

when an input signal is present. Due to gating, filter output signal $y_s(k\Delta)$ will have time-varying phases, including a build-up just after time $k_a\Delta$ and a decay just after time $k_b\Delta$. This nonstationary behavior is included, and exactly accounted for, in the analysis in [3].

The filter output is then squared, scaled by weights w(kΔ), and summed to give system output (decision variable)

$$z = \Sigma_k \, w(k\Delta) \, y^2(k\Delta),$$

where weights w(kΔ) are nonzero only in the limited observation time specified by

$$w(k\Delta) \neq 0 \quad \text{for} \quad k_c \leq k \leq k_d.$$

These weights w(kΔ) need not be uniform; for example, they could be exponential. Output z is compared with a threshold for a declaration of signal absent versus signal present.

RESULTS: The operating characteristics of an equi-weighted energy detector for Gaussian signals in noise, in terms of false alarm and detection probabilities, can be characterized mathematically by a partial exponential expansion [4,5]. However, when the weights are unequal, these results do not apply and can be misleading, especially when the number of samples summed is not large. In the case of an arbitrary number of samples and unequal weights or powers, there is an exact approach [3] in terms of the characteristic function of the decision variable z; this characteristic function is frequently available in closed form and can be utilized for direct evaluation of the exceedance distribution function, which is essential to the accurate determination of the probabilities of detection and false alarm.

The receiver operating characteristics, that is, detection probability P_D versus false alarm probability P_F for a set of signal-to-noise ratios, for the system in figure 1, will be considered in this section. The input signal covariance $R_s(\tau)$ and the input noise covariance $R_n(\tau)$ are taken to be sums of exponentials: $R_s(\tau) = \Sigma_j \alpha_j \exp(-|\tau|/\tau_{sj})$ for all τ, and $R_n(\tau) = \Sigma_m \beta_m \exp(-|\tau|/\tau_{nm})$ for all τ. The impulse response of the filter is given by $h(\tau) = \Sigma_n \gamma_n/\tau_{hn} \exp(-\tau/\tau_{hn})$ for $\tau > 0$. The first example is evaluated for the following set of parameters, where all times are in seconds. All parameters are real.

$$k_a = 4, \ k_b = 11, \ k_c = 2, \ k_d = 16, \ \Delta = .2,$$

$$J=3, \ \{\alpha_j\} = \{11 \ -48 \ \ 75\}, \ \{\tau_{sj}\} = \{1 \ \ 1/8 \ \ 1/15\},$$
$$M=2, \ \{\beta_m\} = \{39 \ \ 60\}, \ \{\tau_{nm}\} = \{.2 \ \ .4\},$$
$$N=4, \ \{\gamma_n\} = \{1 \ 2 \ 3 \ 4\}, \ \{\tau_{hn}\} = \{.3 \ .5 \ .7 \ .9\}.$$

Twenty nonzero values of the system input signal-to-noise ratio, σ_s^2/σ_n^2 = R, were utilized, namely R = 5(1.2)27.8 dB. The receiver operating characteristics are plotted on normal probability paper in figure 2 and cover a wide range of values, ranging from 1E-10 for P_F to .999 for P_D. Since Gaussian random variables would plot on such paper as a set of parallel straight lines, the curvatures of these results illustrate that the Gaussian assumption for decision variable z in figure 1 is not warranted, at least for this example. The major reason for this behavior is the relatively small number of samples used in the detector. This is also a partial explanation for the rather large values of the per-sample signal-to-noise ratio R required at the high-quality region at the top left of figure 2. Another factor to notice is that the input signal duration is only $k_b \Delta - k_a \Delta = 1.4$ seconds, meaning that the filter output signal never reaches steady state before the input signal is turned off; all these time-varying signal effects and their limitations on performance have been included exactly in the analysis and numerical results presented here.

The effect of halving the sampling increment Δ is considered in the next example, but done in such a manner as to keep the total input signal duration the same. That is, the parameter values are kept the same except for the following changes:

$$k_a = 8, \ k_b = 22, \ k_c = 4, \ k_d = 32, \ \Delta = .1.$$

Notice that all absolute times, such as $k_a \Delta$, are kept fixed. The corresponding receiver operating characteristics are plotted in figure 3. Performance has been degraded by a couple of decibels. For example, to realize P_F = 1E-3 and P_D = .5, the per-sample input signal-to-noise ratio R must now be 16 dB, whereas it was 13.8 dB in the previous figure. This is a degradation of 2.2 dB. The main reason for this behavior is again the inability of the filter output signal to reach steady state and thereby contribute significantly to the summer output. By contrast, the filter output noise is in steady state (by assumption) for both examples.

We now return to the original sampling increment Δ = .2 seconds. When

347

Receiver Operating Characteristics

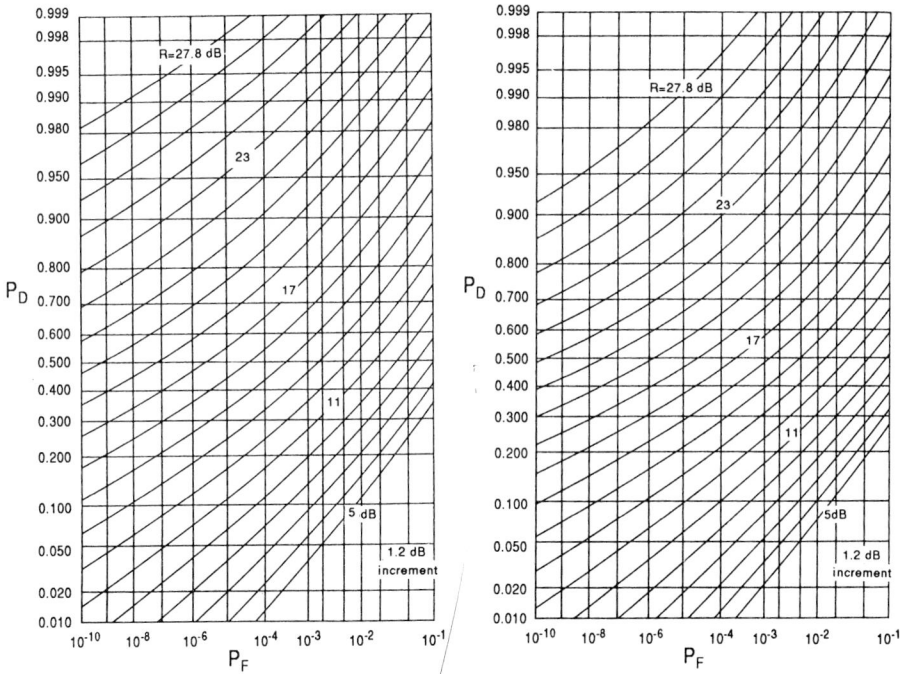

Figure 2. Example A

Figure 3. Example B

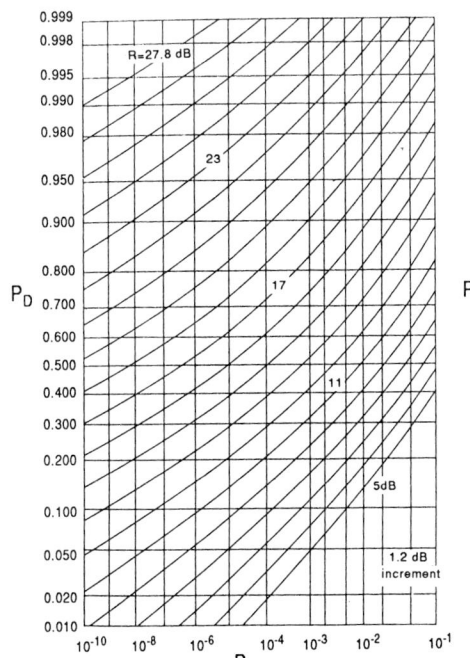

Figure 4. Example C

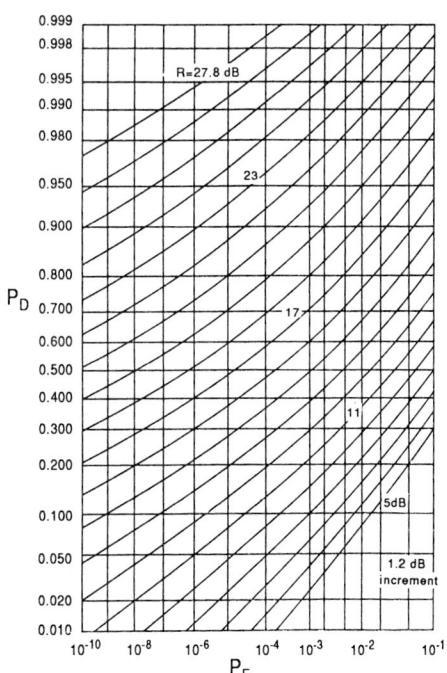

Figure 5. Example D

the observation interval coincides with the input signal nonzero excitation duration, that is, $k_a = k_c = 4$, $k_b = k_d = 11$, the receiver operating characteristics in figure 4 illustrate additional improvement. For example, probabilities P_F = 1E-3 and P_D = .5 can now be realized with R = 13.1 dB, an improvement of .7 dB relative to the signal-to-noise ratio required for the broader observation interval, k_c=2, k_d=16.

Finally, an example with a wide observation interval, namely k_c = 0, k_d = 25, is displayed in figure 5. The desired probabilities can now be achieved only if the input signal-to-noise ratio is increased to 14.7 dB, a degradation of 1.6 dB relative to the best case above.

These examples illustrate the utility of being able to investigate quantitatively and accurately the effects of nonstationarity in a system, without having to make questionable assumptions about, for example, how closely steady state was or was not realized. They also afford a dependable verification or rejection of the Gaussian approximation for the decision variable; in a related work [5], the Gaussian approximation was found to be too optimistic for most working ranges of this detection system.

CONCLUSION: Two programs have been written which enable exact analysis of performance of the mismatched bandpass energy detector, both for the white noise input case as well as the colored noise input case [3]. These results allow for arbitrary sampling time increment Δ and for arbitrary input signal spectra, input noise spectra, and filter transfer functions. By these means, exact quantitative evaluations and degradations can be determined for various combinations of uncertainty regarding the input signal time location and duration, as well as its center frequency, bandwidth, and spectrum. Included in the analysis and programs are the build-up and decay (or any portion thereof) of the nonstationary output signal from the filter, when excited by a burst-like input signal, regardless of the lengths or locations of the excitation and observation intervals. Both programs have been compared with simulation results and confirmed down to as low a probability level as possible, consistent with the number of trials used.

REFERENCES:
[1] R. J. Urick, **Principles of Underwater Sound**, McGraw-Hill Book Company, New York, NY, 1983.
[2] W. S. Burdic, **Underwater Acoustic Systems Analysis**, Prentice-Hall, Inc., Englewood Cliffs, NJ, 1984.
[3] A. H. Nuttall, **Exact Performance of Filtered and Weighted Energy Detector with Mismatched Frequency and Time Locations and Characteristics**, Technical Report 8913, NUSC, New London, CT, 29 July 1991.
[4] A. H. Nuttall and A. F. Magaraci, **Signal-to-Noise Ratios Required for Short-Term Narrowband Detection of Gaussian Processes**, Technical Report 4417, NUSC, New London, CT, 20 October 1972.
[5] A. H. Nuttall, **Operating Characteristics for Weighted Energy Detector with Gaussian Signals**, Technical Report 8753, Naval Underwater Systems Center, New London, CT, 16 July 1990.

Section 8

Applications

A GEOMETRICAL APPROACH TO MEDIUM LOW FREQUENCY REVERBERATION MODELING

A. PLAISANT
Thomson Sintra Activités Sous-Marines
525, route des Dolines B.P 157
06903 SOPHIA ANTIPOLIS CEDEX - France

ABSTRACT
After a quick survey of published data and simple models for reverberation scattering strength in the frequency band 500-2000 Hz, a 3D reverberation model based on the scattering strength concept and geometrical ray propagation with very realistic description of bottom topography is presented. Alternative methods and areas for future work are suggested.

1. INTRODUCTION
Reverberation is known as the part of the acoustic field which is diffracted or scattered in all directions by inhomogeneities of every sort present in the medium and by its boundaries; the phenomenum is therefore relevant to the general theory of diffraction. A general wave solution is very difficult to obtain because of the complexity of practical geometries involved and their size compared to wavelength. The physical effects responsible for reverberation are particularly various and complex, ranging from stochastic effects like scattering by bubbles, by random moving surface, to deterministic effects like scattering by bottom topography, making the acoustic modeling difficult and risky because the necessary input parameters to the models are likely to be widely unknown, leading to large uncertainties in models outputs.

For purposes like systems performance analysis, an empirical approach to characterize the backscattering strength of various inhomogeneities based on measurements rather than physical modeling is satisfactory, because no attempt is made in trying to understand the mechanisms involved. When local scatterers are characterized, the global reverberation effect can be evaluated by taking into account propagation losses and array directivity and summing up local contributions. This is a classical geometrical approach which has been used in various global reverberation models like "NISSM" and also in our 3D variable depth model "REVBAS" presented in the following.

Before presenting our model, we make a quick survey in the next paragraph of existing data and simple models for scattering strength in the band 500 Hz-2000 Hz.

2. LOW - MEDIUM FREQUENCY SCATTERING STRENGTH MEASUREMENTS AND MODELS
The scattering strength is defined as the ratio between the scattered intensity at one meter from the center of the scattering element as if it were a point source, to the incident intensity, the scattering element being one unit surface or volume of the scattering feature.

Since physical effects are different, separation is made between volume, surface and bottom reverberation.

2.1 Volume reverberation

It is due to bubbles and marine organisms present in the ocean, particularly those fish with swimbladders full of air.
SCRIMGER and TURNER [1] made measurements in the Pacific at several depths. In the low frequency band (625 Hz-1250 Hz), their results presented in fig. 1, exhibit a strong variability depending on the time of the day : -85 dB to -105 dB.
The classical deep scattering layer around 500 meters depth during day time does not appear clearly.

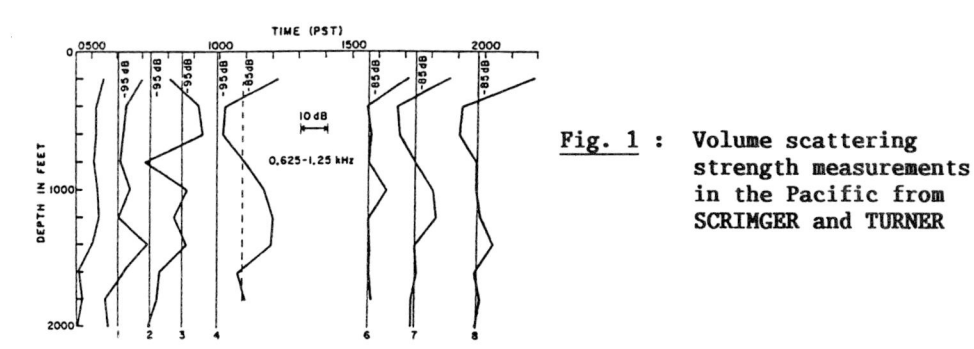

Fig. 1 : Volume scattering strength measurements in the Pacific from SCRIMGER and TURNER

Measurements made by STOCKHAUSEN [2] and DOUTT [3] at Saclantcen are particularly interesting because they allow the identification of a resonant frequency at various depths which in turn allows determination of bubble size, under the hypothesis that reverberation is due to bubble layers. Furthermore, the value of the scattering strength is directly proportional to the number of bubbles per unit volume. Once the bubble size and number per unit volume are determined, it is easy to calculate the scattering strength at other frequencies, even outside the measurements bandwidth. Fig. 2 shows an example of such an extrapolation of DOUTT'S results.

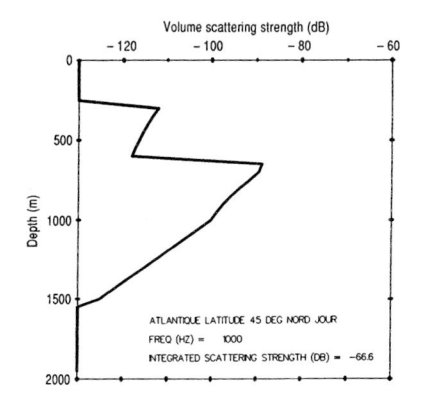

Fig. 2 : Volume scattering strength versus depth modeled from experimental results of J. DOUTT

2.2 Surface reververation

The scattering mechanisms are surface waves, near surface bubbles and marine organisms living close to the surface. Well known measurements were performed by CHAPMAN and HARRIS [4] in the frequency band 400 Hz-6400 Hz. Based on these measurements, the authors have proposed an empirical model (used in NISSM) :

$$S_{sd}B = 3.3 \, \beta \, \text{Log} \left(\frac{\theta}{30}\right) - 42.4 \, \text{Log} (\beta) + 2.6 \quad \text{with} \quad \beta = 158 \left(v.f^{1/3}\right)^{-0.58}$$

f is the frequency in Hz, v is the wind speed in knots, θ is the grazing angle in degrees.

Other authors have also presented experimental results at low frequencies like BROWN and SAENGER [5], their results are close to CHAPMAN-HARRIS's ones for small and high grazing angles but significantly higher for grazing angles between 30 and 60 degrees. An important problem which to our knowledge is not yet solved is the determination of the scattering strength at very low grazing angles, important for long ranges, for which measurements are difficult.

2.3 Bottom reverberation

Not suprizingly, bottom reverberation strength measurements exhibit large variations depending on the areas, as shown for example by ROBISON [6] and in fig. 3.

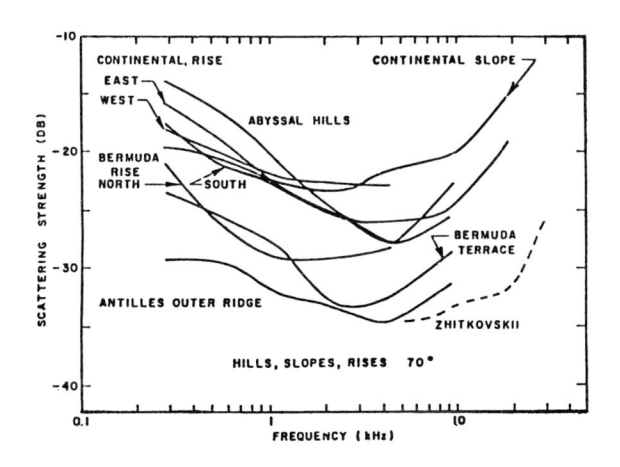

Fig. 3 :

Bottom scattering strength measurements from ROBISON

This is due of course to the variety of bottom types and roughness. Other authors have published measurements in the low frequency band : MACKENZIE [7], URICK and SALING [8], MERKLINGER [9], SCHMIDT [10] and HALL [11]. The frequency dependance is generally found to be small, 1 kHz seems to be the frequency corresponding to minimum scattering strength values, for higher frequencies an increase of 2 to 3 dB per octave is observed.

A popular grazing angle dependance is called "Lambert's" law :
$S_b (\theta) = S_{bo} + 10 \log (\sin \theta_i . \sin \theta_b)$.
θ_i and θ_b are incident and backscattered grazing angles, S_{bo} is the

scattering strength corresponding to normal incidence, it depends on bottom type; from ROBISON'S results, its value ranges from -14 dB on the Azores plateau to -40 dB in the East Antilles Zone.
As for surface reverberation, values of scattering strength are difficult to obtain for very low grazing angles. Lambert's law predicts a minus infinity (in dB) strength for zero grazing angle which is not very realistic at least for rough bottoms.
Once scattering strengths of volume, surface and bottom are known, estimation of the global reverberation can be done as explained in the following paragraph.

3. RAY MODELS FOR GLOBAL REVERBERATION PREDICTIONS

3.1 Principles
- The whole scattering medium is divided in small elements which can be considered homogeneously insonified.
- Each element is treated like a target whose scattering strength is proportional to its surface or volume.
- Each scattering element is insonified by a set of N eigenrays calculated by a ray propagation model.
- For each incident ray the scattering element produces N echoes corresponding to each eigenray back to the receiver.
- Each echo is properly weighted by the transmitter directivity pattern, the propagation losses for the round trip eigenrays and the receiver directivity pattern, and it is also delayed by the round trip travel time.
- All elementary echoes are accumulated. This classical procedure is described in fig. 4.

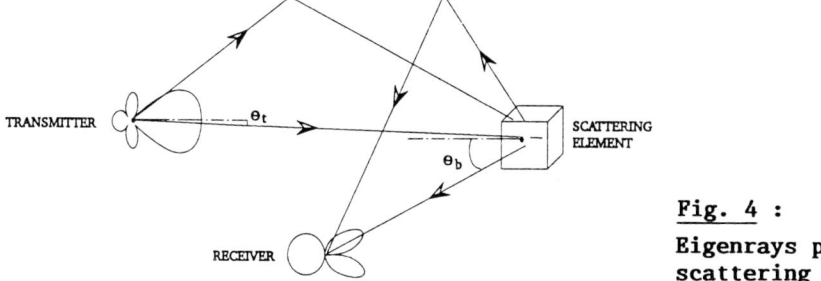

Fig. 4 :
Eigenrays paths to a scattering element

These principles are applied in the NISSM computer program which is commonly used to predict reverberation, but this model is limited to monostatic sonars with arrays having a vertical axis of symetry, therefore, it is not adequate for towed arrays.

3.2 A 3D reverberation model for variable depth bottom = REVBAS
Low frequencies propagate far away therefore bottom topography might have important effects on bottom reverberation like increasing the

reverberation level significantly when the sound reaches the continental slope, coming from deep water. In order to predict such effects, we have started development, under contract from the CERDSM in France, of a reverberation model called "REVBAS" which can take into account the bottom topography in large areas and in much detail. Bottom topography is entered as a disc file containing sampled coast line and iso-depth contours as well as point soundings available from navigation charts. Such files are presently obtained using a digitizing table.
Velocity profiles are taken from a data base, depending on the position of the sonar on the chart and the month of the year.
Transmitter and receiver 3D directivity are described as sampled vertical cuts on the directivity pattern for a set of different azimuths. The whole space is divided in azimuthal sectors centered on the sonar position, they define vertical cuts of the medium. Sector width is adapted to array directivity. Reverberation is calculated in each sector with a variable bottom profile in each vertical plane and summed over all sectors.
In each sector, the bottom is divided into small facets whose depth and orientation (normal vector) have to be known, therefore an algorithm for evaluation of the depth of any point on the chart is required.

Evaluation of the bottom depth of a point given by its coordinates
The depth Z_A of any point A is obtained by considering the depth of a number of neighbour points B_i belonging to isodepth contours or individual soundings. The number N of neighbours is a parameter which has to be adjusted to the level of details of the chart and the accuracy required. Once the neighbour points B_i have been determined, the depth of point A is approximated by a weighted mean of the neighbour's depth, the weighting factor being the inverse of the distance $d(A, B_i)$ from A to B_i :

$$Z_A = \sum_{i=1}^{N} \frac{Z_{B_i}}{d(A,B_i)} \bigg/ \sum_{i=1}^{N} d(A,B_i)$$

Of course, if $d(A, B_i) = 0$ we take $Z_A = Z_{B_i}$
The determination of the neighbour points is as follows :
a) The distances between A and all individual sounding points is computed and the N nearest soundings are retained.
b) The distances between A and the projection of A on every segment making an isodepth line is computed and the N nearest distances retained. If the projection of A on a segment lies outside the segment, the reference point B_i is taken as the closest extremity of the segment; if two consecutive segments have the same reference point, this point is taken only once in the averaging.
c) The N closest points B_i defined in steps a) and b) are finally used in the averaging.

Eigenray determination
Since eigenrays have to be determined for a lot of bottom facets at various depths, this operation has to be done efficiently ; this is why

we have used a procedure based on the "Angle-depth diagram" explained below, see also [12].
In a layered medium of sound velocity C(Z), rays are periodic curves governed by Snell's law :

$$\frac{\cos\theta}{C(Z)} = \frac{\cos\theta_o}{C_o} = \frac{1}{C_V}$$

C_V is called the vertex velocity, it can be regarded as a characteristic constant relative to the ray, independent of the source.
Periods of all possible rays can be represented in the plane [θ, Z] through Snell's law. We thus obtain the angle-depth diagram, an example of which is shown in fig. 5. Each curve of this diagram represents a ray corresponding to a given value of C_V. Curves are symmetrical for negative angles; for large angles they are almost vertical, meaning that rays are straight lines.

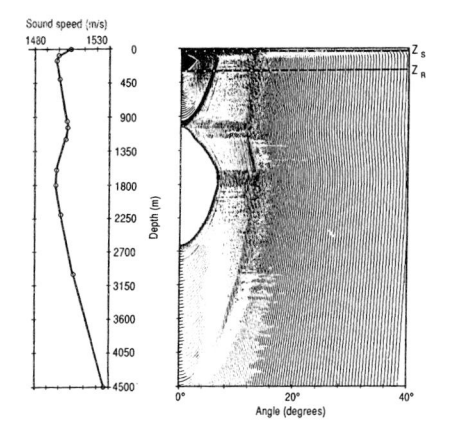

Fig. 5 : Angle-Depth diagram

In order to be useful, these curves are sampled with constant horizontal distance increment between two points, using the classical range increment relation along a ray :

$$x = x_o + \frac{C_V}{g}\left(\sin\theta_o - \sin\theta\right), \text{ g is the velocity gradient.}$$

The depth-angle diagram is only dependant on the velocity profile, it can be computed and stored once for all.
We now consider a source at depth $Z = Z_s$. This point is common to all rays passing by the source at various angles, these points are situated on the line $Z = Z_s$ on the depth-angle diagram, this line is called the source line.
We are interested in finding eigenrays at various ranges and depths, this can be done by first looking to what happens to the source line after

propagation to a given range. The points on the <u>propagated source line</u> will indicate at what depth and what angle a ray starting from the source will be found after propagation to the required range. The determination of the propagated source line is very easy because all curves are sampled with a known range increment, it reduces to a count of the required number of increments and a final interpolation.

The eigenrays at any receiver depth Z_r are finally determined also very easily by finding the intersections of the propagated line source with the horizontal line $Z = Z_r$. This procedure is represented in fig. 6.

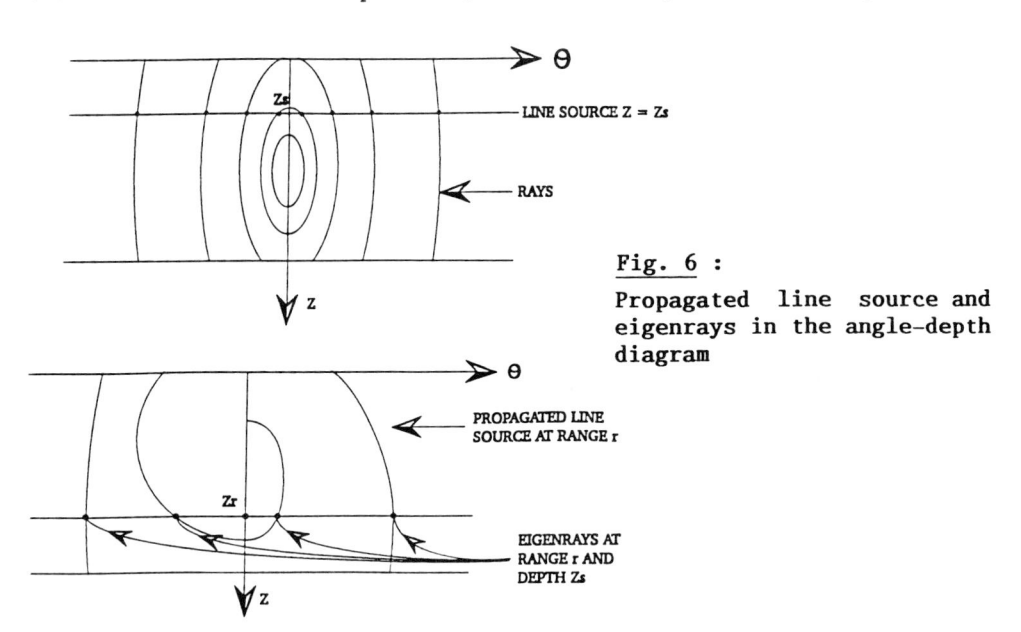

Fig. 6 :
Propagated line source and eigenrays in the angle-depth diagram

Computation of total reverberation
Presently, multiple reflections on the sloping bottom are not permitted in order to take full advantage of ray periodicity, but in order to avoid too severe underestimation of reverberation level due to reduction of the number of eigenrays, particularly in shallow water, a flat piece of bottom is introduced under the source, up to a distance where the bottom depth has changed significantly; multiple reflection is permitted on this flat part of the bottom. All echoes from all facets and sectors are finally accumulated in range boxes along the time axis for FM signals or in frequency boxes for LCW signals as an estimation of the power spectrum of the reverberation after beamforming.

Figure 8 gives an example of level versus time for a situation presented in fig. 7. Effects from the continental slope of Corsica can be identified as being responsible for the level increase observable at delays larger than 75 seconds, which are not present with a flat bottom. Calculation of the spectrum spreading for the same situation is shown on fig. 9.

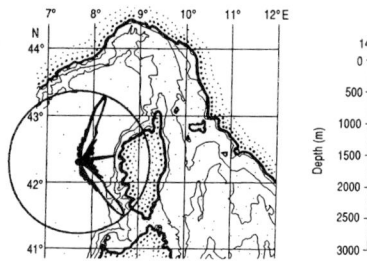

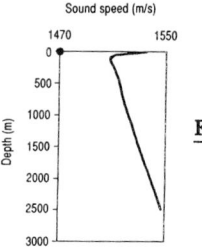

Fig. 7 : Scenario for bottom reverberation estimation by "REVBAS"

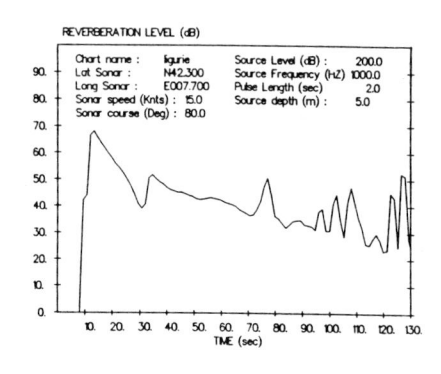

Fig. 8 : Bottom reverberation power level versus time from "REVBAS"

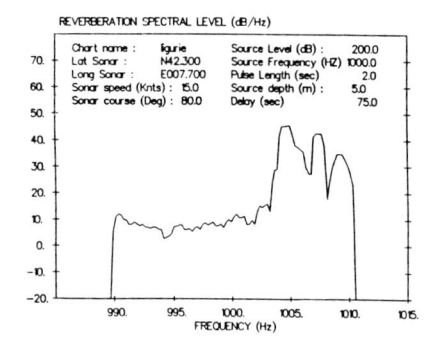

Fig. 9 : Bottom reverberation power spectrum from "REVBAS"

4. ANOTHER POSSIBLE APPROACH FOR LOW FREQUENCY BOTTOM REVERBERATION – COUPLED MODES

For variable depth media, propagation at low frequencies can be conveniently modeled by the "coupled modes" technique in which the medium is split into constant depth zones where normal modes are computed. Transitions between different zones give rise to transmitted and reflected modes. In this case, the reflection is produced by an impedance mismatch between water and sediment and a change in bottom depth, it represents a diffraction effect. Coupled modes can be worked out in 3 dimensions by considering horizontal beams carrying modes with horizontal deviations at each bottom variation encounter, as it is done in the European MAST "SNECOW" project. By keeping only the reflected waves and combining Fourier components to simulate pulsed signals, one would get an evaluation of the monostatic and bistatic reverberation due to bottom topography in a frequency domain where ray approximations are not valid.

5. CONCLUSIONS

Global 3D reverberation can be evaluated using simple concepts like scattering strength and ray propagation. Modeling results have been presented showing the effect of continental slopes in a deep water low frequency active sonar situation with reception on a towed array. This geometrical approach is limited to frequencies for which the ray approximation is valid. We have presented an alternative approach for lower frequencies which could be interesting for evaluation of bistatic reverberation. Modeling scattering strength is difficult, due to the diversity and complexity of the physical processes involved. Work has still to be done in this area but it has to be matched with proper surveys of environmental parameters involved in the scattering processes in order to be able to input proper data into models when they are ready.

REFERENCES

[1] J.A. SCRIMGER AND R.G. TURNER: "Volume scattering strength dependance on depth and frequency in the Pacific Ocean off San Francisco.
JASA 48,5; pp 1266-1274, 1970.

[2] J.H. STOCKHAUSEN: "Volume scattering versus depth and frequency in the eastern Atlantic Ocean and the Western Mediterranean sea" Saclantcen report SM-60, 1975.

[3] J.A. DOUTT: "Broadband measurements of volume scattering strength in the Mediterranean.
Saclantcen report SR 17, 1977.

[4] R.P. CHAPMAN and J.H HARRIS: "Surface backscattering strengths measured with explosive sound sources".
JASA 34,10; pp 1592-1597, 1962.

[5] M.V. BROWN and R.A. SAENGER: "Bistatic backscattering of low frequency underwater sound from the ocean surface".
JASA 52, 3; pp 944-960, 1972.

[6] A. ROBISON: "Bottom reverberation in the North Atlantic.
DREA technical memo 75/B, 1975.

[7] K.V. MACKENZIE: "Bottom reverberation for 530 and 1030 Cps sound in deep water".
JASA 33, 11; pp 1498-1504, 1961.

[8] R.J. URICK and D.S. SALING: "Backscattering of explosive sound from the deep sea bed".
JASA 34, 11; pp 1721-1724, 1962.

[9] H.M. MERKLINGER: "Bottom reverberation measured with explosive charges fired in the deep ocean".
JASA 44,2; pp 508-513, 1968.

[10] P.B. SCHMIDT: "Monostatic and bistatic backscattering measurements from the deep ocean bottom".
JASA 50,1 ; pp 326-331, 1971.

[11] M. HALL: "Bottom backscattering in the Java Trench area and the Tasman sea".
JASA 54,3; pp 730-734, 1973.

[12] S.M. FLATTE: "Angle depth diagram for use in underwater acoustics".
JASA 60,5; pp 1020-1023, 1976.

THE REVERBERATION ARRAY HEADING SURFACE

R. A. Wagstaff, Naval Research Laboratory
SSC Detachment, SSC, MS 39529-5004, USA
and
M.R. Bradley and M.A. Hebert, Planning Systems Incorporated
115 Christian Lane, Slidell, LA 70458, USA

ABSTRACT It is possible to choose a heading for a towed line array that will minimize the ambient noise on the ambiguous beam and achieve a signal-to-noise ratio (S/N) enhancement. The Array Heading Rose (AHR) is a passive sonar tactical decision aid in the form of a polar plot that has been devised and used for that purpose. The AHR presents the improvement in S/N in decibels relative to broadside beam performance as a function of the array heading. The same concept has been applied to reverberation by time-gating the very same analysis techniques that produced the AHR to obtain a similar tactical decision aid for active sonar. It includes an additional dimension of time/range and is called the Reverberation Array Heading Surface (RAHS). The RAHS is described and results for simulated bistatic basin reverberation measurements are discussed. The results demonstrate that significant gains can be achieved by using the RAHS and that the optimum array headings and associated performance gains will vary with time/range.

1. INTRODUCTION

Basin reverberation is not entirely unlike ambient noise. For given oceanographic/acoustic conditions and geometry of source and receiver relative to the basin boundaries (including seamounts), the azimuthal directionality of the reverberation from the basin boundaries should be the same as for other times when the conditions were or are again the same. When this is the case, the measured data can be used to predict active sonar performance in the future. When something is different, the measured data can be used to initialize a basin reverberation model to predict the performance for those changed conditions. Examples include a different bistatic geometry, a different frequency, and a different oceanographic/acoustic season.

Since basin reverberation in active sonar is the active counterpart of ambient noise in passive sonar, some of the techniques used to improve S/N for the passive towed line array can be used to improve the signal-to-reverberation ratio (S/R) of activated towed line arrays. In particular, an active counterpart of the passive Array Heading Rose, for selecting the best towed line array heading to maximize S/N, can be generated from basin reverberation data collected on multiple array headings. The result, called a Reverberation Array Heading Surface (RAHS), is a time/range gated surface that contains at each time/range bin a representation of the reverberation suppression level as a function of array heading.

The development of the RAHS is described in the next section. Following this, the model used to generate synthetic data is described. Next, modeling results are presented to illustrate the RAHS and to quantify the gains that may be achievable. The final section gives the conclusions.

2. REVERBERATION ARRAY HEADING SURFACE DEVELOPMENT

The inputs to the RAHS processing system are measured beam reverberation levels, the source and receiver locations, and a description of the receive array beam pattern. In order to resolve the inherent left-right ambiguity in a towed line array measurement, beam reverberation measurements must be made of the basin reverberation while the array is on several different headings. This is illustrated by the first two columns in Figure 1. The signal return at time t_S denotes the reverberation level received from a particular scatterer during the three different measurements. The measurement geometry is also included in Figure 1. The beam reverberation data corresponding to a particular time/range cell are then processed by the time-gated Wagstaff Iterative Technique (WIT) algorithm [1] to resolve the ambiguities and deconvolve the beam pattern from the data. The result for each time/range cell is an estimate of the azimuthal directionality of the reverberation. When all time/range cells are included on a given polar display, the result is called a Basin Reverberation Rose. The Basin Reverberation Rose is then used for each increment of time/range to produce the RAHS much the same way that the passive AHR [2] is produced for a passive towed array sonar system.

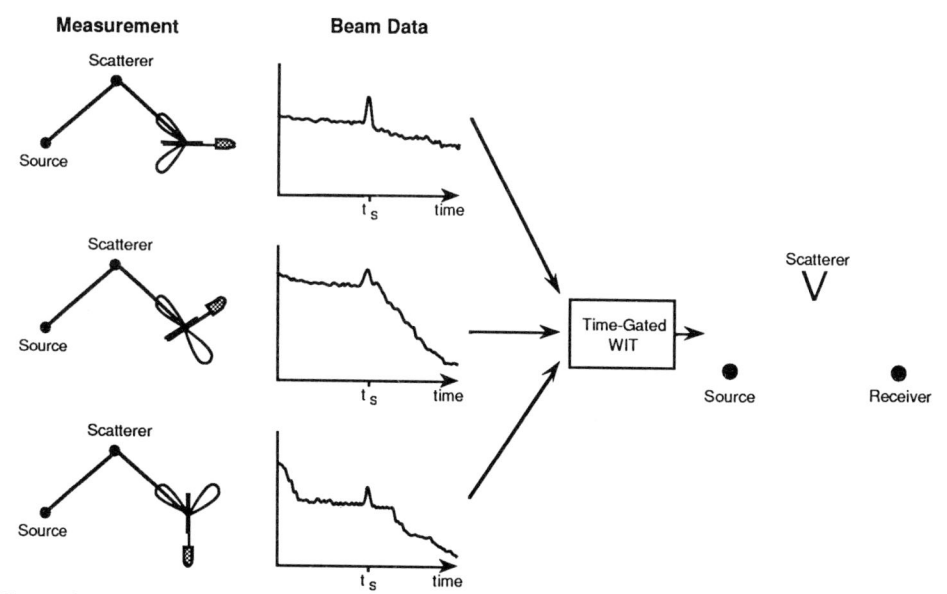

Figure 1. Using WIT to resolve ambiguities in towed array reverberation measurements.

3. BASIN REVERBERATION MODEL

The basin reverberation calculations presented in this paper were made with the Bottom Distributed Active Simulation System (BASS) model [3]. BASS was developed to provide detailed, accurate predictions of the performance of low frequency active sonar systems in noise and reverberation limited environments. The reverberation module in BASS uses range dependent ray theory applied azimuthally about the active source and receiver locations to construct a three-dimensional time dependent reverberation density field applicable at the depth of the receiver. Bottom backscatter is represented in BASS by a Lambert law model in which the level of scatter depends upon the vertical arrival angle of eigenrays at a bottom scattering element and the slope of this element with respect to the horizontal. In setting up a BASS run, the user selects a set of great circle path radials about the source and receiver locations. BASS then makes extractions from oceanographic environmental data bases to determine the variation of bathymetry, bottom type, and sound speed versus depth along these radial paths.

The reverberation density computed by BASS is analogous to noise directionality per steradian. The integration of the BASS reverberation density over a unit sphere at a given instant is simply the omni reverberation level at that time. Beam reverberation levels for a receiver with a directional response, such as a towed line array, are estimated in BASS by convolving the time-azimuth-elevation dependent reverberation density with the receiver beam pattern. This convolution produces time varying beam reverberation levels that depend upon the temporal-directional nature of the reverberation field and the directional response and heading of the receive array.

4. RESULTS AND DISCUSSION

The measurement of reverberation in the Gulf of Alaska was simulated using the BASS model. The source and receiver locations and the major features of the bathymetry within about 450 km of the receiver are given in Figure 2a. The source and receiver were on a north-south line with an approximate separation of 4 km. Several prominent seamounts and guyots are evident in the basin including the Quinn Guyot to the northwest of the receiver and the Surveyor Guyot to the southeast, both at a distances of about 30 km. The Patton Seamounts are located to the southwest of the receiver at a distance of about 360 km. Shallow water off the Alaskan Coast and Aleutian Islands is clearly evident to the northeast and northwest of the receiver location.

Figure 2b gives the unambiguous bistatic basin reverberation as calculated by the BASS model. In this particular example, the source is an unshaded vertical array which emits a 1-sec CW mid-frequency pulse. Both source and receiver are assumed to be located just below the deep sound channel axis, which is at a depth of 100 m near the source-receiver location. The reverberation shown in Figure 2b was plotted in one-degree azimuthal steps out to a maximum time of 600 seconds in one-second increments. A bistatic travel time ellipse with an assumed mean horizontal sound speed of 1.47 km/s has been used to map time and bearing onto a geographic location. The high levels of reverberation near the center of the plot are a result of nearby bottom reverberation. The broad arc of reverberation to the north is a result of a shoaling depths near the Aleutian Islands and Alaskan coast. In addition, the reverberation from other individual seamounts and guyots is evident at a variety of ranges and bearings.

Figure 2c shows the time/range gated beam reverberation directionality for an array heading of 090°. The effects of the shoaling bathymetry to the north are evident at the top and the bottom of the plot. It appears in two places because of the left-right ambiguity of the towed array. Similar results were obtained for five other array headings. Finally, the six sets of beam reverberation data were processed by the time-gated WIT algorithm to produce the unambiguous reverberation rose given in Figure 2d. The fidelity of the ambiguity resolution and deconvolution process can be evaluated by comparing Figures 2b and 2d. If the process were perfect, the two figures would be identical. They are not identical, but they are sufficiently alike to support the conclusion that the basin reverberation azimuthal directionality is being reproduced very well from the ambiguous time/range gated beam reverberation data processed by the time-gated WIT algorithm.

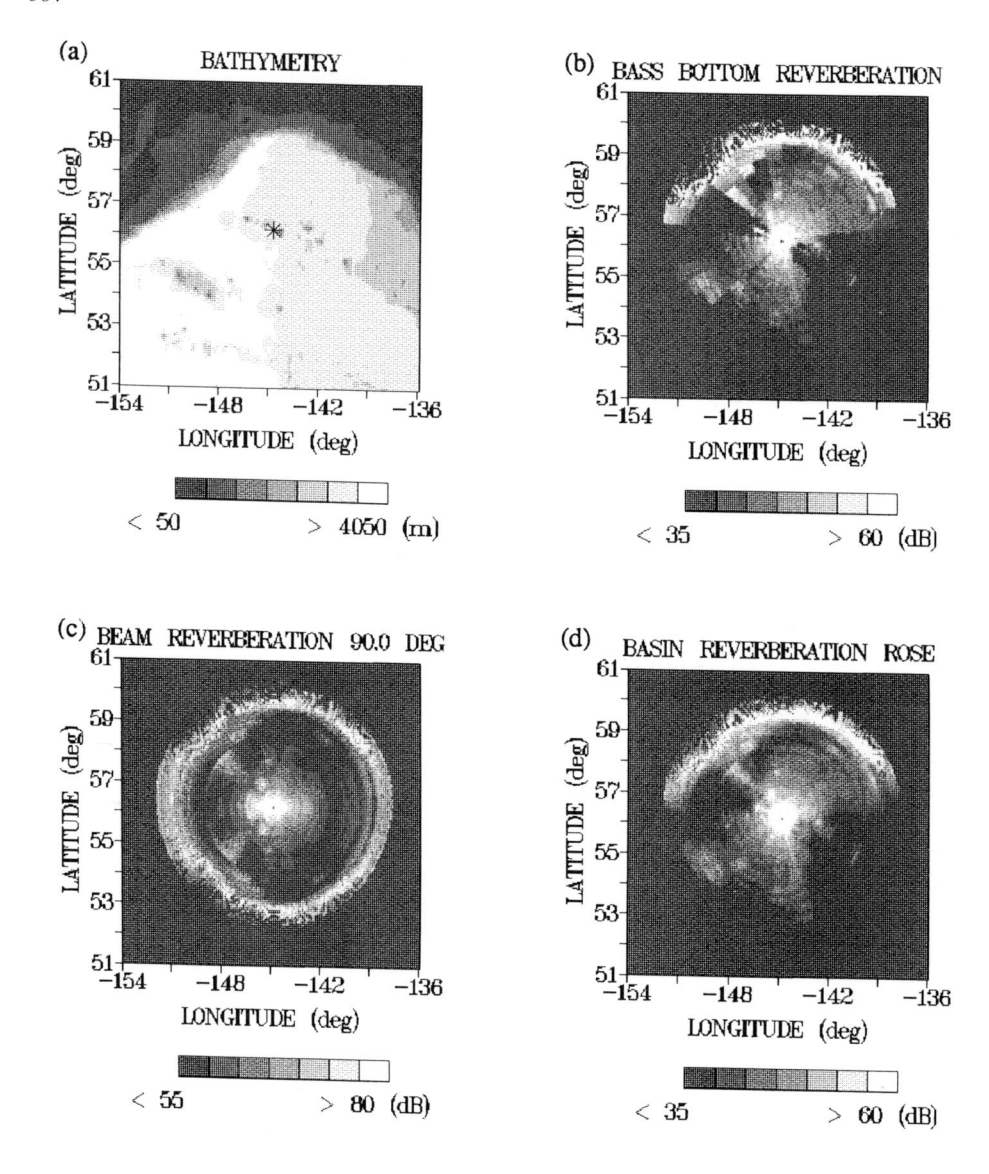

Figure 2. Test of the ambiguity resolution and deconvolution process using synthetic data. (a) Bathymetry about receiver location (*). (b) BASS bottom reverbation prediction. (c) Synthetic beam reverberation data. (d) WIT basin reverberation rose.

REVERBERATION ROSE FOR 450 SEC   ARRAY HEADING ROSE FOR 450 SEC

Figure 3. Reverberation Rose and Array Heading Rose at 450 sec.

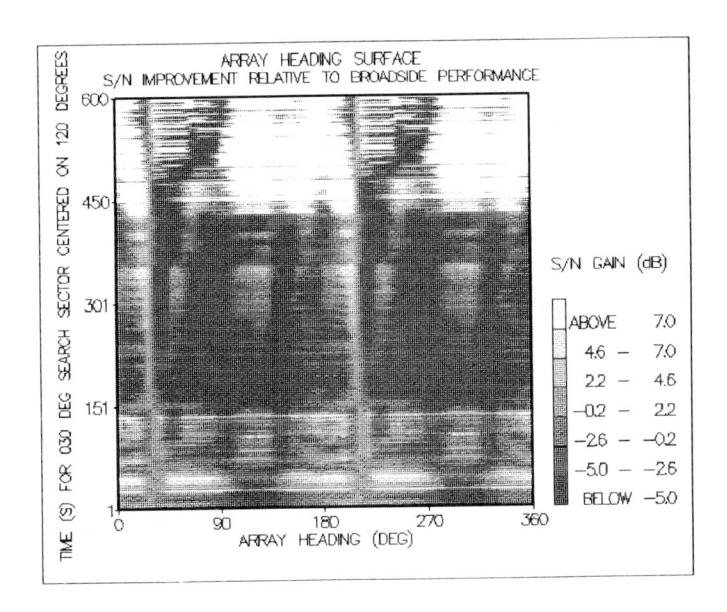

Figure 4. Reverberation Array Heading Surface.

The left hand portion of Figure 3 shows the reverberation rose from Figure 2d at time 450 seconds. Referring back to Figure 2d, there is a large angular region to the southeast of the receiver location with low reverberation levels. That feature is clearly evident in the left-handportion of Figure 3 extending in azimuth from 90 to 150°. In searching for a target in this sector, beam reverberation levels for a towed line array receiver can be minimized by placing the image of the search beam in a low reverberation direction. The right hand portion of Figure 3 shows the gain in S/R improvement over a broadside array heading that can be obtained by properly orienting the array. In this case the gain is 13.6 dB for a 30°

search sector centered on 120°. The optimum array headings at time 450 seconds are 16° and 196°. Because the azimuthal variation in reverberation levels vary in time, so will the optimum headings for searching a given sector.

By examining the variation in reverberation gain as a function of time and receiver azimuth, improved search strategies can be devised. This process is best carried out with the aid of a graphical device called the Reverberation Array Heading Surface. Figure 4 shows the RAHS for a 30° search sector centered on 120°. The right hand portion of Figure 3 is a slice across Figure 4 at a time of 450 seconds. The lightest colored areas in Figure 4 correspond to times and array headings where gains in S/N in excess of 7 dB over a broadside heading can be obtained. The darkest colored areas correspond to times and azimuths where degradations in excess of 5 dB occur. Array headings which produce dark vertical bars are to be avoided, since they correspond to poor performance over extended time periods. Array headings which produce light vertical bars, e.g. an array heading of 210° and time 1-525 seconds, will minimize degradation in array performance due to high reverberation levels.

5. CONCLUSIONS

The WIT algorithm for estimating the horizontal directionality of the ambient noise from towed horizontal line array data can be time-gated and used to produce time/range gated azimuthal directionality estimates (reverberation roses) of the bistatic basin reverberation. Excellent results have been achieved using simulated bistatic basin reverberation data, which demonstrate the ability of WIT to resolve the time-directional reverberation field. The reverberation roses produced by the time-gated WIT algorithm can be used to generate an RAHS from beam reverberation data measured by a towed line array while on three or more different headings. The RAHS is an aid for selecting the best array heading to minimize the masking due to basin reverberation and maximize receiver signal-to-reverberation ratio (S/R).

6. ACKNOWLEDGEMENTS

The authors gratefully acknowledge the funding support of Mr. Edward D. Chaika of the ASW Environmental Acoustic Support (AEAS) Program of the Office of Naval Research (ONR) under program element number 0603785N.

7. REFERENCES

1. Wagstaff, R.A., "Iterative technique for ambient-noise horizontal-directionality estimation from towed line-array data", J. Acoust. Soc. Am., Vol. 63, No.3, March 1978, pp 863-869.

2. Wagstaff, R.A., Bradley, M.R., Gibson, S.M., and B.M. Soroe, "Array Heading Rose User's Guide and Software Documentation", Naval Ocean Research and Development Activity, Technical Report SP 006:245:89, October 1988.

3. Bradley, M.R. and Wagstaff, M.D., "Physics Description of the Bottom Distributed Active Simulation System (BASS)", Planning Systems Inc., Technical Report TR-S518192, May 1992.

BISTATIC OCEAN REVERBERATION EFFECTS
(Reverberation Rejection in a Bistatic Environment)

Sally Sutherland-Pietrzak
Naval Undersea Warfare Center
Code 3314, Bld 122-T
New London, CT 06320

ABSTRACT

Working in a bistatic or multistatic environment creates new and unique ocean reverberation issues as compared with monostatic sonars. For a single line receive array, no longer is there a single area in the frequency domain which represents zero doppler per beam, but two possible reverberation notch frequencies and two possible replica adjustments for match filtering. The bistatic receiver doppler corrections are dependent on target range as well as target bearing. The bistatic/multistatic equations for reverberation rejection are discussed. Also provided are discussions and the relevance to the current Multistatic Sonar System (MSS).

1. INTRODUCTION

Calculation of the waveform effects due to source and receiver ship motion is required for active sonar signal processing. Correcting for the compression or expansion of a waveform allows for better replica correlation, as well as providing a estimate of reverberation regions and target doppler. This paper provides a discussion on calculating the reverberation ridge frequency location, based on own ship and source ship doppler in a bistatic environment, for single line and multiline arrays. Rejecting the CW data in the reverberation regions is required to reduce the clutter on active sonar displays, as well as limiting the amount of data sent to automatic detection algorithms. Included as well, are bistatic reverberation implementation issues for future bistatic active sonar systems.

2. BISTATIC DOPPLER NULLIFICATION (BDN) PROCESS

Bistatic configurations require unique processing methods, as compared with monostatic configurations, for calculating range and doppler. A typical bistatic configuration is shown in figure 1. (A multistatic configuration would have additional sources and/or receivers.) The processor must know or be able to calculate acoustically, source location relative to the receiver in order to compute accurate target ranges. Placing the source directly aft of a receiver ship puts the source in the receiver endfire beams yielding a poor estimate of bearing, hence yielding poor target range estimates. The direct blast time, t3, provides the range to the source and from that, the range of the targets is determined. In a bistatic environment the greatest sources of range errors can generally be traced back to source localization bearing errors.

In a bistatic or multistatic environment the doppler shifts change as a function of ship speed and course, receiver to target angle, as well as source to target angle. Significantly different doppler shifts are normally expected when comparing the detected frequencies on the receiver bearings. However, unlike the monostatic case where there is constant doppler shift over one bearing, there are varying doppler shifts as a function of range on one receive beam.

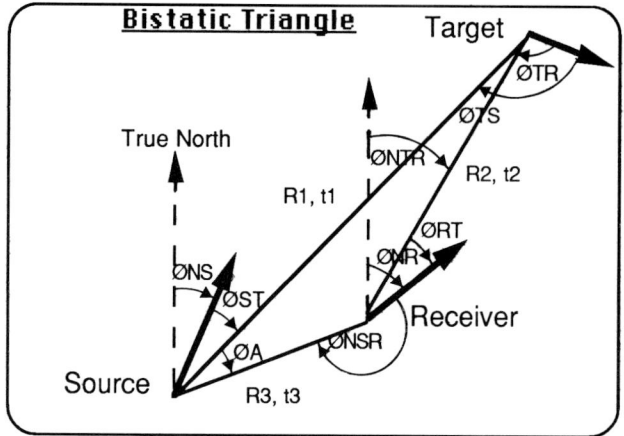

Figure 1 - Bistatic Geometry

Processing active transmissions in a multistatic environment, the effects of the source ship's and receive ship's motion should be compensated for in the sonar processing algorithms, such as the front end baseband coefficients or in the match filter replicas. To compensate for the transmitted ping's frequency compression or expansion, a 'Bistatic Doppler Nullification' (BDN) correction is calculated for all of the beams of interest. For a bistatic environment as shown in figure 1, with a transmit frequency of Fs; the frequency detected Fe, at the receiver is calculated through the following equation[1] :

Fe = Fs * source doppler * target doppler * receiver doppler   (1)
source doppler = 1/(1-(VS/c)*cos (øST))
target doppler = (1+(VT/c)*cos(øTS))/(1-(VT/c)*cos(øTR))
receiver doppler = (1+(VR/c)*cos (øRT))

VS is the source ship velocity
c is the speed of sound in water
øST is the angle between the source ship heading and the target
VT is the target velocity
øTS is the angle between the target heading and the source
øTR is the angle between the target heading and the receiver
VR is the receive ship velocity
øRT is the angle between the receiver heading and the target

The following are observable or known values:
 t1 + t2   - the time of arrival at the receiver
 t3   - the direct blast arrival time
 øNS   - the true heading of the source ship
 øNR   - the true heading of the receiver ship
 øNSR   - the true bearing of the source ship relative to the receiver ship
 øNTR   - the true bearing of the target relative to the receiver ship

---

[1] Cox Henry, Fundamentals of Bistatic Active Sonar, July 1988, NATO UNDERWATER ACOUSTIC DATA PROCESSING, Kingston, Ontario, CANADA

Let us assume for the moment that the target velocity is zero, VT=0. The only unknown to solve for then in equation (1) is øST. To solve for øST, the following steps must be taken. The time delay between the transmission by the source and the arrival of the target echo at the receiver is known to be:

$$t1 + t2 = (R1 + R2)/c. \qquad (2)$$

The direct blast time (t3), allows for the separation angle (∂) and the differential time delay (t) to be measured

$$t = t1 + t2 - t3 = (R1 + R2 - R3)/c \quad \text{and} \qquad (3)$$
$$\partial = |\text{øNTR} - \text{øNSR}| \quad (\text{If } \partial > 180 \text{ then } \partial = 360 - \partial). \qquad (4)$$

Cox states that it is useful to define an equivalent monostatic range, Rm, for the multistatic case. In the monostatic case, R3 is zero, so that R1 and R2 are both equal to c*t/2. For the multistatic case

$$Rm = c*t/2 = (R1 + R2 - R3)/2. \qquad (5)$$

Given that the distance between the source and the receiver (R3) is fixed, the law of cosines implies that equations 2 through 5 may be combined to solve directly for R2.

$$R2 = Rm\,(((1 +(R3/Rm))/(1+(R3/Rm)*(1-\cos\partial)/2)).$$

Once R2 is known, then R1 and øA can be solved, and hence the angle between the source ship and the target, (øST) can be determined.

$$\text{øST} = |(180 - \text{øNSR})| - \text{øNS} - \text{øA}$$

Solving for øST allows for the BDN equation to be calculated, for a specific interval of time, corresponding to a specific range for a specific bearing. As an example the bistatic doppler compensation will be calculated for the Figure 1 geometry.

BDN Calculation for Figure 1

1) Observed or known values:	2) Range Calculations
Direct blast time t3 = 35.376 seconds	R3 = t3*c Range of the source to receiver R3 = 33.5 miles
Echo Arrival Time t1+ t2 = 137.28 seconds	Equivalent monostatic time t = t1 + t2 - t3 = 101.9 seconds
True bearing of the source ship relative to the receiver ship øNSR = 245 degrees True bearing of the target relative to the receiver øNTR = 30 degrees	∂ = Separation angle (source/rec/target) ∂ = |øNTR - øNSR| =145 degrees Rm = c*t/2 Equivalent monostatic range Rm = 48.25 miles
True heading of receive ship øNR = 55 degrees True heading of source ship øNS = 25 degrees	R2 = Range of the receiver to the target R2= Rm*((1+(R3/Rm))/ ((1+(R3/Rm)*(1-cos∂)/2)) R2 = 50.1 miles
Target bearing (relative) øRT = 25 degrees	R1=Rm*2-R2+R3 Range source to target R1= 79.9 miles
VT = 0 knots Target Speed VR = 15 knots Receiver Speed VS = 15 knots Source Speed c= speed of sound = 5000 ft/sec	3) Angle Calculations øA = arccos(-(R2 - R3 - R1)/(2*R1*R3)) øA = 21.1 degrees øST =|(180 - øNSR)| -øA - øNS øST = 18.9 degrees
4. Solving for Fe = Fs*((1-(1-(VS/c)*cos(ST))*(((1+VR/c)*cos(RT)) Fe = Fs*((1-(1-15kts/5000ft/s)*cos(18.9))*((1+15 knts/5000ft/s)*cos (25)) Fe = Fs*.009718	

The center frequency of the waveform will be shifted by .97%.

2.1. BDN as a function of the Constant Time Ellipse

Figure 2 illustrates several of the doppler shifts which are possible given various target ranges. The echo arrival times corresponding to target ranges are defined by the source-receiver-target geometry. For any specified time of arrival (t1+ t2), an ellipse can be computed. The major axis of the ellipse is determined from the distance of the source to the target plus the distance from the receiver to the target (R1+R2). The minor axis is determined by the square root of the major axis distance minus the distance of the source to the receiver $\sqrt{((R1+R2)-R3)}$. Each point on the constant time ellipse represents the possible ranges a target could be located.

For instance in Figure 2, targets 1, 2 and 3 would all be detected at the receiver at the same time. The targets all lie on the same ellipse, however the doppler shifts and ranges for each of the targets is different. For each potential target, at a specified receiver angle øRT, a separate source angle øST, must be calculated to determine doppler and range.

Range can no longer be thought of as simple two-way travel time. This is especially true for distances one convergence zone (CZ) or less from the receiver. As the target to receiver distances increase however, the closer the bistatic equations begin to approximate monostatic equations.

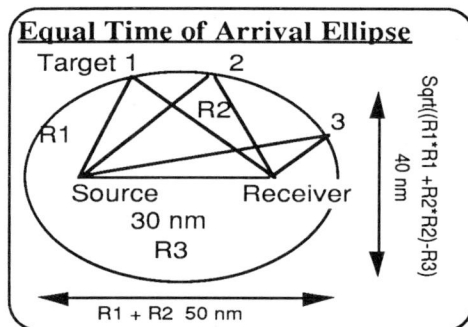

Figure 2 - Possible Target Locations

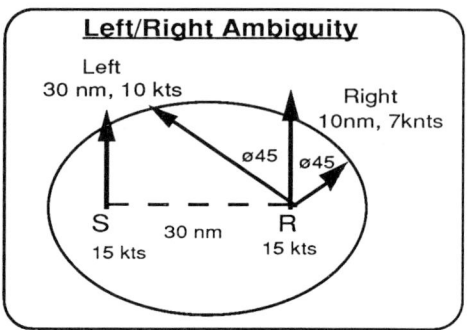

Figure 3 - Left/Right Bearings

2.2. BDN of Single verses Multiline Arrays

Bistatic calculations are more complex than monostatic for computing doppler and range as explained in the earlier sections, and now an additional complication arises; single line arrays verses multiline arrays. A single line array is usually unable to resolve left /right bearing. In a monostatic situation the single line array would simply yield an ambiguous bearing, however in bistatics the left/right bearings correspond to different ranges and different doppler corrections. Figure 3 illustrates a left target estimate at 30 nmiles and right target estimate at 10 nmiles. The source angles are different as well, producing source doppler contributions of 10 and 7 kts, respectively. For accurate match filter processing, two replicas would be required, as well as two reverberation rejection regions.

2.3. BDN as a Function of Range

As the search ranges increase over time, the source to target angles change, as illustrated by figures 4 through 7. Unlike the monostatic case, the doppler correction is no longer constant over range. Notice the receiver to target angles are the same (within each

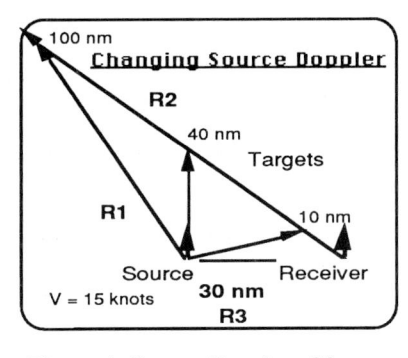

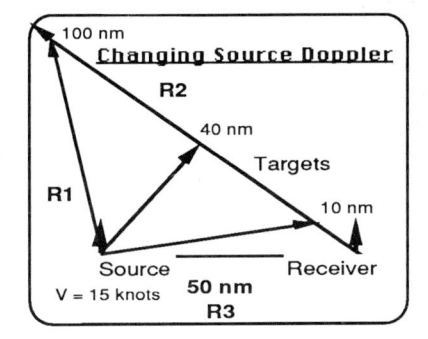

Figure 4 Source-Receiver 30 nm

Figure 5 - Source-Receiver 50 nm

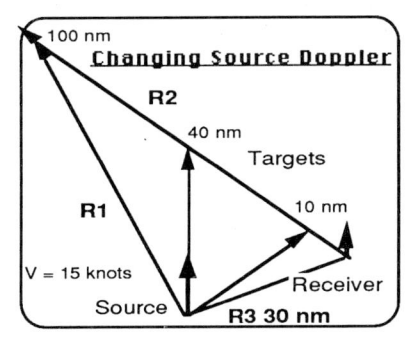

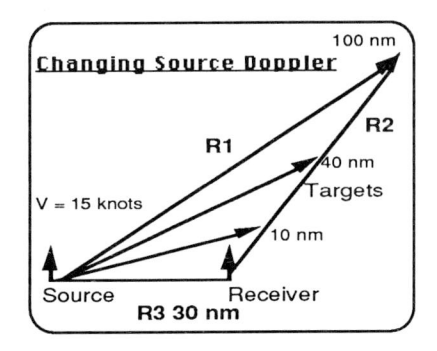

Figure 6 - Increased S/R Angle

Figure 7 - Increased R/T Angle

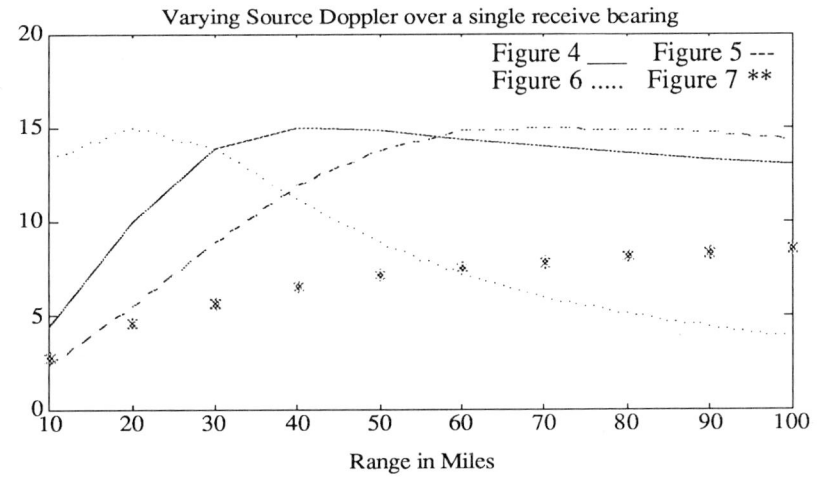

Figure 8 - Varying Doppler as a function of Range

figure) but the source to target angles are changing. Targets at 10, 40 and 100 nautical miles (nm) are shown in each of the four plots, all having varying geometries. Figure 4 shows a source and receiver separation of 30 nm. Figure 5 increases the separation to 50 nm. Figure 6 moves the source 120 degrees port of the receiver and Figure 7 places the targets to the starboard side of the receiver. Figure 8 plots the contribution of source doppler on the target bearings for each of the previous four plots. Notice the doppler changes as function of range or time, as Figure 8 shows it can either be increasing or decreasing with time. The change is greatest in the near field and gradually levels off as the change in source to target angles decreases. The further out in range the less the doppler changes. Not compensating for the source doppler will result in matched filter losses of up to several dBs depending on the configuration and waveform.

3. THE MULTISTATIC SONAR SYSTEM (MSS) PROGRAM

How has this information effected the development of future bistatic sonar systems? A system currently under development is the Multistatic Sonar System (MSS). MSS is designed to be a tactical, multi-purpose, active processor. The MSS system is programmed to operate in both monostatic and multistatic modes. In bistatic operation the source location is required. The processor attempts to locate the source acoustically using the direct blast waveform for range and bearing. A satellite link provides source latitude and longitude as a back-up. From the limited MSS sea test data received to date, it is very apparent high SNR signals are required from the direct blast to obtain accurate bearing information. No high powered bistatic source tests have been run with MSS yet.

After the source is located, the range and bistatic doppler corrections can be computed. It was determined for MSS that the source contribution to doppler, which changes as a function of time (range) need only be computed once every 16 seconds to still insure accurate results.

There is no significant correlation loss due to not processing the data for BDN at every range point, but at 16 second range increments. However, as would be expected, there are large correlation losses by not processing for BDN across receive bearing. There is also no noticeable edge effect due to the 16 second data processing boundaries. For example a data stream was simulated that produced the first half of a coded waveform at the end of a 16 second boundary and the second half of the sweep in the beginning of the next 16 second data segment. Using standard 50 percent overlap match filtering, the amplitude of the data processed with a mismatched replica was within acceptable levels from that of a perfect replica match.

For the problems associated with single line array, MSS opted to utilize a multiline array which better resolves the left/right ambiguity.

4. CONCLUSIONS

The BDN methods described in this paper are currently being employed and will continued to be used for the bistatic, high power source, MSS sea trials scheduled for the summer of 92. Bistatic signal processing, target tracking, clutter reduction as well reverberation issues will be addressed. As an example, new CW reverberation region rejection normalizers, as well as more sophisticated bistatic reverberation displays notches/thresholds will be evaluated for potential future fleet implementation.

MODELING OF WIDE AREA OCEAN REVERBERATION AND NOISE

A.I. Eller, L. Haines, W. Renner, R. Cavanagh
Science Applications International Corporation
McLean, Virginia 22102 USA

and

E.D. Chaika
Office of Naval Research
ASW Environmental Acoustic Support Program (AEAS)
Stennis Space Center, MS 39529 USA

ABSTRACT

Evaluation of acoustic system performance and planning of at-sea acoustic measurements are examples that underline the need for robust, utilitarian models to predict the acoustic background reverberation and noise field that limit performance of active systems. This presentation describes two computer models -- ANDES and the AUAMP Baseline model -- that, in close to real time, predict and display such noise and reverberation fields and the corresponding beam reverberation response. The unique approach features a hands-off, automated data base extraction capability that sets up 3-D range-dependent environmental-input fields and provides a variety of computed files and full azimuthal graphics.

1. INTRODUCTION

The work presented in this paper represents two related long-term projects directed by the US Office of Naval Research through AEAS, the ASW Environmental Acoustic Support program. AEAS is an advanced development program; its products are measurement systems, data bases, and numerical models that provide environmental acoustic information needed by the fleet and by the system acquisition community. In this role the AEAS program provides a bridge connecting scientific principles to application requirements, bringing new science and technology into use, while responding also to the engineering constraints associated with final product reliability.

The two specific products described here are numerical models that combine standard environmental data bases with scientifically based algorithms to predict acoustic ambient noise and reverberation. The first product is ANDES, the Ambient Noise Directionality Estimation System. The second is the AUAMP Baseline model, developed as part of the Advanced Underwater Acoustic Modeling Project under AEAS management. This second model provides performance estimates for active systems; only the part associated with predictions of the reverberation field is discussed here.

The scientific issue represented here is whether numerical implementations of these models can provide predictions with sufficient accuracy and resolution to be useful in their intended support roles, and at the same time can operate in worldwide ocean environments with sufficient dependability and speed on desk top computers. These objectives often are contradictory and force one to make compromises. As a result there necessarily is a different flavor to the evaluation of such products, as compared to the evaluation of purely scientific results.

2. THE AMBIENT NOISE DIRECTIONALITY ESTIMATION SYSTEM

The Ambient Noise Directionality Estimation System (ANDES) is designed to make rapid predictions of average properties of the undersea ambient noise field. The primary outputs of the model are estimates of average omnidirectional and directional (horizontal and vertical) noise levels at any specified depth. The model considers noise generated by the two low-frequency sources that typically dominate in the deep-water open oceans: commercial shipping and wind action on the sea surface. In particular, ANDES is *not* appropriate for modeling noise under ice (or in the marginal ice zone) nor in shallow water (less than 400 meters).

Typically, ANDES predictions are indicative of noise levels averaged over periods of a few hours to several days. For instances in which the noise field is dominated by a few, nearby ships, average noise predictions will belie the temporal and spatial variability of the noise field. For such cases, ANDES provides a submodel that estimates the contributions from a few discrete sources whose location and source level are specified by the user. The discrete noise submodel uses ASTRAL's "convergence zone option" to determine transmission loss.

2.1 KEY ASSUMPTIONS IN APPROACH TO ANDES

The underlying assumptions that allow ANDES to produce noise estimates quickly are that it is valid: 1) to characterize distant shipping as a stationary, continuous shipping density; 2) to estimate transmission loss by range-averaged values determined by the ASTRAL model, without CZ structure; and 3) to support TL calculations with a provinced water mass data base in which new sound speed profiles are encountered along a track no sooner than every 1/2 degree and sometimes only every few hundred miles. The distant shipping densities are specified through a historical data base that gives, on a seasonal basis, the average number of ships in five different size categories in 1-degree by 1-degree regions. Output directional noise estimates are provided with a resolution of approximately 1 degree in the vertical and as small as 5 degrees in the horizontal. The use of ASTRAL with provinced water mass regions is the key to fast execution when noise contributions from distances well over 1000 nmi in all directions are addressed within a matter of a few minutes.

3. THE AUAMP BASELINE MODEL REVERBERATION

The AUAMP Baseline model predicts the detection performance of low frequency active acoustic systems operating with monostatic, bistatic or multi-receiver configurations, in actual real-world, range-varying ocean environments. In representative applications, system performance can be estimated over large ocean areas that measure on the order of 2400 nautical miles in diameter, covering areas on the order of 5 million square nautical miles for a single run.

The model is designed to address a variety of support functions. It has been used to support system concept evaluation studies, to support the planning of experimental sea tests, and to provide at-sea as well post exercise evaluation and interpretation of acoustic observations.

Reverberation predictions are based on Navy standard data bases for sound speed profiles (Provinced GDEM), bathymetry (DBDBC), and bottom loss (LFBL); on adaptations of Lambert's Law and the Chapman-Harris Law for bottom and surface scattering; and on the two computation subroutines ASERT and REVERB. ASERT extracts environmental data needed for acoustic modeling along selected radial directions from the source and receiver positions. ASERT also computes transmission loss and travel time along each radial by means of a modified version of the CZ-ASTRAL model to all scattering boundaries and layers and seamounts. Transmission losses to surface and bottom are modified to account for vertical arrival angles and the angle dependence of the boundary scattering kernels.

The transmission loss values are used next in REVERB to compute the reverberation density field from surface, volume and bottom. The reverberation density indicates a normalized reverberation power resulting from a one-second pulse, as seen at the receiver by an ideal one-degree beam.

3.1 KEY ASSUMPTIONS IN APPROACH TO AUAMP BASELINE

The key assumptions that allow Baseline to make rapid estimates of reverberation are 1) the use of CZ-ASTRAL with a provinced water mass data base to compute propagation between source or receiver to scattering boundaries and layers, 2) the use of a so-called "augmented TL" approach to reverberation in which all arrivals at a point are collapsed to a single time without resolving separate round trip path arrivals, 3) the use of a separate routine for seamounts to guarantee that none are missed because of a coarse azimuthal step size selected for the computation, and 4) the use of factorable scattering kernels. The reverberation field is expressed as a function of time (or range) and bearing angle in steps of 1 sec and 5 degrees or greater.

4. EXAMPLES OF MODEL RESULTS

Representative outputs from ANDES are illustrated in Figure 1. Figures 1a and 1b show the frequency dependence of the vertical and horizontal noise fields at frequencies of 50, 100 and 200 Hz. Omnidirectional spectral noise values computed at these frequencies are 80.0, 73.6 and 66.7 dB/Hz, respectively. The vertical noise field is given in dB/Hz/steradian, and the horizontal noise field in dB/Hz/deg. The horizontal field indicates both a distributed contribution from averaged distant shipping and three angular line components from discrete nearby ships. Figure 1c, a more involved sequence of calculations with a wealth of information, indicates: i) a map with shipping density and a track location, ii) bathymetry and omnidirectional noise level at a bottom sensor along the track, and iii) the horizontal noise directionality along the track.

The set of four displays in Figure 2 illustrate various aspects of the reverberation calculation resulting from a 10-element vertical source at 36 deg North, 18 deg East in the Ionian Basin and a bistatic horizontal line array receiver, with an east-west orientation, located north-east of the source at 36.5 deg North, 17.5 deg East. The source emits a 10 sec, low frequency pulse. Figure 2a is a map of the ensonification of the bottom, with grazing angle taken into account in accordance with Lambert's Law. Figure 2b indicates the reverberation field strength from the bottom. It addresses the propagation from the source to the bottom, the propagation from the bottom to the receiver, and the ensonification area. It include the high intensity forward scattering area between the source and receiver as well. The beam reverberation, found by combining the reverberation field with the array beam pattern, is illustrated in plan view format in Figure 2c and as a waterfall display in 2d.

5. CONCLUSIONS

The noise and reverberation models described here represent unique capabilities in their ability to provide wide area predictions, quickly, sensitive to the range and azimuth dependent environment, and in a format to support system deployment and utilization decisions. Key technical issues are how to improve accuracy and resolution while retaining short execution times. Improvements underway are the planned integration of ASTRAL 4.0, which should improve the handling of complicated ducting environments and the determination of arrival time structure and the integration of a shallow water water mass data base.

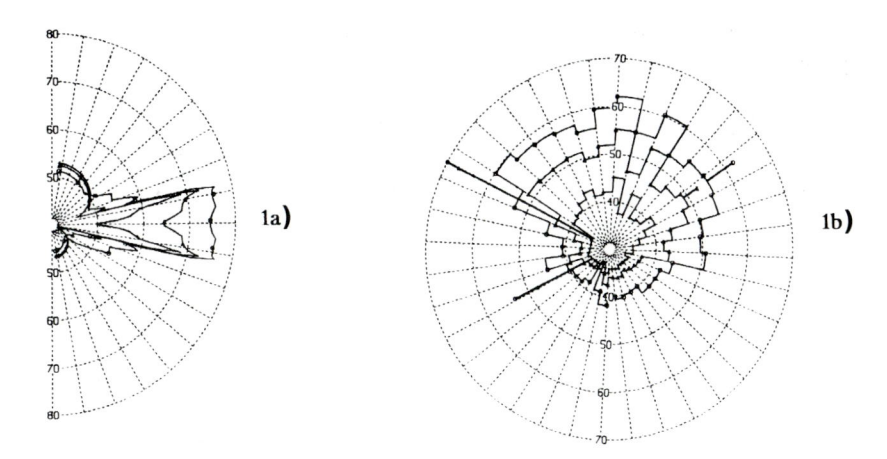

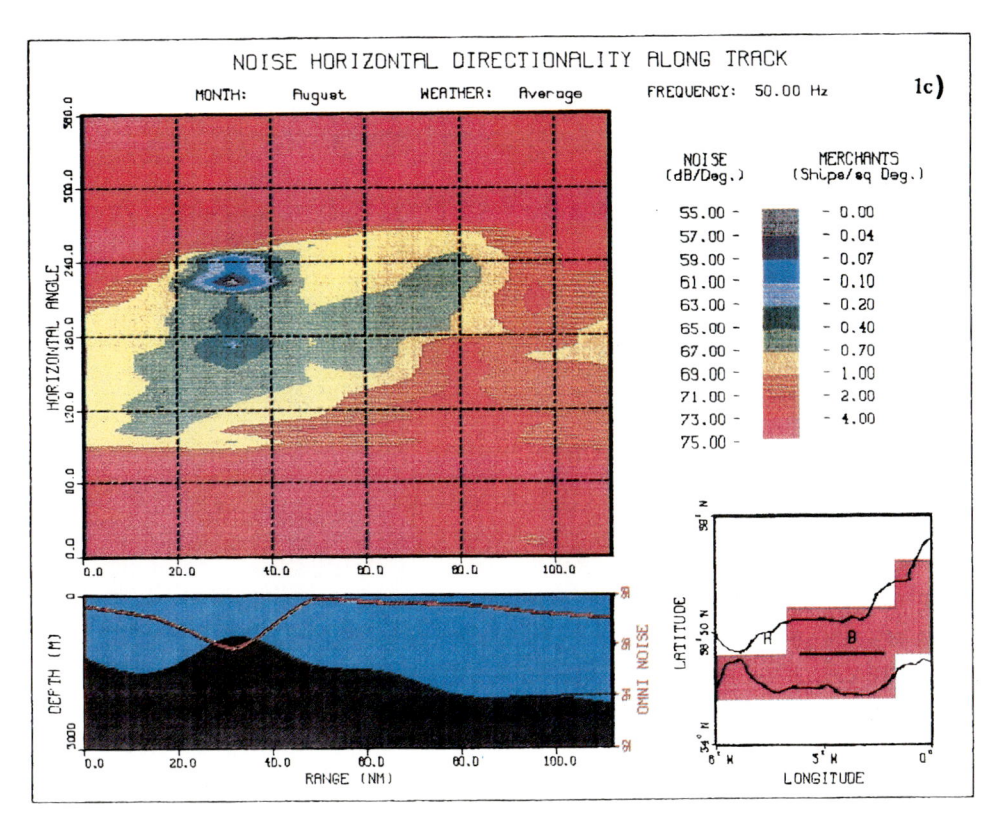

Figure 1. Representative ANDES Output

2 a) Transmission loss from source to bottom

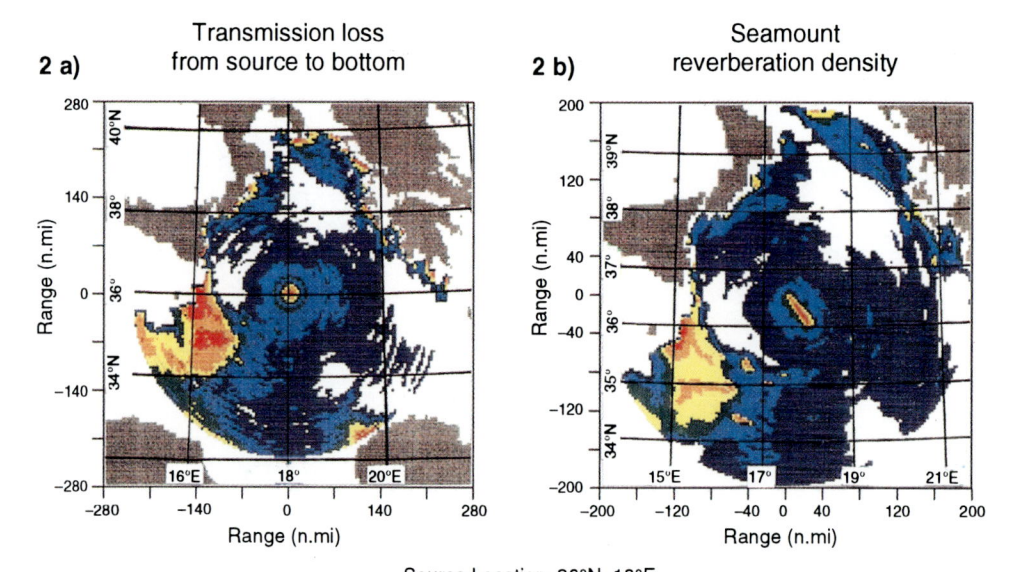

2 b) Seamount reverberation density

Source Location: 36°N, 18°E
Depth: 600 ft
HLA Receiver Location: 36°30'N, 17°30'E
Depth: 600 ft

2 c) Beam reverberation

2 d) Beam reverberation

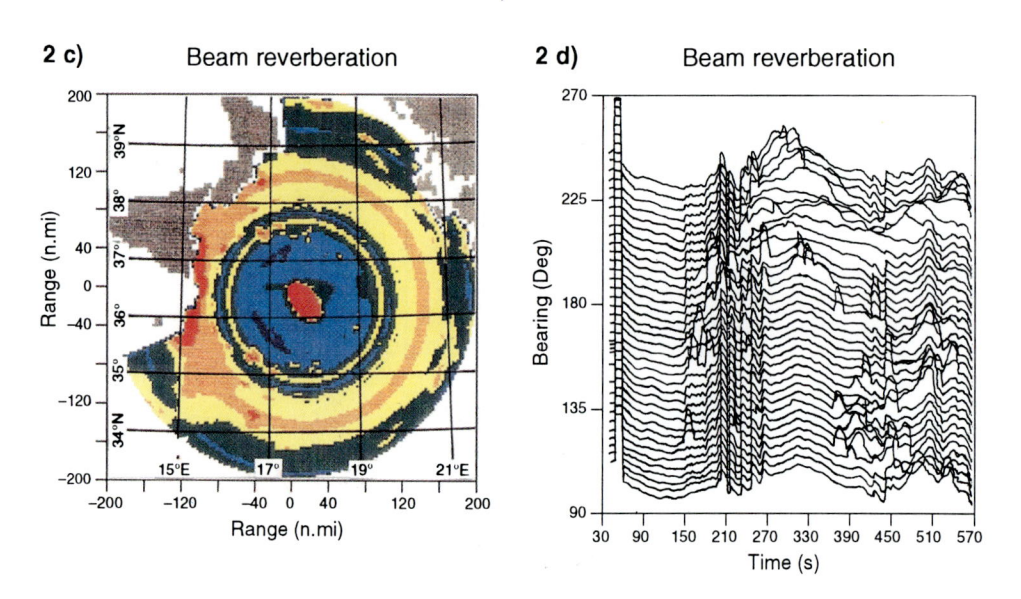

Figure 2. Representative AUAMP Baseline Outputs

RAPID ENVIRONMENTAL ACOUSTIC SURVEY AND MODELING SYSTEM

D. RUBENSTEIN
D. S. HANSEN
Science Applications International Corporation
1710 Goodridge Drive
McLean, VA 22102
USA

ABSTRACT. A two-component acoustic system was developed and deployed during a major Fall 1991 at-sea exercise. The first component is a near-real-time airborne environmental acoustic data acquisition system. This component acquires and processes data, and performs semi-automated quality control and message generation. These messages are transmitted to the second component, an acoustic modeling and validation system. This system is capable of acquiring disparate types of incoming data and messages, performing acoustic performance modeling, and optimizing exercise scenarios based on in situ measurements from several sources, including the airborne system.

1. INTRODUCTION

During harsh-environment concept-evaluation sea tests, we find that many of the emerging active sonar concepts are hindered by nonhomogeneous reverberation-limited environments. To optimize data collection for tactical and surveillance systems during these tests, and to guide post-test analyses, it is mission and cost effective to perform a rapid wide-area Maritime Patrol Aircraft (MPA) EnVironmetal Acoustic (EVA) survey. Such a survey focuses on the reverberation, noise, and transmission loss environment, quantifying the statistics that are important in predicting the performance of candidate systems and processor techniques.

For this purpose, a near-real-time airborne environmental acoustic data acquisition system and a deployable communication and modeling system were developed. The EVA data reduction/analysis system was deployed in an aircraft while the modeling system was deployed shipboard during the previously mentioned ASW exercise. Figure 1 is a schematic overview of the systems. This paper presents the system-design approach and compares results from airborne, shipboard, and fixed arrays.

The requirement for the overall performance monitoring and decision support system was to measure and analyze, in near-real-time, the variable environmental acoustic factors which affect the conduct and outcome of concept-evaluation sea tests. To meet the requirement for rapid data reduction and reporting, messages summarizing EVA and oceanographic ENVironment (ENV) parameters were generated by ships, aircraft and deployable/fixed systems. These messages were automatically inserted into a database system on a central platform over commercial circuits.

The airborne MPA/EVA processing system acquires sonobuoy signals, and performs real-time spectral and matched-filter processing. A high-performance computer workstation system produces near-real-time interactive displays that allow system operators to perform semi-automated quality control checks on the data. The data are then analyzed and their statistics are summarized into EVA messages.

These EVA messages convey information about transmission loss, ambient noise, reverberation, time spread and frequency spread. In parallel, ENV data provide in situ conditions to acoustic performance models. The presence of modeling results and the EVA data allow rapid data/model comparisons and modeling result validation. Time evolving simulations of the exercise that are sensitive to validated model

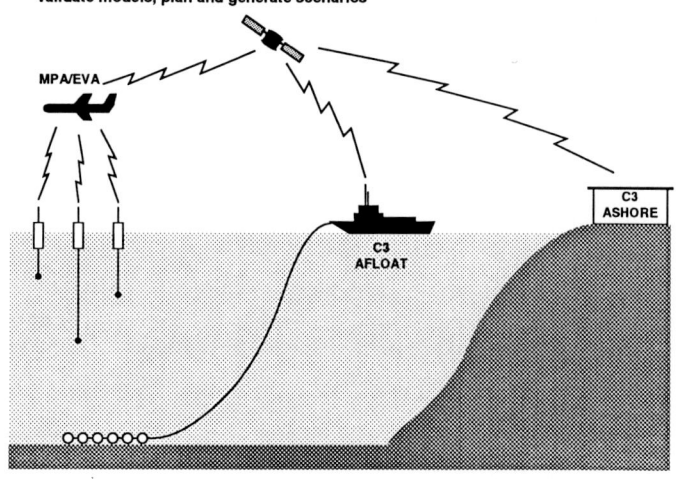

Figure 1. Schematic overview of the MPA/EVA and C3 systems.

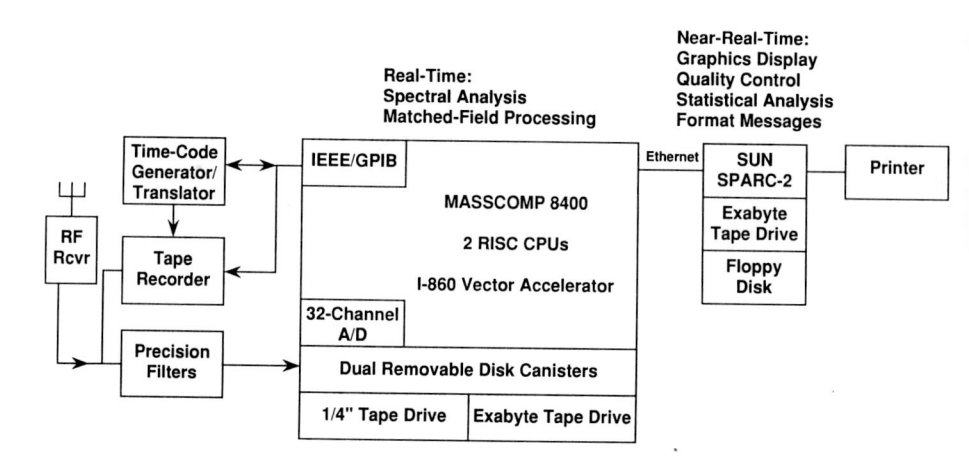

Figure 2. Schematic of Maritime Patrol Aircraft/Environmental Acoustics (MPA/EVA) system architecture.

and measured results and the tactical situation are then used to evaluate various alternatives to exercise execution and focus post-test analyses.

2. MPA/EVA SYSTEM

2.1. System Hardware

The EVA system consists of a Masscomp 8400 and a Sun Sparc 2 workstation. Figure 2 shows a schematic of the system architecture. The Masscomp is used for compute-intensive signal processing operations, while the workstation computer performs user-interactive graphics operations. Both computers use the Unix operating system; the Masscomp uses a real-time version of Unix. This commonality allows the workstation to have transparent access to Masscomp files.

Analog signals are pre-conditioned by a set of Precision Filters, and are passed to the A/D boards, which can handle up to 32 separate channels. An I-860 vector accelerator is hosted by two RISC CPU's. Basic signal processing functions are performed on the Masscomp. The output from these functions are stored onto either of two identical removable disks. This processed data can then be archived onto Exabyte tapes, and simultaneously accessed by the workstation. The workstation has a high-resolution color monitor for rapid, intricate displays. A laser printer produces graphics output for detailed analysis.

2.2. Real-Time Signal Processing

The real-time system performs three basic modes of real-time signal processing; HFM (hyperbolic frequency modulation), CW (continuous wave), and Broadband. The purpose of HFM mode is to acquire calibrated signals that have been projected with a frequency-modulated waveform. In HFM mode, digitized data are de-multiplexed, bandshifted and low-pass filtered. Then the resultant complex-valued time series are correlated with a replica of a projected waveform.

In CW mode, a continuous series of overlapping, windowed FFT's are computed. The output from a preselected set of narrow frequency bands is then saved to disk. Broadband mode entails processing that is similar to that of CW mode, but energy levels are integrated across broad frequency bands before being saved to disk.

2.3. Near-Real-Time Processing

The near-real-time processing portion of the EVA system has two purposes. First, it allows the analyst to view displays of the data, and make judgments concerning data quality. This function is extremely important, because of the various types of noise and signal gaps that accompany sonobuoy data acquisition. Radio frequency interference contaminates the signal with noise for different periods of time. It is a major problem in coastal regions, where proximity to television and radio stations is especially troublesome. Another problem is the generation of data gaps when the sonobuoy antenna is washed over by big surface waves. As the sea test was performed in shallow coastal waters during a stormy winter period, our system was exposed to both types of data problems. Our analyst/operators learned how to discriminate among varying degrees of noise contamination.

The second major purpose of the near-real-time EVA system is to generate messages. Messages are short, user-selected samples or statistical summaries of processed data. The analyst examines a time series, and makes judgments concerning data quality. Using a trackball, he selects good-quality data segments. The segments are then automatically summarized into ASCII-formatted files.

Three types of acoustic messages are generated, corresponding to the three modes of real-time processing; HFM, CW, and Broadband. HFM messages contain averaged time series of statistical quantities; peak energy, mean energy, and standard deviation, in the vicinity of a received pulse. The user specifies various time series parameters, such as duration (on the order of 10-20 seconds) and time sample averaging increment (about 1 second). Sonobuoy models AN/SSQ-57B and AN/SSQ-77A have calibrated frequency response functions, which are used to compute absolute sound pressure levels. Therefore, transmission loss can be derived from the HFM messages, as well as time spread and reverberation measures.

Broadband messages also contain time series of peak, mean, and standard deviation of energy level. Messages are generated covering a selected set of frequencies and sonobuoy channels. They are used for quantifying the reverberation that follows after SUS charges are deployed. They are also useful for describing noise levels, during periods when no artificial sounds are being generated.

CW messages contain time series of frequency spectra. They are useful for describing the frequency spread associated with projected CW waveforms, both during the direct blast, and also during the reverberation period. In addition, they can serve the same purpose as lofar-grams, and aid in the description of passive detections.

Figure 3 shows a comparison between high-quality and low-quality data recorded in HFM mode by the MPA/EVA system (upper and middle panels), and a data record from a single hydrophone in a ship-linked acoustic array (lower panel). The good sonobuoy record is of similar quality to that from the array. The low-quality sonobuoy record clearly shows a ping, but the spiky raw signal shows up as intermittent features that contaminate the matched-filter processed data. An experienced system operator should ideally be present to distinguish these contaminated records, and to make a judgment on their quality, before incorporating their information into messages.

2.4. Operation

The EVA processing system worked well during all of 19 flights. The average flight duration was nine hours. The hardware operated on each flight (a temporary failure at the end of the second flight did not result in the loss of any data). The software worked reliably and provided an adequate Operator-Machine Interface during flight operations. During each flight, we made hard copy graphics of selected interesting records. The availability of these graphics proved invaluable in post-flight briefs and between-flight analyses and discussions. Frequently, the previous experience aided in making decisions concerning subsequent flights and operations.

During each flight, processed data sets were archived onto Exabyte tapes. Messages were generated and accumulated into a set of files, and transferred to floppy disks. These files added up to 40 to 100 KBytes per flight. After landing, the floppy disk would be taken over to a desktop computer and sent through encrypted telephone circuits via COMSAT channels. The messages were received and processed by the C3 system, described below.

The C3 system is composed of three segments. The first segment is the communication segment. It is designed to receive and route message traffic to its final destination. Most messages are typically directed to database parser routines using UNIX EMAIL as the transport mechanism. Tactical C2 is performed by the second segment of the system, a standard Navy Command and Control system that was modified to accept EMAIL messages.

The final segment of the C3 system is the AM&V component. It receives and processes all EVA/ENV data. An interface to the tactical C2 component provides the fused tactical picture to the AM&V to allow experimental simulation and test scenario modification. This segment is described more fully in the following section.

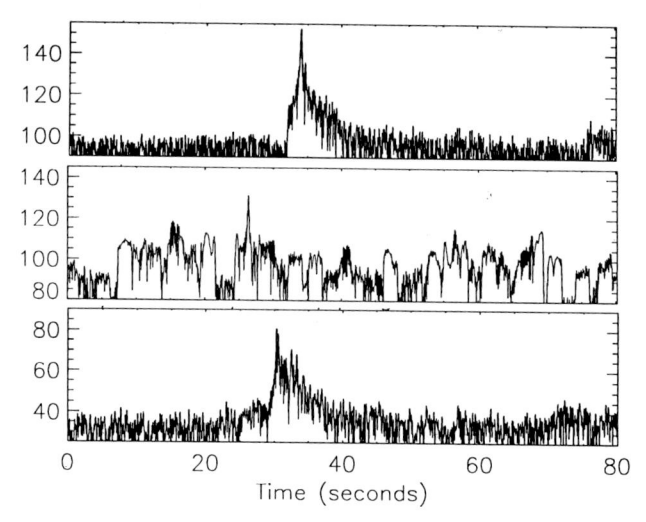

Figure 3. Comparison of matched-filter processed data. Upper two panels show high and low-quality records from sonobuoy data, processed onboard an aircraft. Lower panel shows a ship-linked hydrophone record. Ordinate axes for the sonobuoy records are absolute sound pressure levels in dB re μPa, while the ship-linked record is on a relative scale.

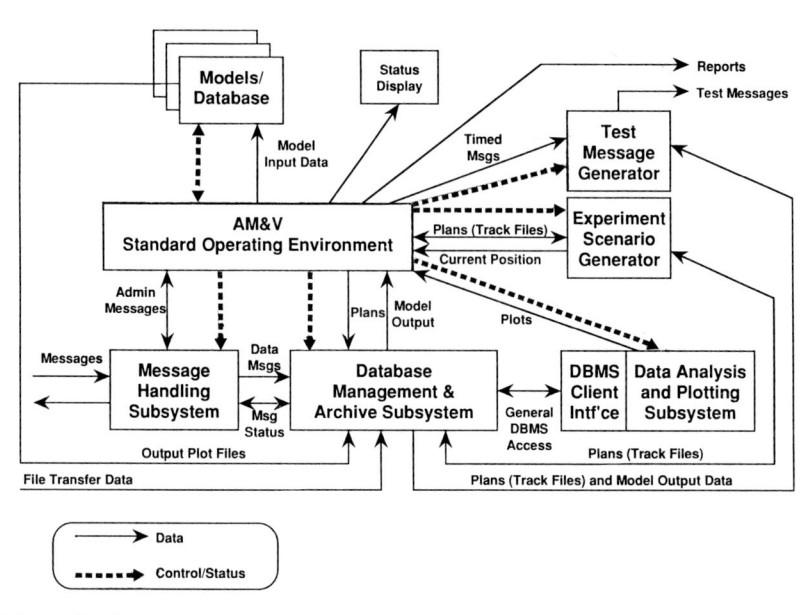

Figure 4. Schematic of the Acoustic Modeling & Validation (AM&V) system.

3. ACOUSTIC MODELING AND VALIDATION (AM&V) SYSTEM SEGMENT

The AM&V system consists of a complex, sophisticated set of software tools residing on a Sun workstation, or on a network of workstations. Its purpose is to aid operators in the acquisition, quality control, and archival of incoming messages, to run and validate models, to perform in situ performance simulations, and to generate experiment scenarios and plans. Figure 4 is a high-level overview of the system.

A message handling system collects and parses ENV and EVA data, signal excess, and source status messages and fused platform position reports. Oceanographic profiles of temperature, salinity, and sound speed contained in ENV messages can be graphically edited by an analyst, to ensure quality control. The various messages are parsed, and stored into a database using Sybase, a state-of-the-art commercial relational distributed database management package.

A data analysis and plotting subsystem is a general tool set for plotting, editing, and analyzing exercise data. The various types of plots include parameter vs. depth profiles, geographic locations of samples, parameter vs. track profiles, and geographic contour plots.

An exercise scenario generator allows an operator to prepare exercise track plans, or to edit previous track plans stored in the database. The planning process uses geographic displays, and is graphically interactive. The process is automated to the extent possible, for example in computing time lines of platform positions.

A suite of acoustic propagation and performance models (AUAMP, PE, MPP, and GSM) is an integral part of the system. The models access bathymetry and bottom data, and recently-acquired sound speed fields from the database. The models produce numerous types of predictions, including transmission loss, noise levels, reverberation levels, and signal excess. Many graphics options are available. An analyst can view contours of predictions overlaid on top of bathymetry and platform tracks. A simulation system displays continuously updated exercise information, both measured and modeled. Predicted and measured signal excess, predicted and measured range, and predicted bistatic angle can be plotted as a function of time.

During the Fall 1991 exercise, the AM&V system proved to be of significant utility. Test direction personnel examined the level and duration of reverberation returns from a wide, environmentally inhomogeneous geographic area. Based on a search strategy, platform laydowns were optimized. Waveform types and wavetrain structures were tuned to the specific environmental conditions.

4. CONCLUSIONS

Field experience with these systems has shown that they are extremely valuable assets for large-scale experiments. As with virtually every other acoustic acquisition system, the MPA/EVA system acquires enormous data sets. What sets the system apart from the others, however, is its ability to perform quality control in near-real-time. Compact messages can effectively convey the essence of the information, and are short enough to be sent over standard data links. The AM&V system has the ability to incorporate these messages, and use them to validate performance models, and to display them in various ways to help gain understanding during the course of the experiment. This near-real-time activity provides test direction personnel with the means to optimize the performance of distributed and mobile ASW systems. In the future, when real-time data links between MPA, surface ships, and shore stations become a reality, this system will be capable of directly integrating large data structures into a sophisticated analysis, modeling, and simulation framework.

LIST OF PARTICIPANTS

Dr. T. Akal
SACLANT Undersea Research Centre
Viale San Bartolomeo, 400
19138 La Spezia
Italy

Dr. R. Ancey
DCN Toulon
GERDSM – Le Brusc
83140 Six-Fours-les-Plages
France

Mr. C. R. Andriani
Chief Engineer, Technical Director
Integrated Undersea Surveillance Project
Department of the Navy
Space and Naval Warfare Systems
    Command
Washington, DC 20363-3700
USA

Dr. A. B. Baggeroer
Department of Ocean Engineering
Room 5-209
Massachusetts Institute of Technology
Cambridge, MA 02139
USA

Dr. J. M. Berkson
Naval Research Laboratory
Acoustic Systems Branch
Code 5160
Washington, DC 20375-5000
USA

Mr. M. Bouvet
CMO - EPSMOM
13 rue le Chatellier
29275 Brest
France

Dr. W. M. Carey
Defence Applied Research Projects Agency
79 Whippoorwill Rd.
Old Lyme, CT 06371
USA

Mr. R. N. Carpenter
Naval Undersea Warfare Center
Code 8212, Bldg 679
Newport, RI 02841-5047
USA

Dr. J. W. Caruthers
Naval Research Laboratory
Code 240
Stennis Space Center
MS 39529-5004
USA

Mr. T. A. Casey
Naval Undersea Warfare Center
Code 382
Newport, RI 02841-5047
USA

Dr. N. P. Chotiros
Applied Research Laboratories
The University of Texas at Austin
P.O. Box 8029
Austin, TX 78713-8029
USA

Dr. J. L. Collé
Groupe d'Etudes et des Recherches en
    Detection Sous-Marine
DCN Toulon
Le Brusc
83140 Six-Fours-les-Plages
France

Dr. M. D. Collins
Naval Research Laboratory
Acoustic Systems Branch
Code 5160
Washington, DC 20375-5000
USA

Dr. F. D. Cotaras
Defence Research Establishment Atlantic
P.O. Box 1012
Dartmouth, NS B2Y 3Z7
Canada

Dr. P. A. Crowther
Marconi Underwater Systems Ltd.
Addlestone
Surrey KT15 2PW
UK

Dr. L. A. Crum
National Center for Physical Acoustics
University of Mississippi
Oxford, MS 38677
USA

Dr. B. De Raigniac
Societé d'Etudes et Conseils
Automation Electronique Recherche
　Opérationelle
3 Avenue de l'Opéra
75001 Paris
France

Dr. O. Diachok
SACLANT Undersea Research Centre
Viale San Bartolomeo, 400
19138 La Spezia
Italy

Dr. J. A. Doutt
Woods Hole Oceanographic Institution
Woods Hole, MA 02543
USA

Dr. L. B. Dozier
Science Applications International Corp.
1710 Goodridge Drive
P.O. Box 1303
McLean, VA 22102
USA

Dr. I. Dyer
Department of Ocean Engineering
Room 5-212
Massachusetts Institute of Technology
Cambridge, MA 02139
USA

Dr. A. I. Eller
Science Applications International Corp.
1710 Goodridge Drive
P.O. Box 1303
McLean, VA 22102
USA

Mr. E. Estalote
Office of Naval Research
Department of the Navy
800 North Quincy Street
Arlington, VA 22217-5000
USA

Dr. D. D. Ellis
SACLANT Undersea Research Centre
Viale San Bartolomeo, 400
19138 La Spezia
Italy

Dr. D. M. Farmer
Institute of Ocean Sciences
P.O. Box 6000, 9860 West Saanich Road
Sidney, BC V8L 4B2
Canada

Dr. D. M. Fromm
Naval Research Laboratory
Code 5160
Washington, DC 20375-5000
USA

Dr. J. K. Fulford
Naval Research Laboratory
Code 223
Stennis Space Center
MS 39529
USA

Dr. R. C. Gauss
Naval Research Laboratory
Code 5160
Washington, DC 20375-5000
USA

Dr. M. Gensane
Thomson-Sintra Activités Sous-Marines
1 av. Aristide-Briand
94117 Arcueil Cedex
France

Dr. R. R. Goodman
Applied Research Laboratory
Pennsylvania State University
P.O. Box 30
State College, PA 16804
USA

Dr. R. F. Gragg
Naval Research Laboratory
Acoustics Systems Branch
Code 5160
Washington, DC 20375-5000
USA

Dr. D. Handschumacher
Naval Research Laboratory
Ocean Sciences Directorate
Bldg. 1005, Code 361
Stennis Space Center
MS 39529-5004
USA

Mr. D. Hanna
Naval Command, Control and Ocean
 Surveillance Center
RDT&E Division (NRaD), Code 702
San Diego, CA 92152-5000
USA

Dr. G. J. Heard
Defence Research Establishment Pacific
FMO Esquimalt
Victoria, BC VOS 1BO
Canada

Dr. P. C. Hines
Defence Research Establishment Atlantic
P.O. Box 1012
Dartmouth, NS B2Y 3Z7
Canada

Dr. A. K. Kalra
Naval Research Laboratory
Tactical, Low Frequency Acoustics Branch
Code 244
Stennis Space Center
MS 39529-5004
USA

Dr. J. G. Kelly
Naval Undersea Warfare Center
Technical Division
Code 8219, Building 679
Newport, RI 02841-5047
USA

Dr. D. B. King
Naval Research Laboratory
Code 223
Stennis Space Center
MS 39529-5004
USA

Dr. R. Kolesar
Naval Command, Control and Ocean
 Surveillance Center
RDT&E Division (NRaD), Code 535
San Diego, CA 92152-5000
USA

Dr. E. Y. T. Kuo
Naval Undersea Warfare Center
New London Detachment
New London, CT 06320
USA

Dr. R. G. Levers
DCN Toulon
GERDSM – Le Brusc
83140 Six-Fours-les-Plages
France

Dr. J. C. Lockwood
Naval Command, Control and Ocean
 Surveillance Center
RDT&E Division (NRaD), Code 734
San Diego, CA 92152-5000
USA

Dr. R. H. Love
Naval Research Laboratory
Code 243
Stennis Space Center
MS 39529-5004
USA

Dr. N. C. Makris
Naval Research Laboratory
Code 5166
Washington, DC 20375-5000
USA

Ing. D. Marandino
Selenia Elsag Sistemi Navali
Via Hermada 6b
16154 Genova
Italy

Mr. J. Marchment
SACLANT Undersea Research Centre
Viale San Bartolomeo, 400
19138 La Spezia
Italy

Dr. R. L. Martin
COMNAVOCEANCOM
Bldg 1020, Code N22
Stennis Space Center
MS 39529
USA

Dr. D. F. McCammon
Applied Research Laboratory
Pennsylvania State University
P.O. Box 30
State College, PA 16804
USA

Dr. S. T. McDaniel
HC-01 Box 62
Spruce Creek, PA 16683
USA

Mr. J. M. Monti
Naval Undersea Warfare Center
Code 3112
New London, CT 06320
USA

Dr. T. G. Muir
Applied Research Laboratories
The University of Texas at Austin
P.O. Box 8029
Austin, TX 78713-8029
USA

Mr. B. Nützel
Forschungsanstalt der Bundeswehr für
    Wasserschall- und Geophysik
Klausdorfer Weg 2-24
2300 Kiel, 14
Germany

Dr. P. M. Ogden
Naval Research Laboratory
Acoustics Systems Branch
Code 5160
Washington, DC 20375-5000
USA

Dr. J. A. Orcutt
Institute of Geophysics and Planetary
    Physics (0225)
Scripps Institution of Oceanography
La Jolla, CA 92093
USA

Dr. G. J. Orris
Naval Research Laboratory
Acoustics Systems Branch
Code 5160
Washington, DC 20375-5000
USA

Dr. F. M. Pestorius
Applied Research Laboratories
The University of Texas at Austin
P.O. Box 8029
Austin, TX 78713-8029
USA

Mr. A. Plaisant
Thomson Sintra ASM
BP 138
06561 Valbonne Cedex
France

Mr. J. R. Preston
SACLANT Undersea Research Centre
Viale San Bartolomeo, 400
19138 La Spezia
Italy

Dr. A. H. Quazi
Naval Undersea Warfare Center
New London, CT 06320
USA

Dr. D. Rauch
USEA S.p.A.
Via Delle Pianazze, 74
19027 Termo, La Spezia
Italy

Dr. J. A. Rice
Naval Command, Control and Ocean
    Surveillance Center
RDT&E Division (NRaD), Code 541
San Diego, CA 92110
USA

Lt. Cdr. R. Rogers
Ministry of Defence
Director of Naval Oceanography and
  Meteorology
Lacon House, Theobalds Road
London WC1X 8RY
UK

Dr. R. A. Roy
The University of Washington
Applied Physics Laboratory
1013 NE 40th Street
Seattle, WA 98105
USA

Dr. D. Rubenstein
Science Applications International Corp.
1710 Goodridge Drive
P.O. Box 1303
McLean, VA 22102
USA

Dr. F. J. Ryan
Naval Command, Control and Ocean
  Surveillance Center
RDT&E Division (NRaD), Code 541
San Diego, CA 92152-5000
USA

Dr. H. Schmidt
Department of Ocean Engineering, 5-204
Massachusetts Institute of Technology
77 Massachusetts Avenue
Cambridge, MA 02139
USA

Ir. P. Schippers
TNO Physics and Electronics Laboratory
Oude Waalsdorperweg, 63
2509 JG The Hague
The Netherlands

Ir. J. G. Schothorst
TNO Physics and Electronics Laboratory
Oude Waalsdorperweg, 63
2509 JG The Hague
The Netherlands

Mr. G. C. Searing
Defence Research Agency
USSA3, Active Sonar Division
ARE Southwell, Portland
Dorset DT5 2JS
UK

Dr. M. M. Sevik
Carderock Division
Department of the Navy
Naval Surface Warfare Center
Bethesda MA 20084-5000
USA

Dr. D. G. Simons
TNO Physics and Electronics Laboratory
Oude Waalsdorperweg, 63
2509 JG The Hague
The Netherlands

Dr. M. Springer
Atlas Elektronik GmbH
P.O. Box 44 85 45
D-2800 Bremen 44
Germany

Dr. R. A. Stephen
Woods Hole Oceanographic Institution
Woods Hole, MA 02543
USA

Dr. S. Sutherland-Pietrzak
Naval Undersea Warfare Center
Code 3314, Bld 122 T
New London, CT 06328
USA

Dr. D. Tielbürger
Forschungsanstalt der Bundeswehr für
  Wasserschall- und Geophysik
Klausdorfer Weg 2-24
2300 Kiel, 14
Germany

Dr. R. Thiele
Forschungsanstalt der Bundeswehr für
  Wasserschall- und Geophysik
Klausdorfer Weg 2-24
2300 Kiel, 14
Germany

Dr. C. H. Thompson
Naval Research Laboratory
Code 243
Stennis Space Center
MS 39529-5004
USA

Dr. D. W. Tufts
Department of Electrical Engineering
Kelley Hall
University of Rhode Island
Kingston, RI 02881
USA

Mr. H. G. Urban
SACLANT Undersea Research Centre
Viale San Bartolomeo, 400
19138 La Spezia
Italy

Dr. R. A. Wagstaff
Naval Research Laboratory
Code 245
Stennis Space Center
MS 39520-5004
USA

Dr. F. Wiekhorst
FGAN
W-5307 Wachtberg-Werthhoven
Germany

Dr. R. B. Williams
Naval Command, Control and Ocean
    Surveillance Center
RDT&E Division (NRaD), Code 7104
San Diego, CA 92152-5000
USA

Dr. J. H. Wilson
Neptune Sciences, Inc.
P.O. Box 1235
San Clemente, CA 92674
USA

Dr. M. A. Wilson
Naval Research Laboratory
Code 243
Stennis Space Center
MS 39529
USA

Dr. T. C. Yang
Naval Research Laboratory
Code 5120
Washington, DC 20375-5000
USA

Dr. J. Ziegenbein
Forschungsgesellschaft für Angewandte
    Naturwissenschaften
Forschunginstitut für Hochfrequenzphysik
Neuenahrer Str. 20
W-5307 Wachtberg-Werthhoven
Germany

AUTHOR INDEX

[**Bold** page numbers refer to authors of papers in this volume; regular typeface corresponds to a bibliographic citation.]

a

Adams, B.B.: 201, 329
Akal, T.: 64, 194, 201, **255**, 278
Akulichev, V.A.: 22
Ali, H.B.: 252
Alpers, W.: 76
Amit, D.J.: 144
Anderson, A.L.: 64
Anderson, C.D.: 84
Anderson, V.: 43, 76
Andreeva, I.B.: 270, 278
ARSRP Project Office: 188
Arvelo, J.I.: 252

b

Bachman, W.: 10, 153
Baer, R.N.: 124
Baggeroer, A.B.: **51**, 57, 112, 166, **183**, 188
Baldy, S.: 22
Banner, M.L.: 70
Barbagelata, A.: 262
Barthel, K.G.: 290
Bass, F.G.: 69
Batchelor, G.K.: 70
Battestin, H.F.: 312
Becker, G.: 64
Belloul, M.: 76
Benthien, G.W.: 312
Bergin, N.T.: 41
Berkson, J.M.: **189**, 194, 195, 200, 201, 278
Berman, D.H.: 153
Bernstein, G.M.: 201
Berrou, J.L.: 194
Berry, M.V.: 137
Blanchard, D.C.: 22
Boehme, H.: 84, 96
Bohn, J.C.: 64
Booth, N.O.: **167**
Boudreaux-Bartels, G.F.: 328
Bowman, J.J.: 166

Bradley, M.R.: **331, 361**, 366
Breitz, N.D.: 22, 23
Brekhovskikh, L.: 69, 160, 208
Briggs, K.B.: 57
Brigham, G.A.: 312
Briscoe, M.G.: 41
Brocher, T.M.: 220
Brooke, G.H.: **247**, 252
Brown, J.R.: 10
Brown, M.V.: 359
Browning, D.G.: 42, 70, 270
Brylow, S.M.: 201
Bucca, P.J.: 290
Bucker, H.P.: **145**, 153, **167**, 174
Buffman, M.: **85**
Buhl, P.: 220
Bulanov, V.A.: 22
Bunchuk, A.V.: 96
Burdic, W.S.: 348
Burns, D.R.: 232
Burns, T.: 137

c

Carey, W.M.: **25**, 41, 42, 70
Carpenter, R.N.: **85**, 90
Carstensen, E.L.: 42
Cartmill, J.: 23
Caruthers, J.W.: 57, **203**, 208, 329
Casey, T.A.: **293**
Castile, B.: 76, 174
Cavanagh, R.: **373**
Červený, V.: 232
Chaika, E.D.: **373**
Chapman, N.R.: 252
Chapman, R.P.: 10, 22, 174, 246, 359
Charnock, H.: 70
Chernov, L.A.: 64
Chertock, G.: 312
Chesterman, W.D.: 174
Chin-Bing, S.A.: **113**, 118, 252
Chotiros, N.P.: **79**, 84, 96
Chou, S.I.: 312

Clarebout, J.F.: 214
Claus, A.: 328
Clay, C.S.: 22, 70
Cline, A.: 180
Codispoti, L.A.: 290
Cole, B.F.: 270
Collins, M.D.: 112, 118, **119**, 124, 201, 252
Commander, K.: 43
Cooke, R.C.: 10, 23
Cox, H.: 174, 304, 369
Craig, D.W.: 252
Crighton, D.G.: 43
Cross, A.: 76
Croswell, W.: 76
Crowe, D.V.: 208, 329
Crowther, P.A.: 10, 23
Curry, T.J.: **293**

d
D'Amico, A.: **145**
Dashen, R.F.: **139**, 144, 153
Davis, J.A.: 252
Davis, N.R.: 201
Dawson, T.W.: 112, 124
Day, S.M.: 232
Deavenport, R.: 112
Deferrare, H.: 41
Deihl, D.T.: 329
DeMary, T.E.: 84, **91**
Detrick, R.S.: 220
Detsch, R.: 76
Dicus, R.: 137
DiNapoli, F.: 112
Ding, H.: 312
Ding, L.: 23
Donelan, M.A.: 10
Dosso, S.E.: 252
Dougherty, M.E.: 130, 232
Doutt, J.A.: 262, 270, **271**, 278, 359
Dozier, L.B.: **313**
Dullea, R.K.: **255**, 278
Dunbar, M.J.: 290
Duplantier, B.: 144
Dyer, I.: **51**, 57, 112, 166, 188, 194

e
Eckart, C.: 208, 329
Efron, A.: 329
Eller, A.I.: **373**
Ellis, D.D.: **125**, 160, 208, 329
Erskine, F.T.: 23, 201, 240, **241**
Estes, L.: 298
Evans, R.B.: 112, 118, 124, 160, 252

f
Fain, G.: 298
Fan, H.: 124
Farmer, D.M.: 10, **11**, 23, 24, 70
Farquhar, G.B.: 290
Fawcett, J.A.: 112, 124
Ferla, C.M.: 252
Feshback, H.: 42, 144
Fialkowski, J.M.: **235**
Figoli, A.: 262, 270, 278
Fitzgerald, J.W.: 42, 70
Flatté, S.M.: 360
Fleischer, P.: 57
Foldy, L.: 42
Foote, K.G.: 270
Fox, C.G.: 153
Franchi, E.R.: 180, 201
Frazer, L.N.: 252
Freese, H.A.: **313**
Fricke, J.R.: **51**
Fromm, D.M.: **155**, 160
Fulford, J.K.: **175**, **209**

g
Garrett, W.: 76
Gauss, R.C.: 201, **235**
Gensane, M.: **59**
Gerstoft, P.: 112
Giannoni, D.: **51**
Gibian, G.L.: **189**, 200
Gibson, S.M.: 366
Gilbert, K.: 43
Gjøsæter, H.: 290
Glass, R.: 76
Goff, J.A.: 220, 226
Goldstein, R.E.: 144
Goodman, R.R.: 41, 43
Gordon, D.F.: **145**

Gorman, A.D.: 10, 76, 90
Gragg, R.F.: **45**, 50
Green, A.W.: 41
Griffin, J.M.: 180
Guerrini, P.: 262
Guidi, G.: **255**, 278
Gulliksen, B.: 290

h
Haines, L.: **373**
Hall, A.: 24
Hall, M.: 360
Haller, D.R.: 160
Halsey, T.C.: 144
Hamilton, E.L.: 96
Hamilton, J.: 10
Hampton, L.D.: 64
Hansen, D.S.: **379**
Hansen, W.J.: 290
Harding, A.J.: 130, **215**, 220, **221**
Harris, I.: 76
Harris, J.H.: 10, 22, 174, 246, 359
Haury, L.R.: 118
Hayek, C.S.: 160, 240, **263**, 270
Hayes, D.E.: 153
Hayward, T.J.: **161**
Heard, G.J.: **247**
Hebert, M.: **361**
Henyey, F.S.: 10, 42, 70, 153
Hersey, J.B.: 284
Herwig, H.: 10, 76, **97**, 102
Hines, P.C.: 64
Huhnerfuss, H.: 76
Hui, W.H.: 10
Humphries, P.N.: 70

i
Ingard, K.U.: 42
Ingenito, F.: 166
Ishimaru, A.: 64, 84
Ivakin, A.N.: 64

j
Jackson, D.R.: 41, 64, 84, 90, 96, 144, 220
Jensen, E.P.: 70
Jensen, F.B.: 112, 118, 252

Jiang, S.-L.: 57
Johnson, B.D.: 10, 23
Johnson, J.: 76
Jones, W.: 76, 76
Jordan, T.H.: 220, 226

k
Kadota, T.: 328
Kalra, A.K.: **209**
Kampanis, N.A.: **125**
Kappus, M.E.: 220
Karplus, H.B.: 43
Keller, W.: 76
Kelly, J.G.: **85**, 90
Kil, D.H.: **319**
King, B.J.: 180
King, D.B.: **113**, 252
Kirsteins, I.P.: 327, 328
Kittappa, R.: 50
Kleinman, R.E.: 50
Koenigs, P.D.: 10, 76, 102
Kolesar, R.: **71**
Kolovayev, P.A.: 23
Krout, T.L.: **189**, 200
Kumaresan, R.: 328
Kuo, E.Y.T.: **65**, 69, 70
Kuperman, W.A.: 70, 112, **119**, 160, 194, 195, 201, 252
Kur'yanov, B.F.: 10

l
La Fond, E.C.: 96
Lackie, K.W.: 270
Ladd, M.E.: 160
Lamarre, E.: 23
Lange, P.: 76
Lee, O.S.: 118
Lemon, D.D.: 23
Levander, A.: 130, **215**, **221**
Levich, V.G.: 23
Levine, E.R.: **85**
Ling, S.C.: 23
Liu, C.: 328
Lockwood, J.C.: **299**
Lønne, O.J.: 290
Love, R.H.: 270, 278, **279**, 284, **285**, 290

Løvik, A.: 23
Lu, M.: 41
Lysanov, Yu.: 64, 69, 160, 208

m

Mac Niocaill, G.: 41
Machlup, S.: 284
Mackenzie, K.V.: 84, 174, 246, 360
Magaraci, A.F.: 348
Makris, N.C.: **189**, 194, 195, 200
Mallick, S.: 252
Marsh, H.W.: 69
McCammon, D.F.: **131**, 137
McCann, K.J.: 160
McCleary, L.: 312
McCoy, K.: 290
McDaniel, S.T.: **3**, 10, 41, 76, 84, 90
McDonald, B.E.: 10, 42, 70
McKee, L.: 252
McKinney, C.M.: 84, 96
Medwin, H.: 22, 23, 70, 84
Mehl, S.: 290
Melbourne, M.G.: 118
Melissinos, C.D.: 328
Melville, W.K.: 23, 70
Menis, R.: **189**, 200
Meredith, R.W.: 290
Merklinger, H.M.: 64, 360
Miller, J.F.: 252
Milne, A.R.: 166
Minster, J.B.: 232
Monahan, E.C.: 41
Monstad, T.: 284, 290
Monti, J.M.: 10, 76, 96, 102, 240, **263**, 270
Morse, P.M.: 42, 144
Mourad, P.D.: 90
Muir, T.G.: 84, **91**, 96
Mulhearn, P.J.: 24
Murphy, J.E.: **113**, 118
Mutter, J.C.: 220

n

Nagl, A.: 252
Nedreaas, K.: 284
Newbold, W.T.: 201
Nolle, A.W.: 84

Nolte, L.W.: 328
Novarini, J.C.: 208
Nuttall, A.H.: 341, **343**, 348
Nützel, B.: 10, 76, 96, **97**, 102

o

Office of Naval Research: 327
Ogden, P.M.: 23, 240, **241**
Orchard, B.J.: **155**
Orcutt, J.A.: 130, **183**, **215**, 220, **221**
Orr, M.H.: 118
Orris, G.J.: **119**, **139**, 144
Owen, M.M.: 304
Ozuler, N.: **51**

p

Palmer, L.B.: 160, 329
Pao, H.P.: 23
Percy, J.L.: 10
Perkins, J.S.: 124, 153, 160, 194, 195, 201
Pesci, A.I.: 144
Phillips, O.: 76
Pierce, A.D.: 112
Pierson, W.J.: 10
Plaisant, A.: **351**
Plant, W.: 76
Popov, M.M.: 232
Porter, M.B.: 153, 160, 174, 194, 201, 252
Potter, D.M.: 270
Preston, J.R.: 194, 201, **271**, 278
Pritchard, R.L.: 312
Prosperetti, A.: 42, 43
Pšenčik, I.: 232

q

Quazi, A.H.: **343**

r

Ray, R.I.: 57
Rees, C.D.: 312
Renner, W.: **373**
Reuter, M.: 304
Rice, J.A.: **167**
Richter, K.: 76
Richter, R.M.: 10

Robison, A.E.: 359
Rogers, P.H.: 118
Rohr, J.: **71**, 76
Romain, D.: 328
Romero-Rochin, V.: 144
Ross, D.: 312
Roy, R.A.: **25**
Rschevkin, S.N.: 42
Rubenstein, D.: **379**
Ruggles, A.: 43
Ryan, F.J.: **71**, **305**, 312
Rynne, E.F.: **305**

s
Saenger, R.A.: 270, 359
Saling, D.S.: 360
Sandy, R.J.: 208
Santaniello, S.R.: 270
Sawyer, W.B.: 57
Schenck, H.A.: 312
Schlichting, H.: 70
Schlude, F.: 76
Schmidt, H.: **51**, **105**, 112, 118, 124
Schmidt, P.B.: 360
Schuler, D.: 76
Scott, J.C.: 76
Scott, M.G.: 290
Scott, W.B.: 290
Scrimger, J.A.: 10, 359
Sen, M.K.: 252
Senior, T.B.A.: 166
Shah, A.A.: 329
Sharman, G.F.: 270
Shooter, J.A.: 84, **91**
Slater, R.R.: **319**
Smailes, I.C.: 96
Smith, S.L.: 290
Smith, W.O.: 290
Sobolev, S.L.: 144
Soroe, B.M.: 366
Soukup, R.J.: **235**
Spitzer, L.: 42
Stanic, S.: 57, 84, 96
Stephen, R.A.: 112, **125**, 130, **227**, 232
Stockhausen, J.H.: **255**, 262, 270, 278, 359

Strasberg, M.: 43
Su, M.-Y.: 23, 41
Sundvik, M.T.: 270
Sutherland-Pietrzak, S.: **367**
Swaszek, P.F.: 328
Szczucka, J.: 24

t
Tague, J.A.: 90
Talham, R.J.: 153
Tappert, F.D.: 112, 124
Tattersall, J.M.: 70
Thiele, R.: 64
Thompson, C.H.: 270, 278, **279**
Thompson, L.A.: 84, **91**
Thomson, D.J.: **247**, 252
Thorpe, S.A.: 24, 41, 70, 76
Thorsos, E.I.: 90, 144, 194, 220, 246
Tucholke, B.E.: 64
Tufts, D.W.: **319**, 327, 328, 329
Turner, R.G.: 10, 359

u
Überall, H.: 252
Updegraff, G.: 76
Urick, R.J.: 76, 84, 102, 130, 180, 278, 290, 348, 360
Uslenghi, P.L.E.: 166

v
Vaccaro, R.J.: 328
Vagle, S.: 10, 23, 24, 70
Van Wijngaarden, L.: 43
Veenkant, R.L.: 312
Vera, E.E.: 220
Vidmar, P.: 341
Virieux, J.: 232

w
Wagstaff, M.D.: **331**, 366
Wagstaff, R.A.: 201, **361**, 366
Walsh, A.L.: 24
Walter, W.: 76
Wang, L.: 43
Ward, J.: 312
Wardale, J.P.: 262
Warfield, J.T.: 329

Weinberg, H.: 90, 174
Wenzel, A.R.: 70
Westwood, E.: 341
Williams, J.E.: 43
Williams, R.B.: **145**
Williams, R.E.: 312
Williams, R.J.: 57, 166
Wilson, D.L.: 290
Wilson, J.H.: **331**, 341
Wilson, M.A.: **285**
Winebrenner, D.P.: 64, 84
Winn, K.: 64
Winokur, R.S.: 64
Wolf, S.N.: **155**
Wong, H.-K.: 174
Wood, A.B.: 24, 43
Woodcock, A.H.: 22
Woolf, D.K.: 24
Wu, J.: 24, 70, 76
Wurmser, D.: **45**, 50, 153
Wyber, R.J.: **91**

y
Yang, T.C.: **161**
Yee, W.: 252
Yoerger, E.J.: **203**, 329

z
Zeskind, R.M.: 304
Zhang, X.Z.: 118
Zhitkovskii, Yu.Yu.: 96
Zhou, J.X.: 118
Zienkiewicz, O.C.: 112
Zittel, J.D.: 57, 166

SUBJECT INDEX

a

Acoustic Reverberation Special Research Program (*see* ARSRP)
Active matched field processing, 313
　clutter rejection, 315
Active sonar, 299
Adaptive beamforming, 300, 321
Adaptive signal detection, 323
Adiabatic processes, 31
Air-sea interaction, 21
Airy functions, 146
Ambient noise,
　directionality, 378
Andreeva's formula, 263
Angle-depth diagram, 355
Arctic (*see also* Scattering, under ice)
Arctic, 161
　Fram Strait, 285
　Marginal ice zone, 285
　Polar front, 285
Array depth, 248
Array tilt, 248
Arrays, 14, 150
　adaptive beamforming, 299
　beam display, 378
　beam intensity, 152
　beam patterns, 168, 354
　　sidelobes, 88
　beam time series,
　　cross correlation, 193
　beamforming,
　　matched field, 313
　bearing ambiguity, 189, 195, 204
　element interactions, 306
　endfire beams, 256
　horizontal line, 155, 185, 235, 271, 378
　　COAMS, 247
　　DTAGS, 209
　　multiline, 370
　inverse beamforming, 333

Arrays (*continued*)
　of sources, 306
　optimum beamforming, 299
　planar, 85, 161, 334
　vertical line, 155, 161, 185, 235, 255, 264, 306, 334, 378
ARSRP, 51, 183, 195, 203, 209, 319
Atlantic Ocean, 241
　Icelandic Basin, 265
　Norwegian Sea, 268, 279

b

Backscattering (*see also* Scattering)
Backscattering, 235
　bubbles, 97
　discrete, 51
　frequency dependence, 93
　internal waves, 117
　low grazing angles, 91, 183
　surface, 98
Basin reverberation, 361
Bathymetry, 248
Beamforming (*see* Arrays)
Binary phase shift keying, 293
Biological respiration, 18
Biological scattering (*see also* Fish, Volume scattering)
Biological scattering,
　zooplankton, 287
Bipolar coordinates, 177
Bistatic (*see also* Models, Scattering, Sonar)
Bistatic geometry, 155
　reverberation, 367
　scattering, 203
Born approximation, 29
Bottom cores, 87
Bottom loss (*see* Reflection loss)
Bottom reverberation, 79, 91, 129, 156, 203, 243, 331, 361
Bottom roughness (*see* Roughness)

Bottom scattering, 66, 89, 227, 317
Bottom slope (*see* Sloping bottom)
Bottom types (*see also* Sediments)
Bottom types,
　basalt, 110, 227
　cobbles, 87
　detritus, 91
　elastic, 105, 215, 225, 227
　grain size, 92
　gravel, 87
　sand, 87, 94
　shell, 87
Boundary conditions,
　impedance, 140
　Dirichlet, 141, 216
　Neumann, 143
Boundary integral method, 45
Boundary element method, 109
Bragg diffraction, 74
Bragg scattering, 5, 149, 237
Bubbles, 97
　clouds, 3, 11, 26, 67, 97, 237, 242
　in sediments, 82
　layers, 89
　plumes, 25
　　ellipsoidal, 46
　saturation limit, 6, 100
　scattering, 25
　size distributions, 5, 11, 25, 74

c

CEAREX 89, 161
Chapman-Harris (*see* Scattering
　　functions)
Characteristic functions, 345
Chernov's theory, 60
Clutter rejection, 315, 319, 331, 367
Coefficient of variation, 99
Command and control systems, 379
Composite model, 145
Composite roughness, 4
Compressibility, 29
Convergence zone scattering, 271
Convergence zones, 189
Correlation,
　space-frequency, 315
Correlation function, 132
　spatial, 59
Covariance matrix, 331
Critical Sea Tests, 11, 235, 241
Cross correlation, 191
CST (*see* Critical Sea Tests)

d

Data,
　acquisition by sonobuoys, 379
　quality control, 379
Data bases, 145, 374
Detectors,
　coherent, 294
　mismatched energy, 347
　zero crossing, 294
Dimensional analysis, 143
Dipole radiators, 38
Dissolved gases, 18
Doppler
　invariance, 300
　nulling, 371
　shift, 14, 238, 305
　　bistatic, 367
　tolerance, 300

e

Echo sounders, 18
Echoes (*see* Target echoes)
Eckart approximation, 132
Eigenvector decomposition, 321
Ensemble average, 140

f

F-K transform, 219, 223
Facets, 188, 193
　reflection, 156, 203
　scattering, 55, 105
　self-selection, 53
False alarms, 294
Fathometer returns, 129
FFT methods, 331
Filter-squarer-integrator, 343
Filtering,
　low grazing angle energy, 212
Fish,
　day/night migration, 274, 283
　swimbladder scattering model, 287

Fish (scattering from)
  blue whiting, 266, 279
  polar cod, 286
  redfish, 284
  salmon, 238, 269
  swimbladders, 263, 272, 283, 286
Fluctuations,
  density, 59
  reverberation, 89
  sound speed, 59, 67
  temporal, 99
Forward scattering, 131
Fourier Integral Method, 331
Fourier-Stieltjes integral, 67
Fractal roughness, 116, 139, 188, 215, 221
Fresnel approximation, 131

g
Gas partial pressures, 12
Gated input signals, 344
Gaussian beams, 145, 168, 227
Gaussian distribution, 131
Gaussian random variables, 346
Geoacoustic models, 92, 126, 157, 177, 250, 340
Geomorphology, 207
Gradient search, 198
Grain sizes, 80
Green's functions, 45, 141, 148

h
Hard limited bandpass correlator, 295
Hard limiters, 295
Head waves, 217
Helmholtz integral, 131
High frequency measurements, 79, 85, 91, 97
Horizontal refraction, 358
Hydrogram, 224
Hyperbolic frequency modulation (*see* Sources)

i
Imaging, 195
  3-D, 87
  high resolution, 88
  sonar, 18

Interface waves, 115
Interference suppression, 320
Internal waves, 113
Inversion, 195
Isothermal processes, 31

k
Kirchhoff approximation, 4, 131, 221

l
Lambert's rule (*see* Scattering functions)
Langmuir circulation, 18
Least mean squares method, 198
Lighthill's method, 28
Linear prediction, 323
Lloyd mirror effect, 250

m
Mackenzie-Lambert scattering (*see* Scattering functions, Lambert's rule)
Matched field processing, 106
  active, 313
  array gain, 316
Matched filters, 236, 319, 368, 383
Measurement systems,
  acoustic, 86, 379
  DTAGS, 209
  environmental, 87, 379
  NORDSEE, 97
  volume scattering, 262
Measurements (*see also* High frequency measurements)
Measurements,
  direct path, 161
  in situ, 80
  laboratory, 80
  long range, 161
  medium range, 161
Mediterranean Sea,
  Ionian Sea, 264
Microbubbles (*see also* Bubbles)
Microbubbles, 3, 11, 25, 237
Mid-Atlantic Ridge (*see also* ARSRP)
Mid-Atlantic Ridge, 51, 184, 189, 195
Minima, global, 198
  local, 198

Minnaert formula, 35
Mode coupling, 314
Model-data comparisons, 161, 178, 247, 275
Models (*see also* Geoacoustic models, Sonar)
Models,
  ambient noise, 373
  ANDES, 374
  ASTRAL, 374
  AUAMP Baseline, 373
  BiRASP, 155
  BIRPS, 45
  BISSM, 203
  boundary element method, 109
  CAPARAY, 331
  elastic, 105, 113, 215, 221, 227
  FFRAME, 113
  finite difference, 125, 215, 221, 227
  finite element, 113
  FOAM, 113
  Gaussian beam, 145, 167
  GSM, 89, 271
  NISSM, 351
  normal mode, 155, 305, 358
  normal mode scattering, 161
  numerical, 45, 105, 113, 119, 125, 145, 167, 227, 305
  parabolic equation,
    two-way, 119, 249
  range dependent, 113, 119, 125, 171, 176, 227, 351
  RASP, 176
  ray theoretic, 90, 176, 351
  REVBAS, 351
  reverberation, 119, 125, 167, 323, 373
    3-D, 105, 145, 155, 176, 351, 373
    bistatic, 145, 155, 167
    Estes-Fain, 296
  RUMBLE, 167
  SAFARI, 105
  SAFE, 113
  system performance, 373, 379
  time dependence, 145
  time domain, 227, 305
  transmission loss, 249

Models (*continued*)
  two-way, 113, 119, 249
  WRAP, 155, 189
Monomolecular films, 71
Monopoles, 30
Monostatic (*see* Backscattering, Scattering)
Multi-scale correlation model, 62
Multiple scatterer correlation, 304
Multipole expansion, 30
Multistatic sonar (*see* Bistatic, Scattering, Sonar)

n
Noise, colored, 343
Non-stationary process, 343
NOREX 85, 8
Normal modes (*see also* Models)
Normal modes
  adiabatic, 164
  attenuation, 158
  cutoff, 157

o
Operator-machine interface, 379

p
Pacific Echo experiment, 132
Pacific Ocean, 241
  Gulf of Alaska, 235, 265
  northeast, 150, 309
  Vancouver Island, 248
Padé approximation, 120, 228
Parabolic equation (*see* Models)
Passive sonar, 343
Pentagonal tracks, 185
Perturbation theory, 65, 140
Plane wave decomposition, 211
Post-test analysis, 379
Principal component inverse method, 321
Probability density functions, 94
Propagation (*see* Transmission loss)
Pulse compression, 302
Pulse distortion, 305
Pulses (*see* Sources)
Pycnocline, 17

r

Ray theory,
  angle-depth diagram, 355
Ray-trace algorithms, 177
Rayleigh method, 149
Rayleigh modes, 220
Rayleigh-Rice approach, 4
Real-time processing, 379
Receiver operating characteristics, 345
Reflection loss,
  bottom
    MGS, 171, 177
  surface, 89
Renormalization group equation, 141
Replica (*see also* Matched filter)
Replica correlator, 305
Resonant
  effects, 47
  frequency, 14, 36, 237, 272
  microbubbles, 3
  scattering, 74, 263, 282
Reverberation (*see* Bottom, Subbottom, Surface, and Volume Reverberation; *also* Models, and Scattering)
Reverberation
  array heading surface, 361
  long range, 189
  low frequency, 161
  mapping, 195
  rejection (*see* Clutter rejection)
  roses, 361
Rough bottom scattering, 51
Rough seafloor, 215
Rough surfaces, 139
Roughness (*see also* Fractal roughness)
Roughness,
  bottom, 82, 92, 116, 355
  small scale, 221

s

S-waves (*see* Shear waves, *also* Bottom, elastic)
SALT tables, 168
Scattering,
  amplitudes, 45
  angle spreading, 131

Scattering *(continued)*
  chamber, 227
  coherent, 27
  cross section, 283
  facets, 188
  frequency dependence, 3
  frequency selective, 188
  from bubbles (*see* Bubbles, scattering)
  from spheres, 32, 88
  grazing angle dependence, 3
  high frequency, 71
  layers, 281
  long range, 183
  low frequency, 11
  low grazing angles, 75, 216, 227, 235
  steep slopes, 188
  time spreading, 131
  under ice, 66, 161
Scattering functions, 48
  bistatic, 156
  bottom, 156, 171, 241
  Brown-Saenger, 353
  Burke-Twersky, 163
  Chapman-Harris, 8, 171, 243, 353
  frequency dependence, 241
  grazing angle dependence, 129
  Lambert's rule, 56, 81, 93, 156, 163, 171, 188, 203, 223, 245
  low frequency, 241
  low grazing angles, 79, 221, 241
  plane-wave, 164
  surface, 241
Scattering strength,
  bottom, 352
  column, 264, 282, 287
  surface, 353
  volume, 351
Sea state, 243
Sediment properties, 61
Sediment types, 245
Sediments (*see also* Bottom types)
Sediments,
  gas bubbles, 82
  gravel, 80
  mud, 80
  sand, 80
  stratification, 81

Sediments *(continued)*
  thin, 132
Seismo-acoustic, 106
Seismograms,
  synthetic, 127, 231
Self affine, 216, 221
Seneca Lake, 34
Sensors,
  acoustic resonator, 19
  backscatter, 19
  bubble trap, 19
  conductivity, 12
  optical, 19
Shallow water, 113, 155
  Block Island Sound, 85, 298
  Coast of Florida, 178
  Gulf of Mexico, 177
  Mission Bay, 91
  Narragansett Bay, 298
  North Sea, 97
  Panama City, 81
Shear waves *(see also* Bottom, elastic)
Shear waves, 116, 127, 213, 227
Signal detection, 323
Signal processing, 299
Signal-to-reverberation ratio, 361
Simulated annealing, 45, 198
Simulations, 145
Singular value decomposition, 321
Slant stacking, 211
Sloping bottom, 133, 157, 248, 355
Solitons, 113
Sonar,
  active, 305
  bistatic, 378
  equation, 128
  multistatic, 367
  side scan, 87
  signal processing, 367
  systems, 367
Sound speed, 11
  profiles, 87, 157, 177, 248
    thermocline, 287
Sources,
  flextensional transducers, 306
  parametric, 34, 92, 98

Sources *(continued)*
  pulses, 86, 127, 227
    broadband, 305
    CW, 235
    HFM, 187, 190, 235, 310, 320, 379
    LFM, 299
    PRN, 235
    SUS charges, 241, 249
Special Research Projects
    (see ARSRP and SRP)
Spectral spreading *(see* Doppler)
Splines, 177
SRP, 11
Stacking, 195
Statistics,
  Rayleigh, 95
  wave breaking events, 13
Subbottom reflections, 129, 338
Subbottom reverberation, 83, 129, 209, 331
Subbottom scattering, 59, 131, 227, 245
Subsurface currents, 11
Surface reflection loss, 89
Surface reverberation, 3, 11, 72, 235, 242
Surface scattering, 89, 139, 236
  frequency dependence, 101
  wind speed dependence, 101

t

Tactical surveillance systems, 379
Target echoes, 88, 172
Target localization, 367
Target scattering, 314
Tau-p processing, 209
Temporal dependence, 240
Temporal variability *(see* Fluctuations)
Tracking signals, 293
Transmission loss, 116, 168
  measured, 251
Travelling acoustician problem, 200
Turbulent diffusion, 20

u

Upslope *(see* Sloping bottom)

V

Void fraction, 13
Volume fraction, 25
Volume inhomogeneities, 61
Volume reverberation, 235, 255, 263, 279, 285
Volume scattering,
   high frequency, 285
   low frequency, 255, 263
   mid frequency, 279
   water column, 271
Volume scattering strength, 260

W

Wave equation,
   elastic, 125, 227
   time domain, 125, 227
Wave rider buoy, 98
Wave spectra,
   Donelan, 5
   Pierson, 5
   Pierson-Moskowitz, 150
Wavebreaking roughness, 68
Waveform design, 305
Waveforms (*see* Sources)
Waves,
   breaking, 12
   directional spectrum, 14
   orbital velocity, 237
   whitecap fraction, 68
Wedge (*see* Sloping bottom)
Wind speed, 2, 15, 89, 101
WIT algorithm, 361